Structural Desig

B Currie

R A Sharpe

Stanley Th rs) Ltd

First published in 1990 by:
Stanley Thornes (Publishers) Ltd
Old Station Drive
Leckhampton
CHELTENHAM GL53 0DN
England

Reprinted 1992

British Library Cataloguing in Publication Data

Currie, Brian
Structural design
1. Structural design
I. Title. II. Sharpe, Robert A.
624.1771

ISBN 0-7487-0417-5

Typeset by Tech-Set, Gateshead, Tyne & Wear.
Printed and bound in Great Britain at The Bath Press, Avon.

Contents

Preface

It is a daunting experience for students to be confronted with the multitude of British Standards which have an input into the design of structures and to be asked to produce design calculations. If they have never been led through these publications their feeling of inadequacy is overwhelming. The authors have combined experience over some thirty-five years in the teaching of structural design and it is perhaps for the reason outlined above, that the question has frequently been asked, 'How is it done?' This book seeks to answer that question.

There are many books available which explain structural theory and allow the question, 'Why is it done?' to be answered. However, it does appear that there is a need to produce a book which illustrates how the available information is used to produce a structurally sound solution. It is hoped that this publication will go some way in meeting that need. The design calculations have been done by hand on calculation sheets with the aim of exposing the student to normal office practice. These sheets have, in the left-hand margin, referred to the various clauses of the British Standards. It is expected that the student will have available either the full standards or the publication *Extracts from British Standards for Students of Structural Design.*

From the above it is obvious that for the successful completion of any structural design there must be the marriage of theory and practice. This book really deals only with the latter, thus its readers are expected to have some competence in the theoretical behaviour of structures. The book should not be used therefore as a layman's guide to structural design. Those who will benefit most from the text are technician and undergraduate students of civil and structural engineering, building and architecture. It may also prove beneficial in updating practising engineers.

Computer Aided Design has been excluded from this book. This has been a conscious, some will say strange, decision. It has been taken for two reasons. The first is the fact that there is such a proliferation of software available that it is totally unnecessary for students to write their own programmes. (This may help their understanding of computer studies but not their understanding of structural design.) It is expected that readers will have such software available to them and can use it to check the designs presented in this book. Secondly, to have done justice to CAD would have taken more space than would have been justifiable economically.

Since a Preface is a personal statement, the authors desire to indicate that while all knowledge is good and serves a purpose, they accept that only one kind is eternal. This has been well put by G.E. Troy:

'What will it profit when life here is o'er,
Though great worldly wisdom I gain,
If, seeking knowledge, I utterly fail,
The wisdom of God to obtain.'

B. Currie, R.A. Sharpe
1990

Acknowledgements

While the authors attract both the praise and the criticism which comes with publishing a book, they wish to thank all who contributed in any way to making this publication possible. They freely acknowledge the part played by many people and, while all could not be named individually, special reference must be made to the following people and institutions: Mr W.H. Crooks, Head of the Department of Civil Engineering and Transport, University of Ulster, for his support and encouragement; the British Standards Institution, the British Constructional Steelwork Association Limited, the Constructional Steel Research and Development Organisation and the Steel Construction Institute, for permission to use extracts from their various publications.

The typing associated with this volume has been efficiently and patiently undertaken by Mrs M. Rea, Mrs L. Lyons and Mrs D. Savage, to whom we express our sincere gratitude.

Finally we must acknowledge the forbearance and support of our respective families during the hours of manuscript preparation.

1

Limit state design in structural steelwork to BS 5950

Permissible stress design and plastic design have for many years been permitted in BS 449: Part 2 *The Use of Structural Steel in Building.* In the permissible stress design method, working or actual loads are determined for a structure and, by using an elastic analysis, the resulting forces and moments are found. The elastic stresses in the members must not exceed permissible stresses given in BS 449. In addition to checking these stresses, the deflection of individual members and the overall structure must be limited. Plastic design is particularly economical in weight of steel used compared with that obtained using permissible stress design, and is associated particularly with rigid jointed portal frame construction. Since members tend to be more slender when this method is used, deflection needs more careful investigation.

Limit state design is now used for structural steelwork. Ultimate limit states include strength, stability, fatigue fracture, brittle fracture. When the limits are reached, collapse of a part or whole of a structure takes place. Serviceability limit states include deflection, vibration, repairable damage due to fatigue, corrosion and durability of the material. When limits are reached, the structure becomes unfit for use, although it may not necessarily collapse. The normal process is to design for ultimate limit state and then check for serviceability.

The overall factor of safety in any design has to cover variability of material strength, loading and structural performance. In BS 5950 the material factor γ_m is taken as 1.0. Working or actual loads are multiplied by relevant load factors γ_f in checking the strength and stability of a structure. The deflection under serviceability loads should neither impair the strength or efficiency of the structure or its components nor cause damage to the finishing, e.g. plaster ceilings. Table 5 of BS 5950 gives recommended deflection limits for a range of structural members. Generally the serviceability loads may be taken as the unfactored imposed loads.

Some properties of steel

For the more common types of steel, the design strength p_y values are obtained from Table 6 of BS 5950. These values of p_y depend on the thickness and grade of the material. The Modulus of Elasticity E is taken as 20.5×10^4 N/mm^2 and Poisson's Ratio v as 0.30.

Section properties and deductions for holes

Gross section properties should be determined using the specified size and profile of the member or element, with allowance made for openings larger than required for fasteners. The net area of a section or element should be taken as:

Gross area less deductions for fastener holes.

The area to be deducted where the holes are not staggered should be the maximum sum of the sectional areas of the holes in any cross-section at right angles to the direction of stress in the member. When the holes are staggered, the area to be deducted should be the greater of:

1) the deduction for non-staggered holes;
2) the sum of the sectional areas of all holes in any zig-zag line extending progressively across the member or part of the member, less $S_p{}^2t/4g$ for each gauge space in the chain of holes,

   where S_p = staggered pitch, i.e. the distance measured parallel to the direction of stress in the member, centre-to-centre of holes in consecutive lines.

   t = thickness of material,
   g = gauge, i.e. the distance measured at right angles to the direction of stress in the member, centre-to-centre of holes in consecutive lines.

For angles with holes in both legs, the gauge should be taken as the sum of the back marks to each hole, less the leg thickness.

Effective area at connections

The effective area A_e of each element of a member at a connection where fastener holes occur may be taken as:

K_e times its net area but not more than its gross area.

K_e = 1.2 for grade 43 steel.

Methods of design

The design of a structure must be carried out in accordance with BS 5950, Clause 2.1.2. The design methods presented are:

1) *Simple design.* Connections between members are assumed not to develop moments adversely affecting either the members or the structures as a whole. For analysis the structure is assumed to be pin-jointed. To maintain stability against sway, braced frames may be introduced or alternatively, use made of shear walls or lift cores designed to fulfil necessary functions within the structure.
2) *Rigid design.* The connections are assumed to be capable of developing strength and/or stiffness required by an analysis, assuming full continuity. Elastic or plastic analysis may be used.
3) *Semi-rigid design.* This method takes partial fixity of practical connections into account, but is very rarely used in practice.

Structures generally are designed to either the simple or the rigid methods of design. Examples throughout this book are confined mostly to the simple method of design.

2

Steel beams

Beams are horizontal structural members spanning between stanchions, columns or walls in a building structure, and are required to be of adequate strength and stiffness to resist the applied loading from the roof, floors, walls, etc. Universal beams (UBs) are commonly used and these are designed generally as single span and considered as simply supported at the ends. It is necessary, however, to consider the tendency of a beam to fail prematurely and below its full moment capacity by lateral torsional buckling. Here the compression flange behaves as a strut, becomes unstable, and buckles in the direction of least resistance, laterally or sideways, accompanied by a twisting action (Fig. 2.1(d)).

This form of buckling can be prevented when the floor or roof construction is capable of effectively restraining the compression flange of the beam.

Lateral restraint

Various degrees of lateral restraint exist in practice and for simply supported beams there are three main categories. Refer to Fig. 2.1.

1) Full lateral restraint is said to exist when the friction or positive connection to the compression flange of a member is capable of resisting a lateral force of not less than 1% of the maximum factored force in the compression flange of the member under factored loading. This load should be considered as distributed uniformly along the flange, provided that the dead load of the floor and the imposed load it carries together form the dominant loading on the beam. The floor or roof construction should also be capable of resisting this lateral force. Examples: A concrete slab bearing along the length of the compression flange of a beam normally provides full lateral restraint. A steel beam with the compression flange embedded in the concrete slab thickness and normally encased in concrete again provides full lateral restraint. When full lateral restraint is provided, no reference need be made to lateral torsional buckling resistance (Fig. 2.1(a)).
2) Beams may have lateral restraint provided only at intervals along the span by transverse secondary load-carrying beams, or alternatively by the introduction of 'tie' beams, both forms being connected positively to the main beam being considered. The system should be adequately braced (Fig. 2.1(b)).
3) Some beams lack any form of lateral restraint along the span, with positive connections only at the end supports (Fig. 2.1(c)).

Limiting proportions of cross-sections

Local buckling can be avoided by limiting the width-to-thickness ratio of each element of a cross-section subject to compression due to moment or axial load. Thin projecting flanges and thin webs of beams may buckle prematurely and this requires investigation.

Classification depends not only on the dimensions of the section but also the distribution of

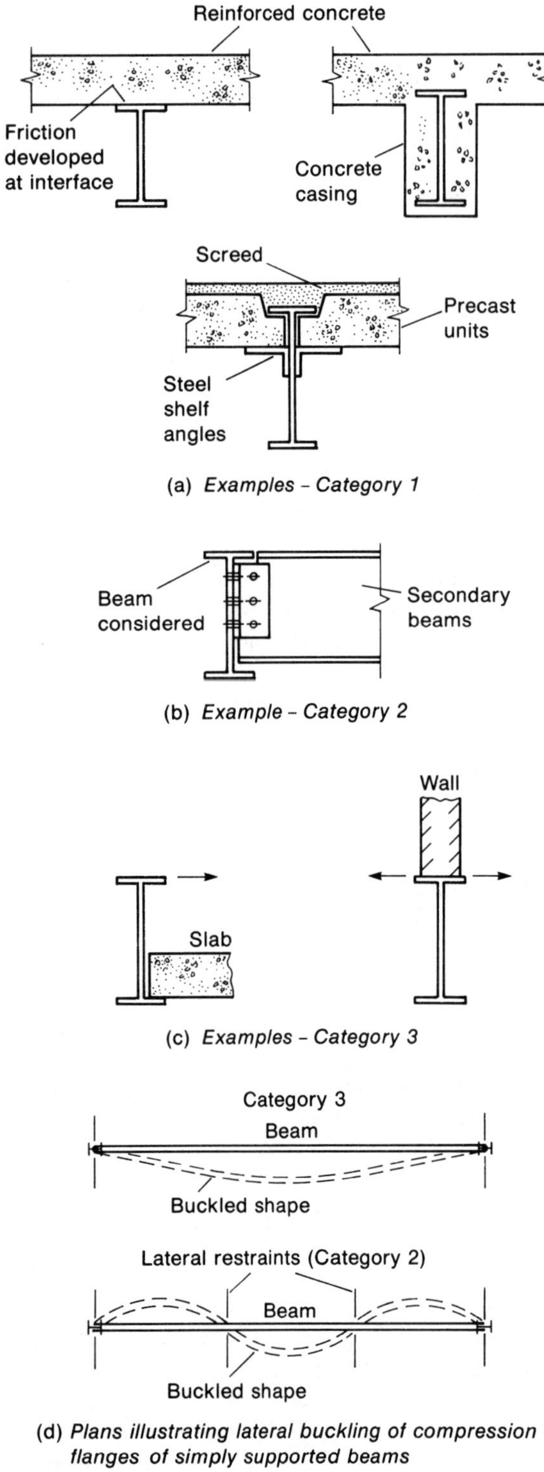

Fig. 2.1

stress across the section and the position of the neutral axis. Beam cross-sections are classified in accordance with their behaviour in bending, as plastic, compact, semi-compact or slender.

Class 1: Plastic. A plastic hinge can be developed with sufficient rotation capacity to allow redistribution of moments within the structure. Only Class 1 sections may be used for plastic design.

Class 2: Compact. The full plastic moment capacity can be developed, but local buckling may prevent development of a plastic hinge with sufficient rotation capacity to permit plastic design. Cannot be used for plastic design.

Class 3: Semi-compact. The stress at the extreme fibres can reach the design strength, but local buckling may prevent the development of the full plastic moment.

Class 4: Slender. Local buckling may prevent the stress in this type from reaching the design strength. The capacity of slender cross-sections is limited.

General conditions

All beams should meet the following conditions:

1) At critical points, the combination of
   a) maximum moment and co-existent shear;
   b) maximum shear and co-existent moment
   should be checked.
2) Deflection limits should not be exceeded.
3) Lateral torsional buckling resistance of the section should be checked if full lateral restraint is not provided.
4) Local buckling should be considered.
5) At end supports and at concentrated loads where loads or reactions are applied through the flange to the web:
   a) web buckling;
   b) web bearing;
   should be checked.

Shear

Two modes of shear failure are considered:

1) shear capacity of the web;
2) shear buckling of relatively thin webs. This

applies particularly to fabricated sections rather than rolled sections.

$$\text{Shear force } F_v \ngtr \text{ shear capacity } P_v$$

$$\text{Shear capacity } P_v = 0.6\,p_y A_v$$

where p_y = design strength (Table 6, BS 5950),
A_v = shear area taken as tD for rolled I, H, [sections, load parallel to web,
D = overall depth of section,
t = web thickness.

When d/t of a web $> 63\varepsilon$ then it should be checked for shear buckling. If the actual maximum shear stress is required then using the well-known formula and referring to Fig. 2.2,

$$\text{Maximum shear stress } f_v = \frac{FA\bar{Y}}{It} \ngtr 0.7\,p_y$$

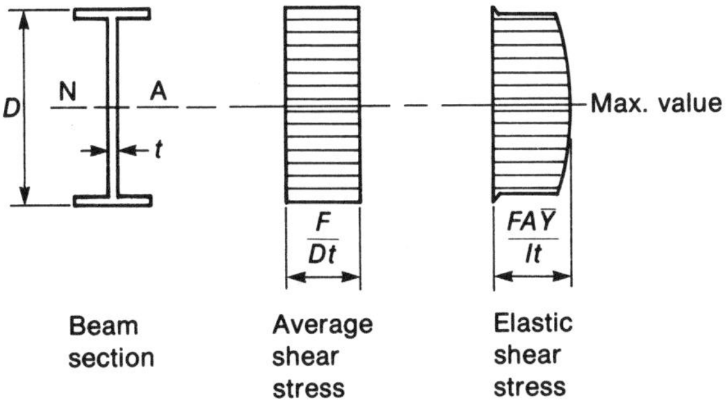

Fig. 2.2 *Shear stress distribution in an I-beam*

Combined bending and shear

Two conditions are considered, depending on whether the shear load is 'low' or 'high'. The moment capacity M_c of a beam section is reduced by the presence of a 'high' shear force.

1) *Moment capacity with low shear load.* If $F_v \leqslant 0.6\,P_v$ then the load is 'low' and the effect of shear on the moment capacity is small and no reduction in moment capacity M_c is required. For plastic or compact sections, plastic stress distribution is used and:

$$M_c = p_y S \ngtr 1.2\,p_y Z \qquad \text{(Fig. 2.3)}$$

For semi-compact and slender sections, elastic stress distribution is used and:

$$M_c = p_y Z \qquad \text{(Fig. 2.3)}$$

where S = plastic modulus of section about the relevant axis,
Z = elastic modulus of section about the relevant axis,
p_y = design strength (reduced for slender sections).

2) *Moment capacity with high shear load.* If $F_v > 0.6\,P_v$ then the load is 'high' and the effect of shear should be taken into account. See 4.2.6, BS 5950.

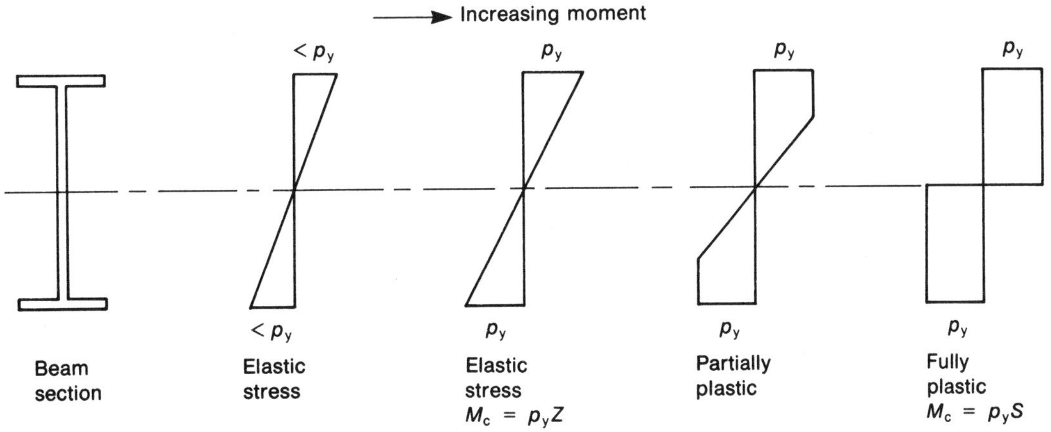

Fig. 2.3 *Beam behaviour in bending with increasing moment*

Lateral torsional buckling

When a beam is not provided with full lateral restraint, it is necessary to check its resistance to lateral torsional buckling. Lateral restraints may exist in the form of secondary load-carrying beams or 'tie' beams at intervals along the span. These prevent sideways movement of the compression flange at the restraint, and should be capable of resisting a lateral force of not less than 1% of the maximum factored force in the compression flange, shared equally between the points of restraint. Where several members share a common restraint, the maximum total lateral force may be taken as the sum of those derived from the largest three members.

Torsional restraints prevent movement of top and bottom flanges of the beam relative to each other.

Effective length

In considering lateral torsional buckling, the effective length L_E for a beam restrained at the ends only with different conditions of support may be obtained from Table 9 of BS 5950. For beams restrained at intervals by the steel members, L_E should be taken as L, the length of the beam between restraints.

Lateral torsional buckling resistance

For members, or a portion of a member, between adjacent lateral restraints, subject to bending about their major axis, the following condition should be satisfied:

$$\overline{M} \not> M_b$$

$$\overline{M} = mM_A$$

where $\overline{M}$ = equivalent uniform moment,
m = equivalent uniform moment factor,
M_A = maximum moment on the member, or portion of the member under consideration,
M_b = lateral torsional buckling resistance moment = $S_x P_b$
S_x = plastic modulus of the section about major axis,
p_b = bending strength.

Bending strength p_b is related to the equivalent slenderness λ_{LT} the design strength of the material p_y and the member type, i.e. rolled or fabricated by welding, and is obtained from Tables 11 or 12 of BS 5950, as appropriate.

The equivalent slenderness is given by:

$$\lambda_{LT} = nuv\lambda$$

where $\lambda = \dfrac{L_E}{ry} = \dfrac{\text{effective length}}{\text{radius of gyration minor axis}}$,
u = buckling parameter. For a rolled I, H, [section taken from section tables or conservatively as 0.9,
v = slenderness factor from Table 14 of BS 5950, using N and $\dfrac{\lambda}{X}$,
η = slenderness correction factor.

(X = torsional index from section tables,
N = 0.5 for members with equal flanges.)

Equivalent uniform moment factor, *m*

Formerly, beams were designed for uniform moment loading, a process which is unnecessarily conservative. Cases of moment gradient, i.e. unequal end moments are less prone to lateral instability, and this is recognized in BS 5950 by permitting the use of an equivalent uniform moment previously given as:

$$M = mM_A$$

For beams not loaded in the length between adjacent lateral restraints, the values of m are obtained from Table 18 of BS 5950, and depend on the ratio and the direction of the end moments M and βM. Values of β may be positive or negative, ranging from 1.0 to −1.0. Values of m range from 1.0 to 0.43 with the lower values where β is negative. Some cases with factors are shown in Fig. 2.4(a).

Correspondingly, the slenderness correction factor n is taken as 1.0 from Table 13 of BS 5950.

Slenderness correction factor *n*

For beams loaded in the length between adjacent lateral restraints, a more accurate method of

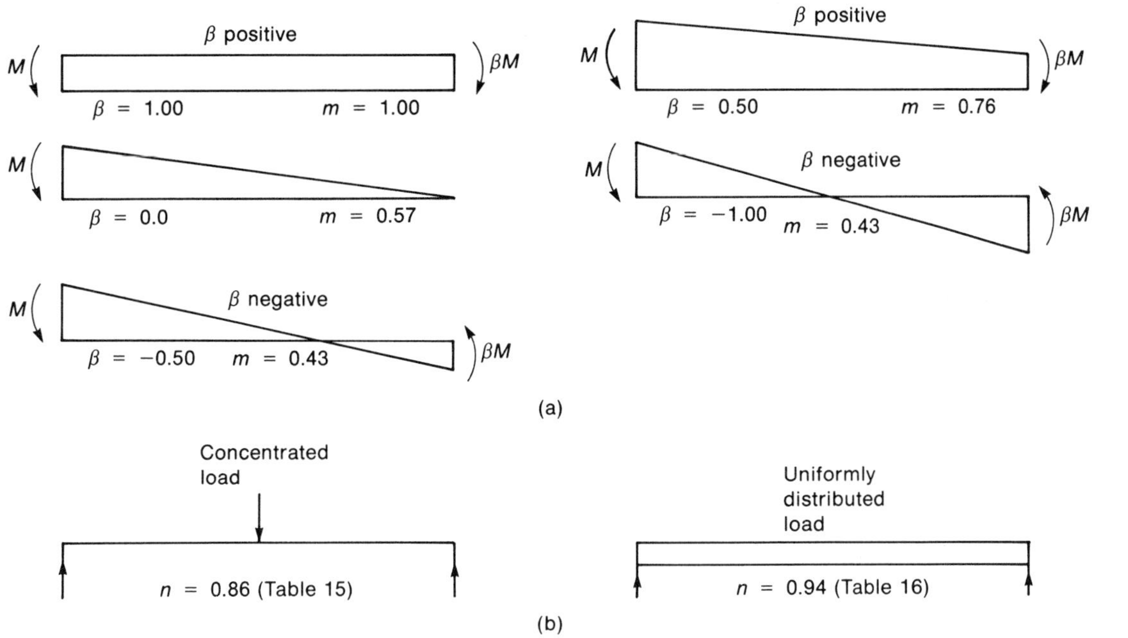

Fig. 2.4 *(a) Examples of values of m from Table 18 of BS 5950. (b) Examples of values of n for common cases of simply supported beams with end restraints only*

correcting for non-uniform moment may be used, by implementing the slenderness correction factor n rather than m. Values of n are given in Tables 15 and 16 of BS 5950, and depend on:

1) how the load is applied, i.e. with loading substantially concentrated within the middle fifth of the unrestrained length, or otherwise;
2) the ratio of the end moments at the points of restraint and the ratio of the larger moment to the mid-span simply supported moment.

Some common cases with factors are shown in Fig. 2.4(b). Correspondingly, the equivalent uniform moment factor m is taken as 1.0 from Table 13 of BS 5950.

Web buckling

The application of heavy concentrated loads on the top flange of a beam will produce a region of very high stress in the web directly under the load, causing possible buckling, with the web acting as a strut between the beam's flanges. This situation exists also at the end supports with the reaction applied now to the bottom flange. The load of reaction spreads or disperses into the web and is considered as carried by a vertical strut of width $b_1 + n_1$, where:

b_1 = stiff bearing length (Fig. 2.5)
n_1 = length obtained by dispersion at 45° through half the depth of the beam (Fig. 2.6(a)).

The buckling resistance is given by:

$$P_w = (b_1 + n_1)\, t p_c$$

where t = web thickness,
p_c = compressive strength using Table 27(c) and λ as follows.

In determining p_c the slenderness λ of an unstiffened web should be taken as $2.5d/t$ (where d is the depth of the web), provided that the flange through which the load or reaction is applied is restrained effectively against:

1) rotation relative to the web,
2) lateral movement relative to the other flange.

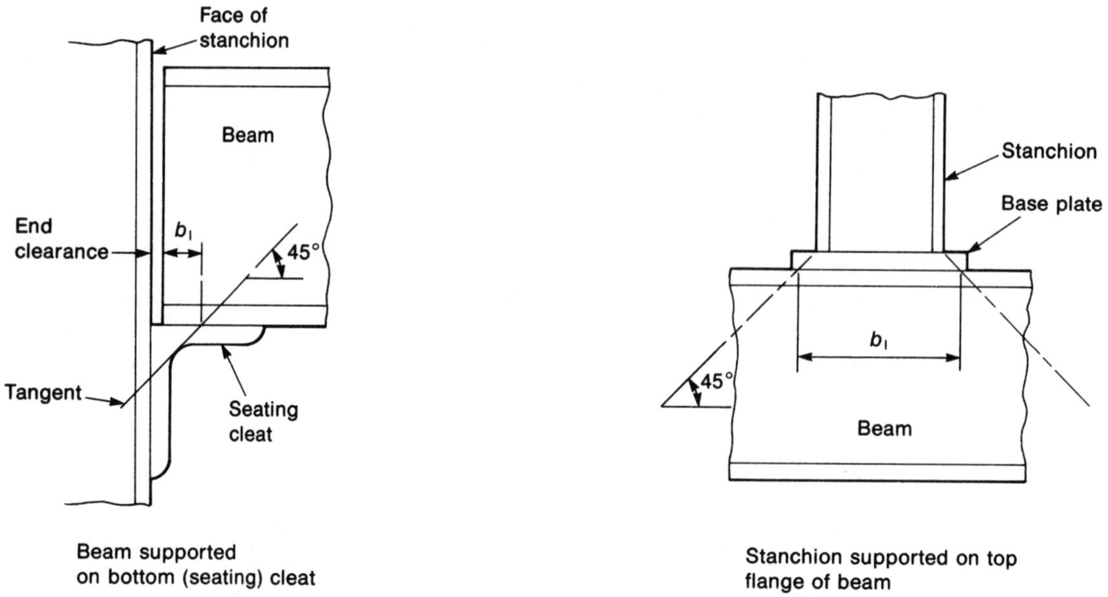

Fig. 2.5 *Examples of stiff bearing length* b_l

In this case the web acts as a fixed end strut with $L_E = 0.7d$.

$$\lambda = \frac{L_E}{ry} = \frac{0.7d}{ry}$$

$$ry = \sqrt{\left(\frac{I_y}{A}\right)} = \sqrt{\left(\frac{t^3}{12t}\right)} = \frac{t}{3.46}$$

$$\frac{0.7d}{t} \times 3.464 = \frac{2.43d}{t}, \text{ taken as } 2.5\,\frac{d}{t}$$

If the web is not restrained as required above, then a value of L_E appropriate to the conditions of web/flange restraint can be chosen. The applied load or reaction must not exceed P_w. If the load or reaction exceeds P_w then vertical stiffeners should be provided.

Web bearing

The web of the beam can fail also by bearing or crushing of the material when the whole end reaction or concentrated load has to be carried on a small, highly stressed horizontal section of the web area. The load or reaction is dispersed into the web and the level to be considered is at the junction of the web and the root fillet.

The bearing capacity is given by

$$(b_l + n_2)tp_{yw}$$

where b_l = stiff bearing length (Fig. 2.5),
n_2 = length obtained by dispersion through the flange to the junction of the web and the roof fillet at a slope of 1 in 2.5 (Fig. 2.6(b)),
t = web thickness,
p_{yw} = design strength of the web.

If the bearing capacity is exceeded by the load or reaction, then vertical stiffeners should be provided.

Deflection

Deflection under serviceability loads should not impair the strength or efficiency of a structure or its components to cause damage to the finishings, e.g. plastered ceilings. Generally the serviceability loads may be taken as the unfactored imposed loads. Table 5 of BS 5950 gives recommended limitations for certain structural members.

Formulae for calculating the deflection of beams under various conditions of loading are

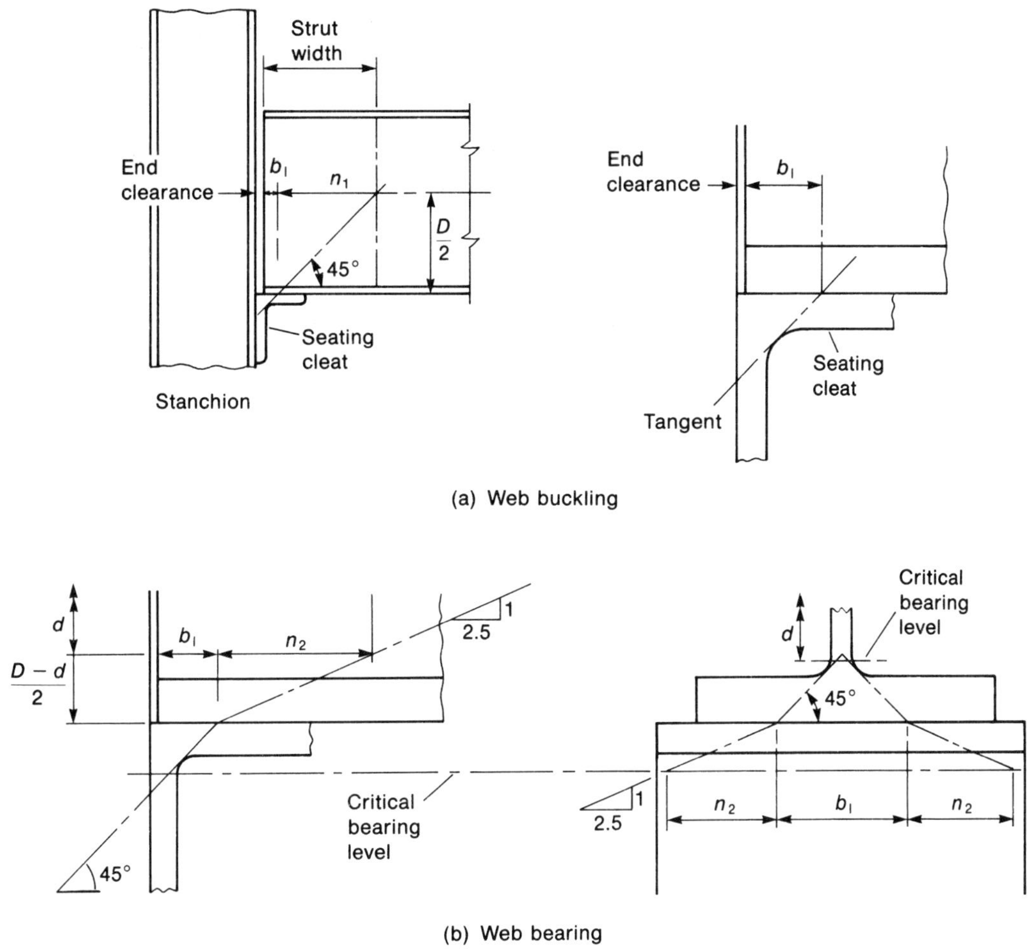

Fig. 2.6 *Web buckling and bearing at end seating cleat*

given in structural handbooks. Two of the most commonly used formulae are for simply supported beams carrying:

1) a uniformly distributed load,
2) a point load at the mid span.

For case (1) the deflection is given by:

$$\frac{5}{384}\,\frac{WL^3}{EI}$$

For case (2) the deflection is given by:

$$\frac{1}{48}\,\frac{WL^3}{EI}$$

where W = total load,
L = span
E = 20.5×10^4 N/mm^2
I = relevant second moment of area of the beam.

Lattice beams

Lattice beams are designed to support various forms of roofing or flooring applied as a uniformly distributed loading to the upper surface of the top chord. For example, suitable steel metal decking positively fixed to the top chord and capable of acting as a stiff horizontal diaphragm provides full lateral restraint to the top chord. Lattice beams may be fabricated from rectangular hollow sections or tee and angle sections, welded together at the node points. Camber is provided to allow for deflection of the beams under loading.

Beams may be connected together and restrained horizontally at intervals along their span by transverse members and bracing systems, to provide temporary stability until the roof or floor decking is laid and fixed permanently in position.

Castellated beams

Castellated beams are fabricated from UB, UC and joist sections by flame cutting the web to form castellations, separating the two parts and then butt welding the teeth together to produce a new section of greater depth than the original section. The regular pattern of holes thus formed in the web allows for easy access and route of service ducts and pipes in a building. These beams are suitable for long spans with light loading and where heavy concentrated loads are not present. It is normal practice to fill in the web openings at and near to the ends of the beams, to prevent shear buckling or bearing failure, and to facilitate the provision of suitable end connections to supporting members.

Design of castellated beams is quite complex, including investigation of combined stresses, local buckling, web buckling and bearing deflection. Analysis of these sections is outside the scope of this book.

Excavations – sheet piling – walings and struts

Design of a typical waling, i.e. a laterally unrestrained, simply supported beam is shown for a deep sheet piled excavation. The pressure given is assumed for design purposes only, and actual soil conditions dictate the actual pressure values used in calculations.

CURPE CONSULTANTS
46 Orburn Road
Dunfield

Project: Steel Beams | Job ref:
Part of structure: Beam with full lateral restraint | Calc sheet no: C2/ 1 / rev
Drawing ref: | Calc by: R.A.S. | Date: | Check by: | Date:

Ref	Calculations	Output
BS5950	Ex.1. Beam with full lateral restraint	

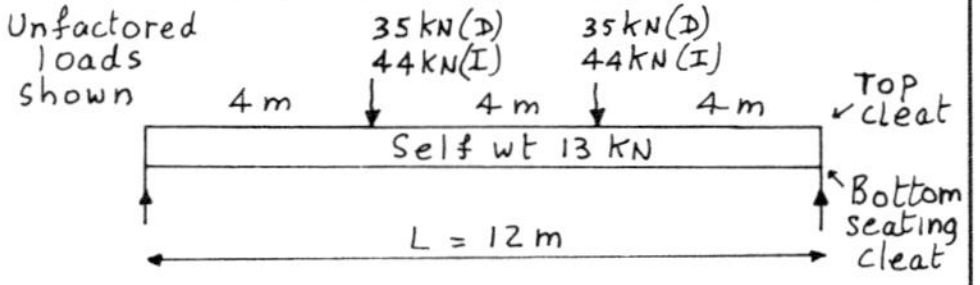

Table 2 — Factored loads

Point load $= 1{\cdot}4(35) + 1{\cdot}6(44) = 119{\cdot}4$ kN

Self wt U.D.L $= 1{\cdot}4(13) = 18{\cdot}2$ kN

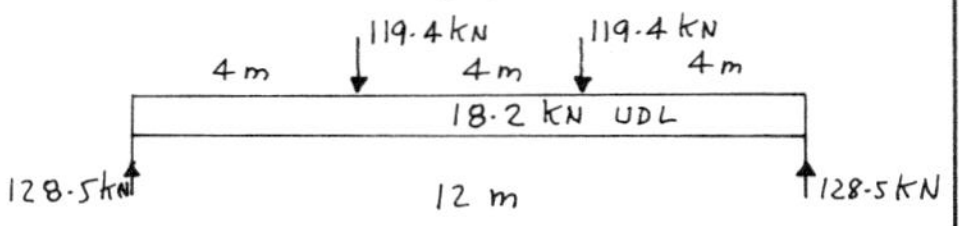

Output: Reactions 128·5 kN

At mid span $M_x = (128{\cdot}5\times6) - (119{\cdot}4\times2) - \left(\frac{18{\cdot}2}{2}\times3\right)$

$M_x = 771 - 238{\cdot}8 - 27{\cdot}3 = 504{\cdot}9$ kNm — Output: $M_x = 504{\cdot}9$ kNm

Cl.4.2.5 — $M_c = p_y S \not> 1{\cdot}2\, p_y Z$ for plastic or compact sections.

S_x reqd $= \frac{M_x}{p_y} = \frac{504{\cdot}9\times10^6}{275} = 1836000\ \text{mm}^3$

Try 533 × 210 × 82 UB ($S_x = 2060\ \text{cm}^3$)

Table 6 — For $T = 13{\cdot}2 < 16$, $p_y = 275\ \text{N/mm}^2$ — Output: $p_y = 275\ \text{N/mm}^2$

Table 7 — Section classification

$\frac{b}{T} = \frac{B}{2T} = \frac{208{\cdot}7}{2\times13{\cdot}2} = 7{\cdot}9 < 8{\cdot}5\varepsilon$

$\frac{d}{t} = \frac{476{\cdot}5}{9{\cdot}6} = 49{\cdot}6 < 79\varepsilon$ Plastic. — Output: Plastic

Shear

F_v at supports $= 128{\cdot}5$ kN — Output: $F_v = 128{\cdot}5$ kN

Cl.4.2.3 — Shear capacity $P_v = 0{\cdot}6\, p_y A_v$.

CURPE CONSULTANTS
46 Orburn Road
Dunfield

Project: Steel Beams. | Job ref:
Part of structure: Beam with full lateral restraint | Calc sheet no: C2 / 2 / rev
Drawing ref: | Calc by: R.A.S. | Date: | Check by: | Date:

Ref	Calculations	Output

$P_v = 0{\cdot}6\, p_y D t$.

$= \frac{0{\cdot}6\times275\times528{\cdot}3\times9{\cdot}6}{10^3} = 837$ kN — Output: $P_v = 837$ kN

Checks with value given in Design Guide Vol 1 Page 140

Since $F_v < P_v$ Shear O.K. — Output: Shear O.K.

Also since $F_v < 0{\cdot}6 P_v$ shear force is 'low.'

Cl. 4.2.5. — Moment Capacity M_c

$M_c = p_y . S_x = \frac{275\times2060\times10^3}{10^6} = 566{\cdot}5$ kNm

$\not> 1{\cdot}2 . p_y . Z = \frac{1{\cdot}2\times275\times1800\times10^3}{10^6} = 594$ kNm

∴ $M_c = 566$ kNm, checks with value given in Design Guide Vol 1, Page 130. — Output: $M_c = 566$ kNm

Since $M_c > M_x$ 566 > 504·9 O.K — Output: Moment O.K.

Table 5 — Check deflection

Use unfactored imposed loads

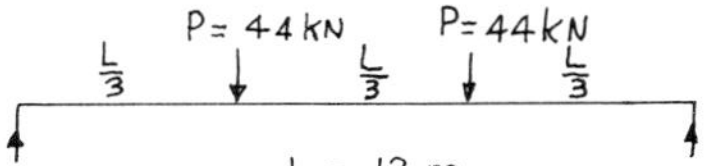

Concentrated loads at third points

Actual $d = \frac{23}{648}\frac{PL^3}{EI}$

$d = \frac{23\times44\times10^3\times12^3\times10^9}{648\times20{\cdot}5\times10^4\times47500\times10^4} = 27{\cdot}7$ mm

Allowable $= \frac{L}{360} = \frac{12\times10^3}{360} = 33{\cdot}3$ mm

Since $27{\cdot}7 < 33{\cdot}3$ O.K. — Output: Deflection O.K.

CURPE CONSULTANTS
46 Orburn Road
Dunfield

Project: Steel Beams			Job ref	
Part of structure: Web buckling & bearing			Calc sheet no: C2 / 3	rev: 1
Drawing ref	Calc by: R.A.S.	Date	Check by	Date

Ref	Calculations	Output

Check for web buckling at supports

End reaction = 128·5 kN — *Output:* Reaction 128·5 kN

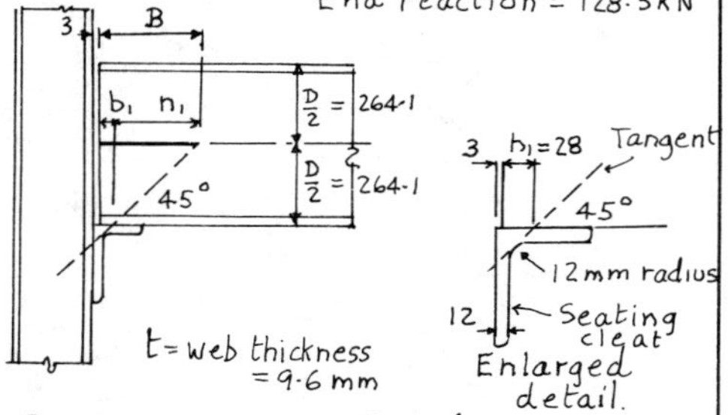

Cl. 4.5.2.1 — Buckling resistance $P_w = (b_1 + n_1)t\,p_c$

Web slenderness $\lambda = \frac{2{\cdot}5d}{t} = \frac{2{\cdot}5 \times 476{\cdot}5}{9{\cdot}6} = 124$

Table 27c — For $\lambda = 124$ and $p_y = 275$, $p_c = 92\ N/mm^2$

$P_w = \frac{(28 + 264{\cdot}1)\,9{\cdot}6 \times 92}{10^3} = 258\,kN$ — *Output:* P_w = 258 kN

Since reaction < P_w, 128·5 < 258 O.K. — *Output:* Web buckling O.K.

Check for web bearing at supports — *Output:* Reaction 128·5 kN

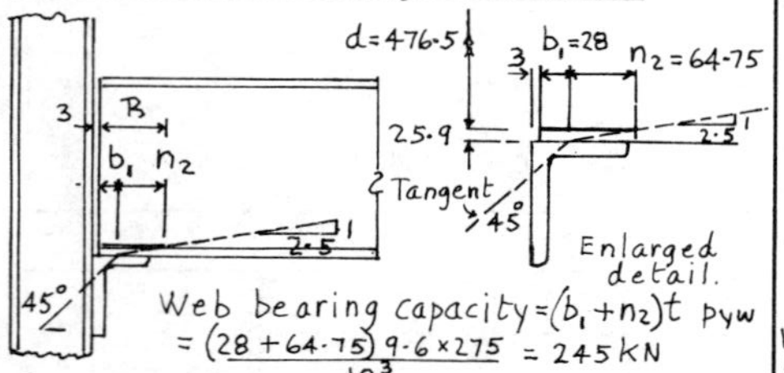

Cl. 4.5.3 — Web bearing capacity $= (b_1 + n_2)t\,p_{yw}$

$= \frac{(28 + 64{\cdot}75)\,9{\cdot}6 \times 275}{10^3} = 245\,kN$

Since reaction < Bearing capacity O.K. — *Output:* Web bearing O.K. Capacity 245 kN. USE 533×210×82UB

CURPE CONSULTANTS
46 Orburn Road
Dunfield

Project: Steel Beams			Job ref	
Part of structure: Beam restrained at points			Calc sheet no: C2 / 4	rev: 1
Drawing ref	Calc by: R.A.S.	Date	Check by	Date

Ref: BS5950	Calculations	Output

Ex. 2. **Beam restrained at A, B, C, D only**

Unfactored loads: Dead 9 kN/m U.D.L; Imposed 10 kN/m UDL

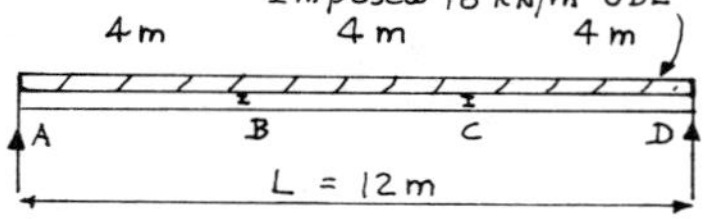

Table 2 — **Factored loads.**

Dead = 1·4 (9 × 12) = 151·2 kN

Imposed = 1·6 (10 × 12) = 192·0 kN

Total UDL = 343·2 kN

End reactions = 171·6 kN

Applied moments

M_x at mid span $= \frac{343{\cdot}2 \times 12}{8} = 514{\cdot}8\,kNm$ — *Output:* M_x = 514·8 kNm

M at B and C $= (171{\cdot}6 \times 4) - (114{\cdot}4 \times 2) = 457{\cdot}6\,kNm$

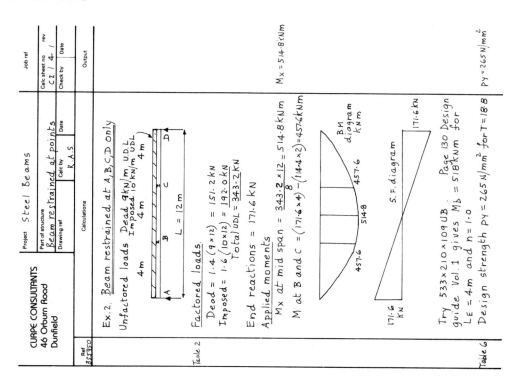

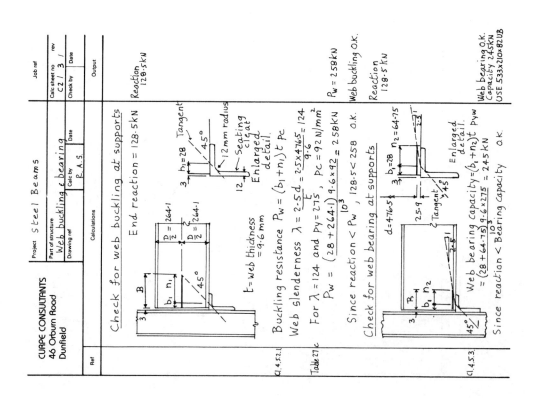

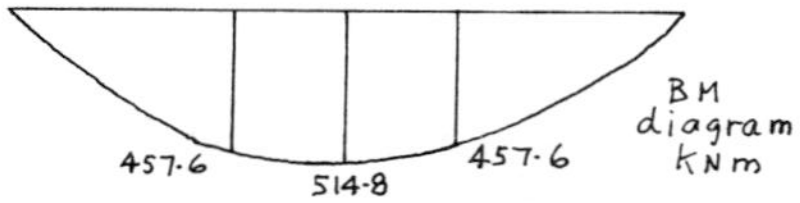

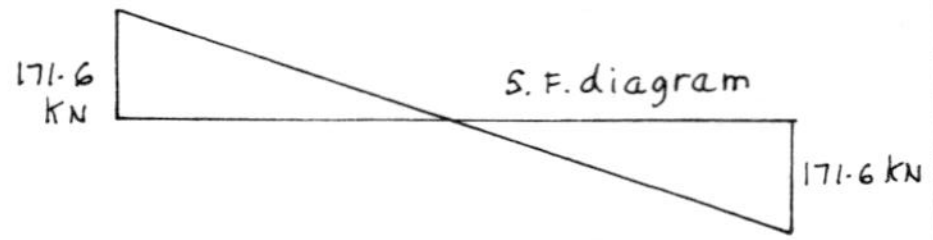

Try 533 × 210 × 109 UB. Page 130 Design guide Vol. 1 gives M_b = 518 kNm for L_E = 4 m and n = 1·0

Table 6 — Design strength $p_y = 265\ N/mm^2$ for T = 18·8 — *Output:* p_y = 265 N/mm²

CURPE CONSULTANTS
46 Orburn Road
Dunfield

Project: Steel Beams. | Job ref
Part of structure: Beam restrained at points | Calc sheet no: C2 / 5 | rev: 1
Drawing ref | Calc by: R. A. S. | Date | Check by | Date

Ref BS5950	Calculations	Output
Table 7	Section classification	
	$\frac{b}{T} = \frac{105.35}{18.8} = 5.6 < 8.5\varepsilon$	
	$\frac{d}{t} = \frac{476.5}{11.6} = 41.1 < 79\varepsilon$	
	$\varepsilon = \sqrt{\left(\frac{275}{py}\right)} = 1.02$ Section plastic	Plastic.
	Shear F_v at supports = 171.6kN	$F_v = 171.6\,kN$
	Shear area $A_v = t\,D = 539.5 \times 11.6$	
Cl.4.2.3	Shear capacity $P_v = 0.6\,p_y\,A_v$	
	$P_v = \frac{0.6 \times 265 \times 539.5 \times 11.6}{10^3} = 995\,kN$	$P_v = 995\,kN$
	$P_v > F_v$ $995 > 171.6$ O.K.	Shear O.K.
Cl.4.2.5	Moment capacity	
	$M_{cx} = p_y\,S_x \not> 1.2\,p_y\,Z$	
	$M_{cx} = \frac{265 \times 2820 \times 10^3}{10^6} = 747\,kNm$	
	$\not> \frac{1.2 \times 265 \times 2470 \times 10^3}{10^6} = 785\,kNm$	
	$\therefore M_{cx} = 747\,kNm$	$M_{cx} = 747\,kNm$
	$0.6\,P_v = 0.6 \times 995 = 597\,kN$	
	Since $F_v < 0.6\,P_v$ shear force is 'low'	
	and M_{cx} need not be reduced.	
	$M_{cx} > M_x$ $747 > 514.8$ O.K.	Moments O.K. so far but need further checks
Cl 4.2.1.3(a)	Combined shear and bending	
	At mid span $F_v = 0$ $M = 514.8\,kNm$	
	$F_v < 0.6\,P_v$ $\therefore M_{cx} = 747$	
	At supports $F_v = 171.6$ $M = 0$	
	$F_v < 0.6\,P_v$ $\therefore M_{cx} = 747$	
	$M_x < M_{cx}$ O.K.	Combined shear + bending O.K.
CL.4.3.	Check lateral torsional buckling	
4.3.7.1.	$\bar{M} \not> M_b$	

CURPE CONSULTANTS
46 Orburn Road
Dunfield

Project: Steel Beams. | Job ref
Part of structure: Beam restrained at points | Calc sheet no: C2 / 6 | rev: 1
Drawing ref | Calc by: R. A. S. | Date | Check by | Date

Ref BS5950	Calculations	Output
4.3.7.2.	$\bar{M} = m\,M_A$	
Table 13	$m = 1.0$ for member lengths AB, BC, CD loaded between adjacent lateral restraints	
	n to be determined	
4.3.7.3.	$M_b = S_x\,p_b$	
	Bending strength p_b depends on equivalent	
4.3.7.5	slenderness $\lambda_{LT} = n\,u\,v\,\lambda$	
	Length AB	
	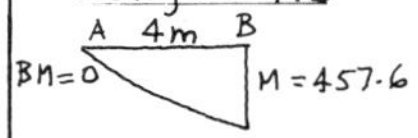	
	U.D.L on AB = $\frac{343.2}{3} = 114.4\,kN$	
	$M_o = \frac{114.4 \times 4}{8} = 57.2\,kNm$	
	$\gamma = \frac{\text{larger end mom}}{\text{mid length mom}} = \frac{M}{M_o} = \frac{457.6}{57.2} = 8.0$	
Table 16	$\beta = \frac{\text{Smaller end mom}}{\text{Larger end mom}} = \frac{0}{457.6} = 0$ $\therefore n = 0.826$	
4.3.5.	$L_E = 4\,m$	
4.3.7.5	Slenderness $\lambda = \frac{L_E}{r_y} = \frac{4 \times 10^3}{4.61 \times 10} = 86.7$	
	$\frac{\lambda}{x} = \frac{86.7}{30.9} = 2.8$	
Table 14	For $\frac{\lambda}{x} = 2.8$ and $N = 0.5$ $v = 0.92$	
	$\lambda_{LT} = n\,u\,v\,\lambda = 0.826 \times 0.875 \times 0.92 \times 86.7 = 57.5$	
Table 11	For $\lambda_{LT} = 57.5$ and $p_y = 265$ $P_b = 213\,N/mm^2$	$P_b = 213\,N/mm^2$
4.3.7.3	$M_b = S_x\,p_b$	
	$= \frac{2820 \times 10^3 \times 213}{10^6} = 600\,kNm$	$M_b = 600\,kNm$
	Since $\bar{M} < M_b$ $457.6 < 600$, **Length AB** O.K.	Length AB O.K. lat tors buckg
	Length BC	
	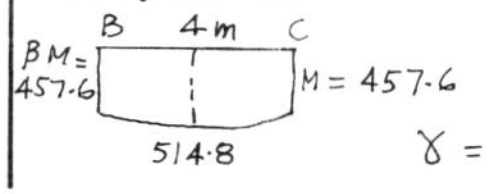	
	Design moment = 514.8kNm	
	$M_o = \frac{114.4 \times 4}{8} = 57.2\,kNm$	
	$\gamma = \frac{M}{M_o} = \frac{457.6}{57.2} = 8.0$	

CURPE CONSULTANTS
46 Orburn Road
Dunfield

Project: Steel Beams	Job ref
Part of structure: Beam restrained at points	Calc sheet no C2 / 7 / rev
Drawing ref / Calc by R.A.S. / Date	Check by / Date

Ref BS5950	Calculations	Output
	$\beta = \frac{457.6}{457.6} = 1.0$	
Table 16	$\therefore n = 0.99$	
	$\lambda_{LT} = 0.99 \times 0.875 \times 0.92 \times 86.7 = 69$	
Table 11	$P_b = 186.4\ N/mm^2$	$P_b = 186.4\ N/mm^2$
	$M_b = \frac{2820 \times 10^3 \times 186.4}{10^6} = 525.6\ kNm$	$M_b = 525.6\ kNm$
	Since $\bar{M} < M_b$ 514.8 < 525.6 BC O.K.	Length BC O.K. lat. tors. buckg
	Length CD	
	Calculations as for length AB	Beam O.K. lat. tors. buckg
Table 5	Check deflection	
	Total imposed load = 10×12 = 120kN U.D.L.	
	$d = \frac{5}{384}\frac{WL^3}{EI} = \frac{5 \times 120 \times 10^3 \times 12^3 \times 10^9}{384 \times 20.5 \times 10^4 \times 66700 \times 10^4} = 19.7\ mm$	
	Allowable $= \frac{L}{360} = \frac{12 \times 10^3}{360} = 33.3\ mm$	
	Actual < Allowable 19.7 < 33.3 O.K.	Deflection O.K.
	Web buckling and bearing capacities checked as shown in previous example for seating and top cleat type connection.	USE 533×210×109 UB

CURPE CONSULTANTS
46 Orburn Road
Dunfield

Project: Steel Beams	Job ref
Part of structure: Beam restrained at points	Calc sheet no C2 / 8 / rev
Drawing ref / Calc by R.A.S. / Date	Check by / Date

Ref BS5950	Calculations	Output

Ex.3. Beam restrained at A, B, C, D only

Member not loaded in length between lateral restraints except for self weight

Unfactored loads

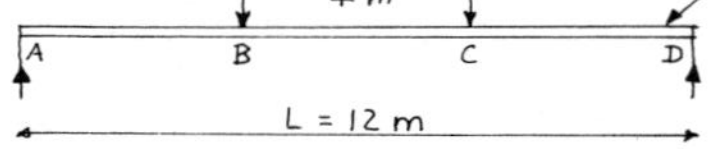

Table 2 — Factored loads

Point loads = 1.4(35) + 1.6(44) = 119.4 kN

Self weight U.D.L = 1.4(13) = 18.2 kN

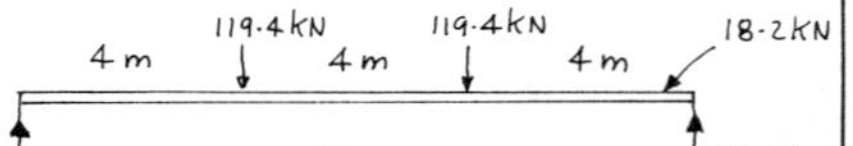

Output: Reactions 128.5kN

Applied moments

M_x at mid span $= (128.5 \times 6) - (119.4 \times 2) - (\frac{18.2}{2} \times 3)$

$= 771 - 238.8 - 27.3 = 504.9\ kNm$

Output: $M_x = 504.9\ kNm$

M at B and C $= (128.5 \times 4) - (\frac{18.2}{3} \times 2)$

$= 514 - 12.1 = 501.9\ kN\ m$

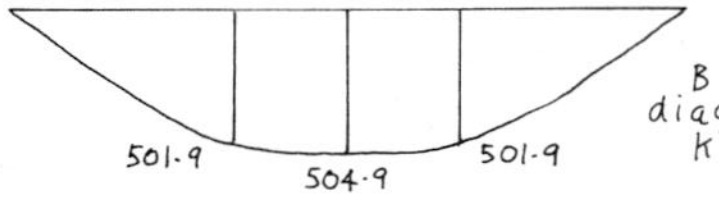

BM diagram kNm

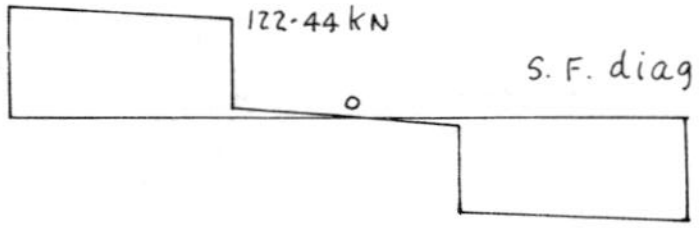

S. F. diagram.

CURPE CONSULTANTS 46 Orburn Road Dunfield	Project: Steel Beams Part of structure: Beam restrained at points Drawing ref: Calc by: R.A.S. Date:	Job ref: Calc sheet no: C2 / 9 / rev Check by: Date:
Ref BS5950	**Calculations**	**Output**
	Try 533 × 210 × 109 UB. Page 130 Design Guide Vol 1 gives $M_b = 518$ kNm for $n = 1.0$ and $L_E = 4$ m.	
Table 7	<u>Section Classification</u> From previous calculations section is plastic	Plastic
4.2.3.	<u>Shear</u> $F_v = 128.5$ kN $P_v = 995$ kN $F_v < P_v$ $128.5 < 995$ O.K.	$F_v = 128.5$ kN $P_v = 995$ kN Shear O.K.
4.2.5.	<u>Moment Capacity</u> $M_{cx} = 747$ kNm Since $F_v < 0.6P_v$ shear force is low M_{cx} need not be reduced. $M_{cx} > M_x$ $747 > 504.9$	$M_{cx} = 747$ kNm Moments O.K. so far but need further checks
4.2.1.3(a)	<u>Combined shear and bending</u> (I) At B. $F_v = 122.44$ kN $M = 501.9$ kN $F_v < 0.6\,P_v$ ∴ $M_{cx} = 747$ kNm Since $M < M_{cx}$ $501.9 < 747$ O.K. (II) At A. $F_v = 128.5$ kN $M = 0$ $F_v < 0.6\,P_v$ ∴ $M_{cx} = 747$ kNm Since $M < M_{cx}$ $0 < 747$ O.K.	Combined shear + bending O.K.
	<u>Check lateral torsional buckling</u> $\bar{M} \not> M_b$ $\bar{M} = m.\,M_A$	
Table 13	Member not loaded between restraints since self weight is insignificant then $n = 1.0$ but m must be determined.	
Table 18	<u>Length AB</u> 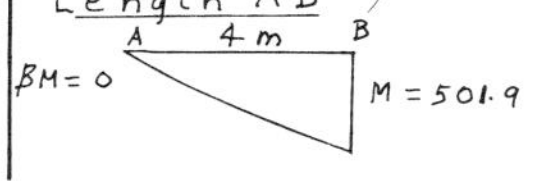 $\beta = \frac{0}{501.9} = 0$ ∴ $m = 0.57$ $\bar{M} = 0.57 \times 501.9$ $= 286.08$ kNm	Length AB $\bar{M} = 286.08$ kNm

CURPE CONSULTANTS 46 Orburn Road Dunfield	Project: Steel Beams Part of structure: Beam restrained at points Drawing ref: Calc by: R.A.S. Date:	Job ref: Calc sheet no: C2 / 10 / rev Check by: Date:
Ref BS5950	**Calculations**	**Output**
Table 18	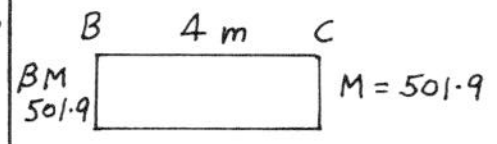 $\beta = \frac{501.9}{501.9} = 1.0$ ∴ $m = 1.0$ $\bar{M} = 1.0 \times 501.9 = 504.9$ kNm	Length BC $\bar{M} = 504.9$ kNm
	∴ Length BC is the critical span Page 130 Design Guide Vol 1 for $n = 1.0$ and $L_E = 4$ m, a 533 × 210 × 109 UB has an $M_b = 518$ kNm $M_b = S_x.p_b.$ $\lambda_{LT} = nuv\lambda$ $\lambda = \frac{L_E}{r_y} = \frac{4 \times 10^3}{4.61 \times 10} = 86.7$ $\frac{\lambda}{x} = \frac{86.7}{30.9} = 2.8$	
Table 14	For $\frac{\lambda}{x} = 2.8$ and $N = 0.5$ $v = 0.92$ $\lambda_{LT} = 1.0 \times 0.875 \times 0.92 \times 86.7 = 69.7$	
Table 11	For $\lambda_{LT} = 69.7$ and $p_y = 265$ $P_b = 184.72\ \text{N/mm}^2$ $M_b = \frac{2820 \times 10^3 \times 184.72}{10^6} = 520.9$ kNm Since $\bar{M} < M_b$ $504.9 < 520.9$ OK	$P_b = 184.7\ \text{kN/mm}^2$ $M_b = 520.9$ kNm Lateral torsional buckling O.K.
	Web buckling and bearing capacities checked as shown previously for seating and top cleat type connection.	USE 533 × 210 × 109 UB

CURPE CONSULTANTS 46 Orburn Road Dunfield	Project: Steel Beams	Job ref
	Part of structure: Lattice Beam	Calc sheet no: C2 / 11 / rev
	Drawing ref: — Calc by: R.A.S. Date: —	Check by: — Date: —

Ref	Calculations	Output

Ex.4. Design of Lattice Beam

Beams are fully restrained laterally at top chord level by direct fixing of steel decking & roofing system.

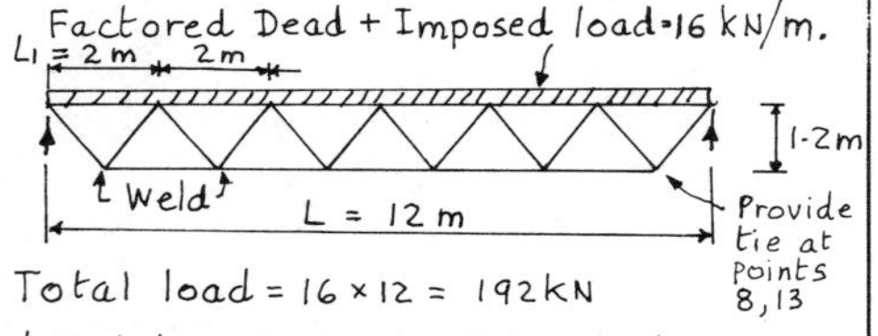

Camber beams 50 mm. at mid span.

Total load = 16 × 12 = 192 kN

Load to intermediate top node points $= \frac{192}{6} = 32$ kN

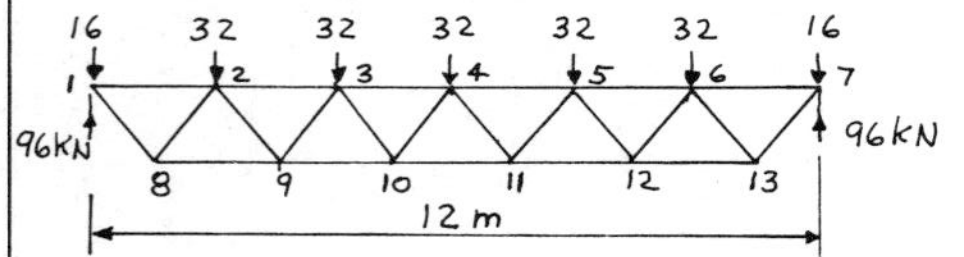

Top Boom. Max. force, members 3-4, 4-5

By method of sections, moments about point 10. Force $F_c = \frac{M}{\text{Depth}}$

$$F_c = \frac{(96-16)5 - (32\times3) - (32\times1)}{1\cdot2}$$

$$= \frac{400 - 96 - 32}{1\cdot2} = 226\cdot6\text{ kN Comp.}$$

Output: $F_c = 226\cdot6$ kN

$$M_x \text{ due to U.D.L} = \frac{FL_1}{12} = \frac{32\times2}{12} = 5\cdot33\text{ kNm}$$

Output: $M_x = 5\cdot33$ kNm

Try 146 × 127 × 22 Structural Tee.

CURPE CONSULTANTS 46 Orburn Road Dunfield	Project: Steel Beams	Job ref
	Part of structure: Lattice Beam	Calc sheet no: C2 / 12 / rev
	Drawing ref: — Calc by: R.A.S. Date: —	Check by: — Date: —

Ref	Calculations	Output

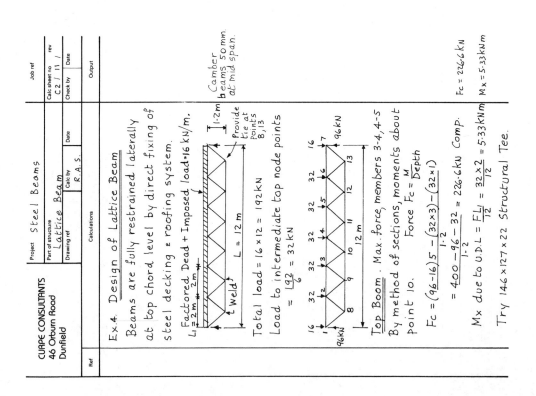

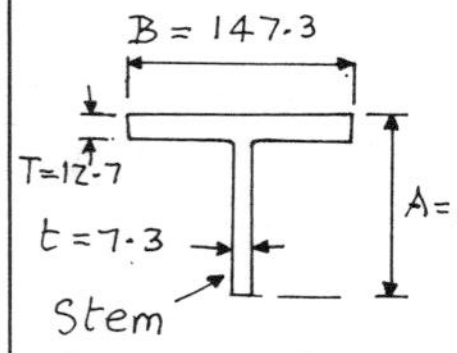

$Ag = 27\cdot6\text{ cm}^2$

Section classification

Table 6 — For T = 12·7 $p_y = 275\text{ N/mm}^2$

$$\varepsilon = \sqrt{\frac{275}{p_y}} = 1\cdot0$$

$$b = \frac{B}{2} = \frac{147\cdot3}{2} = 73\cdot65$$

Table 7 — $\frac{b}{T} = \frac{73\cdot65}{12\cdot7} = 5\cdot8 < 9\cdot5\varepsilon$, flange compact

$$\frac{d}{t} = \frac{129\cdot8}{7\cdot3} = 17\cdot78 < 19\varepsilon\text{, stem semi-compact}$$

Hence section is semi-compact

Output: Semi-compact

Axial capacity $Ag.p_y =$

$$\frac{27\cdot6\times10^2\times275}{10^3} = 759\text{ kN}$$

Output: Axial cap = 759 kN

Compression resistance $P_c = Ag.p_c$

To calculate p_c,

$$\lambda_x = \frac{L_{EX}}{r_x} = \frac{0\cdot85L_1}{r_x} = \frac{0\cdot85\times2\times10^3}{3\cdot56\times10} = 47\cdot75$$

Table 27(c) — For $p_y = 275$ then $p_c = 224\cdot5\text{ N/mm}^2$

$$P_c = Ag.p_c = \frac{27\cdot6\times10^2\times224\cdot5}{10^3} = 619\cdot6\text{ kN}$$

Output: $P_c = 619\cdot6$ kN

Since top chord continuously restrained laterally $M_b = M_{cx}$

CURPE CONSULTANTS 46 Orburn Road Dunfield	Project: Steel Beams Part of structure: Lattice Beam Drawing ref: — Calc by: R.A.S. Date: —	Job ref Calc sheet no: C2 / 13 / rev Check by: — Date: —

Ref	Calculations	Output
	$M_{cx} = p_y . Z = \frac{275 \times 33.9 \times 10^3}{10^6} = 9.32$ kNm	$M_{cx} = 9.32$ kNm
	Local capacity check:	
	$\frac{F}{A_g . p_y} + \frac{M_x}{M_{cx}} + \frac{M_y}{M_{cy}} \not> 1.0$	
	$\frac{226.6}{759} + \frac{5.33}{9.32} + 0 = 0.87 < 1.0$ O.K.	
	Overall buckling check.	
	$\frac{F}{A_g . p_c} + \frac{m\,M_x}{M_b} + \frac{m\,M_y}{p_y . Z_y} \not> 1.0$	
	$\frac{226.6}{619.6} + \frac{5.33}{9.32} + 0 = 0.938 < 1.0$ O.K.	
	Use 146×127×22 T	Use 146×127×22 Struct. Tee.
	Bottom Boom Max. force, member 10-11	
	Moments about point 4 $F_t = \frac{M}{Depth}$	
	$F_t = \frac{(96-16)6 - (32 \times 4) - (32 \times 2)}{1.2} = 240$ kN, Tension	$F_t = 240$ kN
	Tension capacity $P_t = A_e . p_y$	
	With ⊥ section and welded connections to stem then $A_e = A_g$.	
	A_e reqd $= \frac{F_t}{p_y} = \frac{240 \times 10^3}{275} = 872.7$ mm²	A_e reqd = 872.7 mm²
	For 102×127×11 Struct. Tee, $A_e = 1420$ mm²	Use 102×127×11 Struct. Tee.
	Diagonal Members	
	Assume for construction 50×50×6 L minimum size.	
	Forces in members 1-8 and 2-8	
	Force $= \frac{(96-16)\,1.562}{1.2} = 104.1$ kN	$F = 104.1$ kN

CURPE CONSULTANTS 46 Orburn Road Dunfield	Project: Steel Beams Part of structure: Lattice Beam Drawing ref: — Calc by: R.A.S. Date: —	Job ref Calc sheet no: C2 / 14 / rev Check by: — Date: —

Ref	Calculations	Output
	Member 1-8 $F_t = 104.1$ kN Tension	$F_t = 104.1$ kN
	A_e reqd $= \frac{F_t}{p_y} = \frac{104.1 \times 10^3}{275} = 378.5$ mm²	A_e reqd = 378.5 mm²
	Try 50×50×6 L	
	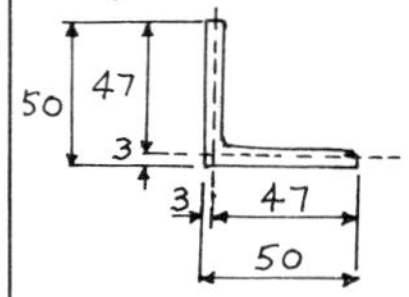	
	$\frac{t}{2} = 3$ mm	
	$a_1 = 47 \times 6 = 282$ mm²	
	$a_2 = 47 \times 6 = 282$ mm²	
	$A_e = a_1 + \left[a_2 \left(\frac{3a_1}{3a_1 + a_2}\right)\right]$	
	$= 282 + \left[282 . \left(\frac{3 \times 282}{3 \times 282 + 282}\right)\right]$	
	$A_e = 493$ mm²	$A_e = 493$ mm²
	$P_t = A_e . p_y = \frac{493 \times 275}{10^3} = 136$ kN	$P_t = 136$ kN
	Since $F_t < P_t$ section satisfactory	Use 50×50×6 L
	Member 2-8 $F_c = 104.1$ kN Compression	$F_c = 104.1$ kN
	Try 70×70×8 L $L = 1.562$ m.	
	$\frac{b}{t} = \frac{d}{t} = \frac{70}{8} = 8.75 < 9.5\varepsilon$, Compact	
	$\lambda_{vv} = \frac{0.85L}{r_{vv}} = \frac{0.85 \times 1.562 \times 10^3}{1.36 \times 10} = 97.6$	
	$\lambda_{a\text{-}a} = \frac{0.7L}{r_{a\text{-}a}} + 30 = \frac{0.7 \times 1.562 \times 10^3}{2.11 \times 10} + 30 = 81.8$	
	p_c for $\lambda = 97.6 = 129.6$ N/mm²	
	$P_c = A_g . p_c = \frac{10.6 \times 10^2 \times 129.6}{10^3} = 137.3$ kN	$P_c = 137.3$ kN
	Since $F_c < P_c$ section satisfactory	Use 70×70×8 L
	Other Diagonals Design similar to above members 1-8 and 2-8.	
	Node points	
	Welded joint 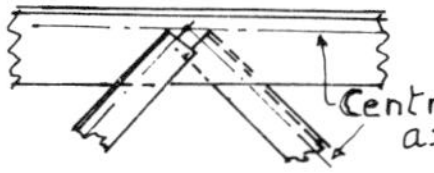	Use 6mm fillet welds

CURPE CONSULTANTS 46 Orburn Road Dunfield	Project: Steel Beams	Job ref
	Part of structure: Excavation- Struts & Walings	Calc sheet no: C2 / 15 / rev
	Drawing ref: — Calc by: R.A.S. Date: —	Check by: — Date: —

Ref	Calculations	Output
BS5950	Ex.5. Braced Cofferdam for excavation in medium dense sand	
	3.5 m Long Struts, 3 m centres along excavation	

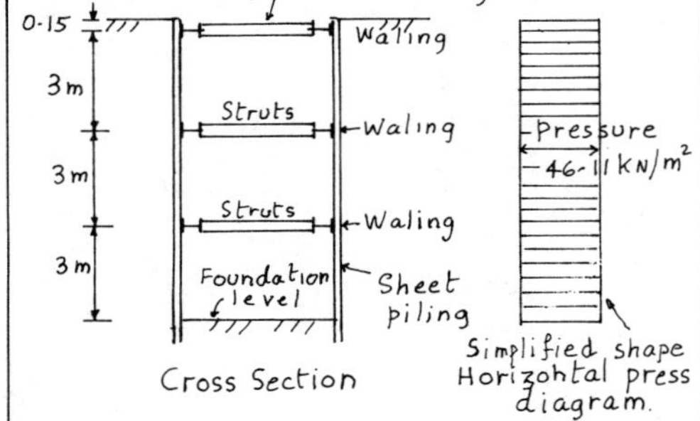

Ref	Calculations	Output
	For 1 m length of excavation press = $46{\cdot}11\,kN/m^2$	
	Walings 3 m span simply supported	
	W to intermediate walings $= 46{\cdot}11 \times 3 \times 3 = 415\,kN$	
	Factored load $= 1{\cdot}4 \times 415 = 581\,kN$	
	Design moment $M_x = \frac{WL}{8} = \frac{581 \times 3}{8} = 217{\cdot}8\,kNm$	$M_x = 217{\cdot}8\,kNm$
	Waling laterally unrestrained so consider lateral torsional buckling	
	Effective length $L_E = 1{\cdot}0L$	
	$\bar{M} \not> M_b$	
	$m = 1{\cdot}0$	
	$\bar{M} = 1{\cdot}0 \times 217{\cdot}8 = 217{\cdot}8\,kNm$	$\bar{M} = 217{\cdot}8\,kNm$
	M_b for $L_E = 3\,m$ and $n = 1{\cdot}0$ is given as 208 kNm Page 132 Design Guide Vol 1 for 356 × 171 × 57 UB	

CURPE CONSULTANTS 46 Orburn Road Dunfield	Project: Steel Beams	Job ref
	Part of structure: Excavation- Struts & Walings.	Calc sheet no: C2 / 16 / rev
	Drawing ref: — Calc by: R.A.S. Date: —	Check by: — Date: —

Ref	Calculations	Output
BS5950	Try 356 × 171 × 57 UB Plastic section.	Plastic
	$M_b = S_x\,p_b$	
	$\lambda_{LT} = n\,u\,v\,\lambda$	
	$\lambda = \frac{L_E}{r_y} = \frac{3 \times 10^3}{3{\cdot}92 \times 10} = 76{\cdot}5$	
	$\frac{\lambda}{x} = \frac{76{\cdot}5}{28{\cdot}9} = 2{\cdot}65$	
Table 14	For $\frac{\lambda}{x} = 2{\cdot}65$ and $N = 0{\cdot}5$ $v = 0{\cdot}925$	
	To find n	
	$\gamma = \frac{M}{M_o} = \frac{0}{217{\cdot}8} = 0$	
Table 16	$\beta = \frac{0}{0} = 0$ $\therefore n = 0{\cdot}94$	
	$\lambda_{LT} = n\,u\,v\,\lambda$ $= 0{\cdot}94 \times 0{\cdot}884 \times 0{\cdot}925 \times 76{\cdot}5 = 58{\cdot}8$	$\lambda_{LT} = 58{\cdot}8$
Table 11	For $p_y = 275\,N/mm^2$ $p_b = 216{\cdot}1\,N/mm^2$	$P_b = 216{\cdot}1\,N/mm^2$
	$M_b = S_x . p_b$ $= \frac{1010 \times 10^3 \times 216{\cdot}1}{10^6} = 218{\cdot}2\,kNm$	$M_b = 218{\cdot}2\,kNm$
	Since $\bar{M} < M_b$ $217{\cdot}8 < 218{\cdot}2$ Section adequate for lateral torsional buckling.	Lateral tors. buckling O.K. use 356 × 171 × 57 UB walings.
	Shear will be found satisfactory	
	Struts $L = 3{\cdot}5\,m$ $L_E = 1{\cdot}0L = 3{\cdot}5\,m$	
	W to inter. strut $= 46{\cdot}11 \times 3 \times 3 = 415\,kN$	
	Factored load $= 1{\cdot}4 \times 415 = 581\,kN$	$F_c = 581\,kN$
	Try 152 × 152 × 37 UC Section is not slender	
Table 27b	$\lambda_x = \frac{L_{EX}}{r_x} = \frac{3{\cdot}5 \times 10^3}{6{\cdot}84 \times 10} = 51{\cdot}2$ $p_{cx} = 235{\cdot}5\,N/mm^2$	
" 27c	$\lambda_y = \frac{L_{EY}}{r_y} = \frac{3{\cdot}5 \times 10^3}{3{\cdot}87 \times 10} = 90{\cdot}4$ $p_{cx} = 140{\cdot}5$ "	
	$P_{cx} = \frac{47{\cdot}4 \times 10^2 \times 235{\cdot}5}{10^3} = 1116\,kN$	
	$P_{cy} = \frac{47{\cdot}4 \times 10^2 \times 140{\cdot}5}{10^3} = 666\,kN > 581\,kN$ Section adequate	$P_c = 666\,kN$ Use 152 × 152 × 37 UC struts

3

Plate girders

Plate girders are used over large spans carrying heavy loads associated with high bending moment and shear forces which exceed the capacity of any of the range of standard rolled steel universal beam sections. Plates added to flanges of U.B. may be used as an alternative. Modern construction consists of steel plates normally welded together to form an I-section (Fig. 3.1(b) and (c)). In the past, plate girders consisted generally of double angle steel section flanges riveted to a steel web plate. Additional steel plates were riveted to the horizontal legs of the flange angles if required for design purposes (Fig. 3.1(a).

Proportions of plate girders are limited usually by web buckling. The buckling capacity of slender webs can be improved by providing an arrangement of vertical plate stiffeners welded to the girder at intervals along the span. Generally, closer spacing proves more effective. Slender flanges are avoided where possible to prevent occurrence of local buckling. Lateral torsional buckling needs to be considered if adequate lateral restraint is not provided.

Minimum web thickness should satisfy two conditions:

1) Serviceability.
2) Avoidance of flange buckling into a thin web.

Capacity of section

To determine the capacity of the section, several methods are available. The method used in the example is that the moment is assumed to be resisted by the flanges alone, and the web resisting the shear only.

Slender web carrying shear only

Webs with intermediate stiffeners may be designed with or without utilizing tension field action. Different panels between vertical stiffeners and flanges may be designed by either method. It is more conservative to design without utilizing

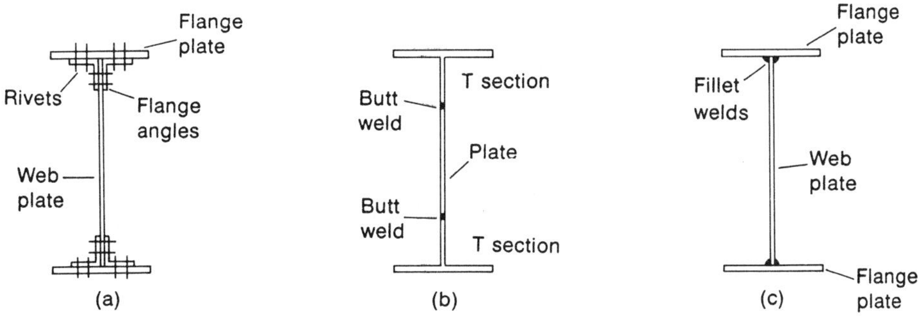

Fig. 3.1 *Types of plate girder. (a) Riveted, (b) welded, (c) welded–most common*

tension field action. Tension field action occurs in the web because of stress redistribution after web buckling occurs as a result of shear. The web is divided into panels. In each panel a diagonal strip of web acts as a tie and the stiffeners as struts, analogous to the vertical and diagonal members in a lattice girder.

Design without using tension field action – elastic critical method
Tension field action is necessary if:

Shear stress f_v in a panel
> critical shear strength q_{cr}

or shear force F_v in a panel
> shear buckling resistance V_{cr}

$$V_{cr} = q_{cr}dt$$

where q_{cr} is obtained from Table 21 (a)–(d) of BS 5950, as appropriate

Values of q_{cr} depend on $\frac{a}{d}$ and $\frac{d}{t}$

where d = depth of web,
t = thickness of web,
a = spacing of stiffeners.

For girders with no intermediate stiffeners take the spacing as infinity.

Design using tension field action
Shear buckling resistance $V_b = q_b dt$

where q_b = basic shear strength from Table 22(a)–(d) of BS 5950 as appropriate.

Values of q_b depend on $\frac{a}{d}$ and $\frac{d}{t}$.

Stiffeners

Vertical stiffeners are positioned at concentrated loads and at intervals along the length of the girder to divide the web into panels, each bounded by the stiffeners and the girder flanges.

1) Intermediate stiffeners

Their primary function is to prevent buckling of a slender web. They may be on one or both sides of the web, and their spacing is dependent on the web thickness. The outstand from the face of the web should not exceed $19t_s\varepsilon$, where t_s = thickness of stiffener.

For stiffeners not subject to external loads or moments, a minimum second moment of area I_s is specified about the centre line of the web.

$$I_s \geqslant 0.75\,dt^3 \text{ for } a \geqslant \sqrt{2}\,d$$

$$I_s \geqslant \frac{1.5\,d^3t^3}{a^2} \text{ for } a < \sqrt{2}\,d$$

where d = depth of web,
t = minimum required web thickness for spacing a, using tension field action,
a = stiffener spacing.

Designing for tension field action imposes additional loading and induced compression forces on the stiffener, and this must be designed as a strut for buckling, for webs designed using tension field action.

If F_q = stiffener force,
P_q = buckling resistance of the stiffener,
F_x = external load or reaction,
P_x = buckling resistance of a load-carrying stiffener,
M_s = moment on stiffener due to eccentric applied load,
M_{ys} = moment capacity of stiffener based on its elastic modulus,

then, for stiffeners not subject to external loads or moments,

$$F_q = V - V_s < P_q$$

where V = maximum shear adjacent to stiffener,
V_s = shear buckling resistance of web panel designed without using tension field action.

Stiffeners subject to external loads and moments should meet the conditions for load-carrying web stiffeners and

$$\frac{F_q - F_x}{P_q} + \frac{F_x}{P_x} + \frac{M_s}{M_{ys}} \leqslant 1$$

If $F_q < F_x$ then $(F_q - F_x)$ should be taken as zero.

Stiffeners not subject to external loading should be connected to the web to withstand a shear force between each stiffener and the web (in kN/mm^2) of not less than

$$\frac{t^2}{8b_s}$$

where b_s = stiffener outstand in mm,
t = web thickness in mm.

Shear between web and stiffener due to external loading should be added to the above value.

2) Load-carrying stiffeners

These are provided to prevent local buckling of the web due to concentrated loading or reactions applied through a flange.

The buckling resistance of the unstiffened web is given by

$$P_w = (b_1 + n_1)\,tp_c$$

where b_1 = stiff bearing length,
n_1 = length obtained by 45° dispersion through half the depth of the section,
t = web thickness,
p_c = compression strength from Table 27(c), BS 5950.

Load-carrying stiffeners must be checked for both buckling and bearing.

Buckling

$$F_x \ngtr P_x$$

where F_x = external load or reaction on a stiffener,
P_x = buckling resistance of the stiffener based on the compressive strength p_c of a strut using Table 27(c), BS 5950, the radius of gyration being taken about the axis parallel to the web.

The effective section is the full area of the stiffener together with an effective length of web on each side of the centreline of stiffeners limited to 20 times the web thickness. The design strength used should be the minimum value obtained for the web or the stiffener. A reduction of 20 N/mm^2 is made for the stiffeners attached to a welded section. Effective length L_E of stiffeners used in calculating P_x is taken as $0.7L$ if the loaded flange is effectively restrained by other members, or L_E taken as L where the flange is not so restrained.

Bearing

When checking for bearing, only that area of the load-carrying stiffener in contact with the flange is taken into account.

$$A > \frac{0.8\,F_x}{P_{ys}}$$

where F_x = external load or reaction,
A = contact area of stiffener with flange,
P_{ys} = design strength of stiffener.

Bearing stiffeners

These are provided to prevent local crushing of the web due to concentrated loads or reactions applied through the flange. Local capacity of the web is given by:

$$(b_1 + n_2)tp_{yw}$$

where b_1 = stiff bearing length,
n_2 = length obtained by dispersion through the flange to the flange/web connection at a slope of 1:2.5 to the plane of the flange,
p_{yw} = design strength of web,
t = thickness of web.

If the buckling resistance of the web is not exceeded, the load-carrying stiffeners are not needed. If the local bearing resistance is exceeded, then bearing stiffeners should be designed to provide sufficient bearing area.

Design example

In the design example given, two cases are considered for the same girder:

1) Girder fully restrained.
2) Girder restrained at ends and at point loads only. Calculations are therefore necessary for lateral torsional buckling.

CURPE CONSULTANTS 46 Orburn Road Dunfield	Project: Plate Girders	Job ref
	Part of structure: Design example - Full lateral restraint	Calc sheet no: C3/1 rev: 1
	Drawing ref: — Calc by: R.A.S. Date: —	Check by: — Date: —

Ref	Calculations	Output

Ex. 1. Plate Girder design

Two cases are considered:

Case(I) - Girder with full lateral restraint

Case(2) - Girder restrained at ends and point loads only

Case(I) Factored loads are shown.

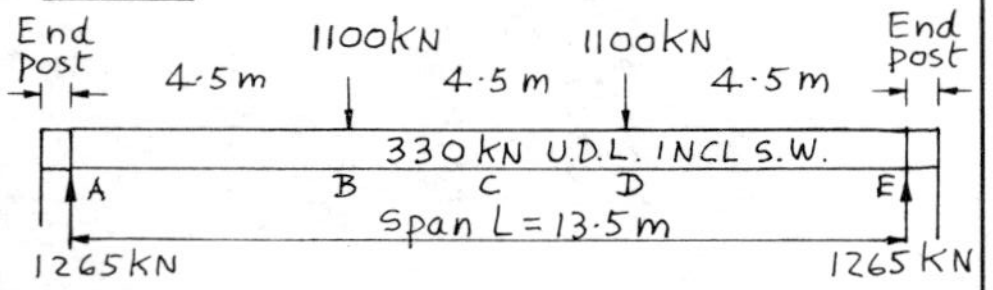

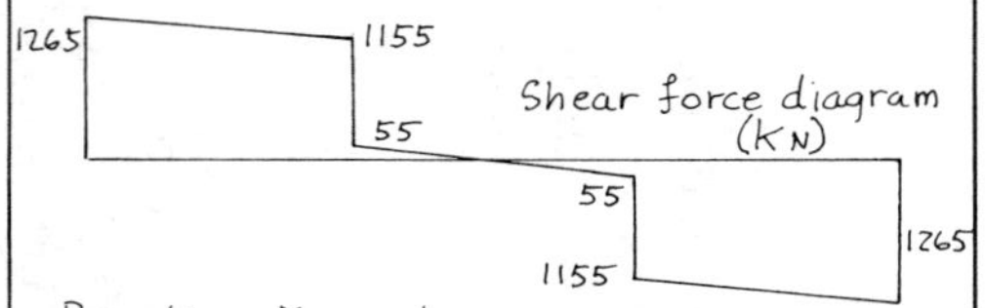

Output: Max. shear force 1265 kN

Bending Moments

M at $B = (1265 \times 4{\cdot}5) - (110 \times \frac{4{\cdot}5}{2}) = 5445$ kNm

M at $C = (1265 \times \frac{13{\cdot}5}{2}) - (1100 \times \frac{4{\cdot}5}{2}) - (\frac{330}{2} \times \frac{13{\cdot}5}{4})$

$= 5507$ kNm

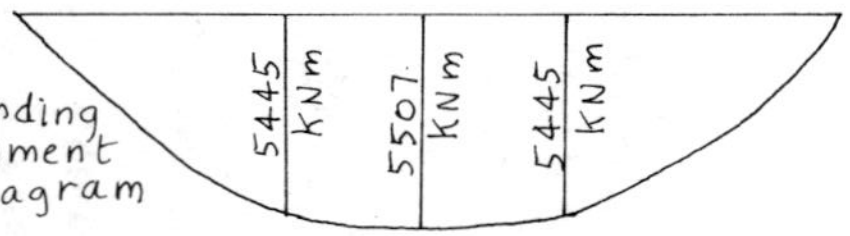

Bending Moment Diagram

Output: Max. moment 5507 kNm

Initial sizing Depth say $\frac{span}{10} = \frac{13{\cdot}5}{10} = 1{\cdot}35$ m

Flange area $A_f = \frac{M_{max}}{Depth\ p_y}$

Assume $p_y = 265$ N/mm²

CURPE CONSULTANTS 46 Orburn Road Dunfield	Project: Plate Girders	Job ref
	Part of structure: Design example - fully restrained	Calc sheet no: C3/2 rev: 1
	Drawing ref: — Calc by: R.A.S. Date: —	Check by: — Date: —

Ref	Calculations	Output
BS 5950		
	$A_f = \frac{5507 \times 10^6}{1350 \times 265} = 15393$ mm²	
	Flange width say $\frac{Depth}{3{\cdot}5} = \frac{1350}{3{\cdot}5} = 385$ mm	
	Flange $400 \times 40 = 16000$ mm²	Flanges 400×40
	Web thickness say $\frac{Depth}{150} = \frac{1350}{150} = 9$ say 10mm	Web 10mm
	Trial Section 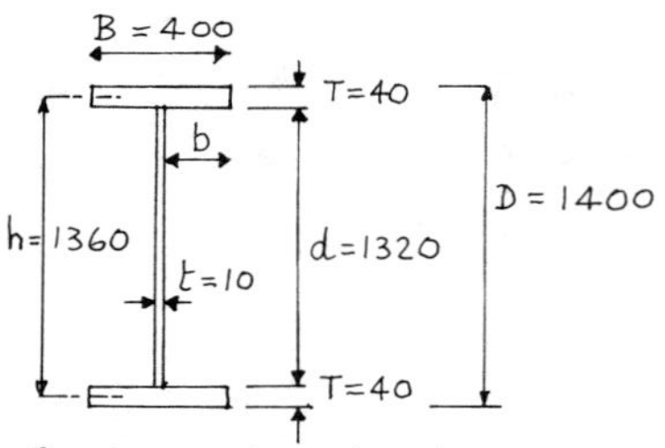	
	Section classification	
Table 6	Flange $p_y = p_{yf}$ For $T = 40$, $p_{yf} = 265$	$p_{yf} = 265$ N/mm²
	$\varepsilon = \sqrt{\frac{275}{p_y}} = \sqrt{\frac{275}{265}} = 1{\cdot}02$	
Table 7	Outstand $b = \frac{B - t}{2} = \frac{400 - 10}{2} = 195$ mm	
	$\frac{b}{T} = \frac{195}{40} = 4{\cdot}875 < 7{\cdot}5\varepsilon$. Welded section	
	∴ Flange plastic.	
	Web $p_y = p_{yw}$	
Table 6	For $t = 10$, $p_{yw} = 275$	$p_{yw} = 275$ N/mm²
Table 7	$\frac{d}{t} = \frac{1320}{10} = 132 > 120\varepsilon$ ∴ Web slender	
4.4.4.2	Since $\frac{d}{t} > 63\varepsilon$ check web; shear buckling	

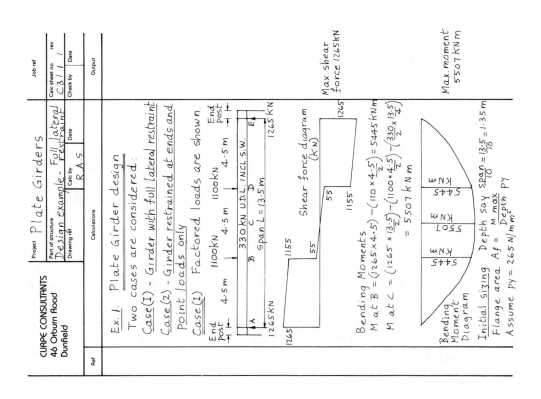

CURPE CONSULTANTS
46 Orburn Road
Dunfield

Project: Plate Girders | Job ref:
Part of structure: Design example - fully restrained | Calc sheet no: C3 / 3 | rev: 1
Drawing ref: | Calc by: R.A.S. | Date: | Check by: | Date:

Ref BS5950	Calculations	Output
4.4.2.	<u>Web and Flange dimensions</u>	
4.4.2.2(b)	Stiffener spacing $a = 1500\,mm > d$	
	then $t \geqslant \frac{d}{250}$ $\quad \frac{d}{250} = \frac{1320}{250} = 5{\cdot}28$	
	Since $t > \frac{d}{250}$ $\quad 10 > 5{\cdot}28$	
	Web t is O.K. for serviceability	web thickness O.K. serviceability
4.4.2.3(b)	With $a < 1{\cdot}5d$ then $t \geqslant \frac{d}{250}\left(\frac{p_{yf}}{455}\right)^{\frac{1}{2}}$	
	$\frac{d}{250}\left(\frac{p_{yf}}{455}\right)^{\frac{1}{2}} = \frac{1320}{250}\left(\frac{265}{455}\right)^{\frac{1}{2}} = 4{\cdot}03\,mm$	
	Since $t > \frac{d}{250}\left(\frac{p_{yf}}{455}\right)^{\frac{1}{2}}$ $\quad 10 > 4{\cdot}03$	
	Web t is O.K. for avoiding flange buckling into web	web thickness OK web buckling by flange
4.4.4.2(a)	<u>Moment resisted by flanges alone and web designed for shear</u>	
	Moment capacity of section Mc	
	$M_c = p_{yf}.A_f.h$	
	$= \frac{265 \times 400 \times 40 \times 1360}{10^6} = 5766\,kNm$	
	Since $M_x < M_c$ $\quad 5507 < 5766$	
	Section O.K. for moment	Section O.K for moment
4.4.5.4	Webs with inter. stiffs designed in this example using tension field action.	
	Try stiffener spacing as shown.	Stiffeners 1·5 m centres
	Usually the end post and the end panel are found to be most critical.	
	M at 1 $= (1265 \times 1{\cdot}5) - \left(\frac{330}{9} \times \frac{1{\cdot}5}{2}\right) = 1870\,kNm$	
	M at 2 $= (1265 \times 3) - \left(\frac{330}{9} \times 2 \times 1{\cdot}5\right) = 3685$ "	
	M at 3 $= (1265 \times 6) - \left(\frac{330}{9} \times 4 \times 3\right) - (1100 \times 1{\cdot}5) = 5500$ "	

CURPE CONSULTANTS
46 Orburn Road
Dunfield

Project: Plate Girders | Job ref:
Part of structure: Design example - fully restrained | Calc sheet no: C3 / 4 | rev: 1
Drawing ref: | Calc by: R.A.S. | Date: | Check by: | Date:

Ref BS5950	Calculations	Output
	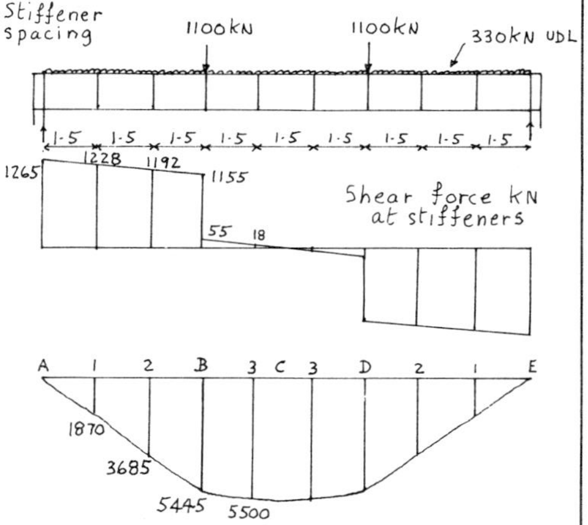	Proposed Stiffener spacing a; Shear forces at stiffeners
	Bending moments at stiffeners. kNm	Bending moments
	<u>Consider end panel A-1</u>	
	$d = 1320\,mm$ $\quad t = 10\,mm$	
	Shear stress in panel $f_v = \frac{F_v}{d.t}$	
	$f_v = \frac{1265 \times 10^3}{1320 \times 10} = 95{\cdot}83\,N/mm^2$	
4.4.5.3	Shear buckling resistance V_{cr} of a stiffened or unstiffened panel is given by:	
	$V_{cr} = q_{cr}.d.t.$	
	To obtain q_{cr} from Table 21	
	$\frac{a}{d} = \frac{1500}{1320} = 1{\cdot}14$ $\quad \frac{d}{t} = \frac{1320}{10} = 132$	

CURPE CONSULTANTS 46 Orburn Road Dunfield	Project: Plate Girders.	Job ref
	Part of structure: Design example - fully restrained	Calc sheet no: C3 / 5 rev: /
	Drawing ref: — Calc by: R.A.S. Date: —	Check by: — Date: —

Ref BS5950	Calculations	Output
Table 21b	For $p_y = 275\ N/mm^2$ Critical shear strength $q_{cr} = 91.96\ N/mm^2$ Since $f_v > q_{cr}$ $\quad 95.83 > 91.96$ Tension field action must be used Also $V_{cr} = \frac{91.96 \times 1320 \times 10}{10^3} = 1213\ kN < 1265$	Tension field action is necessary
4.4.5.4.1	Using tension field action Shear buckling resistance of a stiffened panel is given by: $V_b = q_b \cdot d \cdot t.$	
Table 22b	For $\frac{a}{d} = 1.14$ $\quad \frac{d}{t} = 132$ $\quad p_y = 275\ N/mm^2$ Basic shear strength $q_b = 119.9\ N/mm^2$ $V_b = \frac{119.9 \times 1320 \times 10}{10^3} = 1582\ kN$ Since $F_v < V_b$ $\quad 1265 < 1582$ O.K.	Panel A-1 O.K shear buckling
4.4.5.4.3 b.	Design of End Post - with double stiffeners	

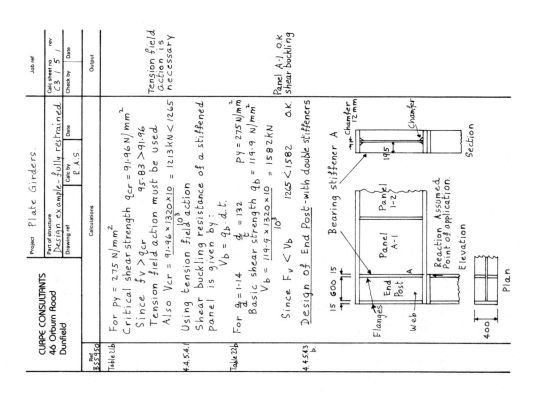

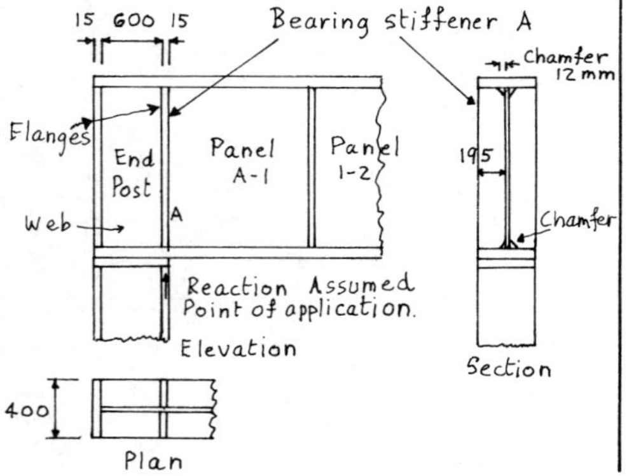

CURPE CONSULTANTS 46 Orburn Road Dunfield	Project: Plate Girders	Job ref
	Part of structure: Design example fully restrained	Calc sheet no: C3 / 6 rev: /
	Drawing ref: — Calc by: R.A.S. Date: —	Check by: — Date: —

Ref	Calculations	Output
	Shear and Moment from tension field	
4.4.5.4.3 b	With double stiffener. Check end post as a beam spanning between flanges of the girder capable of resisting: (1) a shear force R_{tf} (2) a moment M_{tf} due to anchor forces For end panel A-1 $f_v = \frac{F_v}{d\,t} = \frac{1265 \times 10^3}{1320 \times 10} = 95.83\ N/mm^2$	
Table 21b $p_y = 275$	Critical shear strength $q_{cr} = 91.96\ N/mm^2$ for $\frac{a}{d} = \frac{1500}{1320} = 1.14$ and $\frac{d}{t} = \frac{1320}{10} = 132$	
Table 22b	Basic shear strength $q_b = 119.9\ N/mm^2$ for $\frac{a}{d} = 1.14$ and $\frac{d}{t} = 132$	
4.4.5.4.4	Shear force $R_{tf} = \frac{H_q}{2}$ Moment $M_{tf} = \frac{H_q\, d}{10}$ $H_q = 0.75\, d\, t\, p_y \left(1 - \frac{q_{cr}}{0.6 p_y}\right)^{\frac{1}{2}}$ $H_q = \frac{0.75 \times 1320 \times 10 \times 275}{10^3} \left(1 - \frac{91.96}{0.6 \times 275}\right)^{\frac{1}{2}}$ $H_q = 1811\ kN$ $R_{tf} = \frac{H_q}{2} = \frac{1811}{2} = 905.5\ kN$	

CURPE CONSULTANTS
46 Orburn Road
Dunfield

Project: Plate Girders | Job ref:
Part of structure: Design example - fully restrained | Calc sheet no: C3 / 7 | rev: 1
Drawing ref: | Calc by: R.A.S | Date: | Check by: | Date:

Ref	Calculations	Output
	$M_{tf} = \frac{H_q . d}{10} = \frac{1811 \times 1320}{10 \times 10^3} = 239$ kNm	
	Shear capacity of end post	
	Web = 600 mm × 10 mm	
	$\frac{d}{t} = \frac{600}{10} = 60 < 79\varepsilon$ ∴ Plastic	
	Shear capacity $P_v = 0.6\ p_{yw}\ A_v$	
	$P_v = \frac{0.6 \times 275 \times 10 \times 600}{10^3} = 990$ kN	
	Since $R_{tf} < P_v$ $905.5 < 990$ kN	End post O.K. shear
	Note. If $f_v < q_b$ and $R_{tf} > P_v$	
	then H_q may be reduced by ratio:	
	$\frac{f_v - q_{cr}}{q_b - q_{cr}}$	
	Moment Capacity	
	Flange proportions of stiffener at A	
Table 6	$\frac{b}{T} = \frac{195}{15} = 13$ $p_y = 275$ N/mm^2	
Table 7	$\frac{b}{T} < 13\varepsilon$ semi-compact	
	$M_{cx} = \frac{275 \times 400 \times 15 \times 615}{10^6} = 1014$ kNm	
	Since $M_{tf} < M_{cx}$ $239 < 1014$	End post OK. moment
4.5.4.2	Bearing check Stiffener at A	
4.5.15.	$F_x = 1265$ kN comp. $p_{ys} = 275 - 20 = 255$ N/mm^2	
	Contact with flange area $A > \frac{0.8 F_x}{p_{ys}}$	
	Min. A reqd = $\frac{0.8 \times 1265 \times 10^3}{255} = 3968$ mm^2	
	Try 2 flats 195×15 A given = $2(195-12)15 = 5490$	
	Since $3968 < 5490$ contact area O.K.	
	Outstand of stiffener = 195 mm	

CURPE CONSULTANTS
46 Orburn Road
Dunfield

Project: Plate Girders | Job ref:
Part of structure: Design example - fully restrained | Calc sheet no: C3 / 8 | rev: 1
Drawing ref: | Calc by: R.A.S. | Date: | Check by: | Date:

Ref BS 5950	Calculations	Output
	$13\ t_s . \varepsilon = 13 \times 15 \times 1.0 = 195$ mm O.K	Bearing O.K.
4.5.1.5.	Buckling check - Stiffener at A	
4.5.4.1.	Reaction $F_x \ngtr$ Buckling resistance P_x	
4.5.1.5	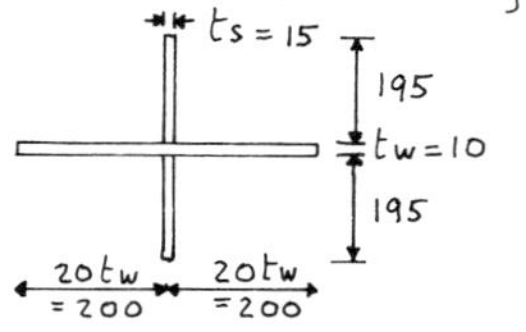	
	Effective section of strut	
	$F_x = 1265$ kN	
	Area $A = (2 \times 195 \times 15) + (400 \times 10)$	
	$= 5850 + 4000 = 9850$ mm^2	
	$I = \frac{15 \times 400^3}{12} + \frac{400 \times 10^3}{12} =$	
	$= 80000000 + 33333 = 80033333$ mm^4	
	$r = \sqrt{\frac{I}{A}} = \sqrt{\frac{80033333}{9850}} = 90$	
4.5.1.5 a	$L_E = 0.7 L = 0.7 \times 1320 = 924$ mm	
	Slenderness $\lambda = \frac{L_E}{r} = \frac{924}{90} = 10.2$	
4.7.5 4.5.1.5	$p_{ys} = 275 - 20 = 255$ N/mm^2	
Table 27c	For $p_y = 255$ $\lambda = 10.2$ $p_c = 255$ N/mm^2	
	∴ Buckling resistance of stiffener $P_x = p_c . A$.	
	$P_x = \frac{255 \times 9850}{10^3} = 2512$ kN	
	Since $F_x < P_x$ $1265 < 2512$ O.K.	Buckling O.K.
	Check local bearing capacity of girder web and bearing stiffener at A	
	Local capacity of web assuming conservatively stiff bearing length $b_1 = 0$.	
4.5.3.	Capacity $= (b_1 + n_2)\ t . p_{yw}$	
	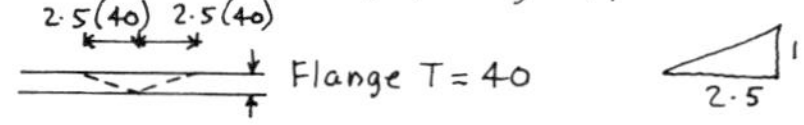	

CURPE CONSULTANTS
46 Orburn Road
Dunfield

Project	Plate Girders	Job ref	
Part of structure	Design example - fully restrained	Calc sheet no	C3 / 9 / rev
Drawing ref	Calc by RAS	Check by	Date

Ref BS5950	Calculations	Output
	Web capacity $= \frac{[0+(2\times100)]\,10\times275}{10^3} = 550$ kN	
4.5.5.	Bearing stiffener to be designed for: applied load − local capacity of web $= 1265 - 550 = 715$ kN	
	Bearing capacity of stiffener alone $= p_{ys}.A$ $= \frac{255\times5490}{10^3} = 1400$ kN	
	Since $715 < 1400$ O.K.	End Post Stiffeners at A 195 × 15 mm
	<u>Outside stiffener of End Post</u>	
	Compressive force $= \frac{Mtf}{\text{stiff. spacing}} = \frac{239\times10^6}{615\times10^3} = 388$ kN	
	Area reqd $= \frac{388\times10^3}{275} = 1410\ mm^2$	
	Area given $= 400\times15 = 6000\ mm^2$ O.K.	End Post Outside stiff. 400 mm × 15 mm
4.4.6	<u>Intermediate stiffener at 1</u>	
	Spacing $a = 1500$ mm	
	Web depth $d = 1320$ mm	
	$I_s \geq \frac{1.5d^3t^3}{a^2}$ for $a < \sqrt{2}d$	
	$\sqrt{2}d = \sqrt{2}\times1320 = 1866 \therefore a < \sqrt{2}d$	
	$I_s \geq \frac{1.5d^3.t^3}{a^2}$. taking t as actual web thickness i.e. conservatively	
	$I_s \geq \frac{1.5\times1320^3\times10^3}{1500^2} = 1533312\ mm^4$	
	Try 2 flats 70 mm × 8 mm	
	$I_s = \frac{8\times150^3}{12} = 2250000\ mm^4 > 1533312$	
4.5.1.2	Outstand $= 70$ mm $< 13\times8$ O.K.	
	web 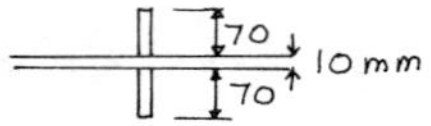	

CURPE CONSULTANTS
46 Orburn Road
Dunfield

Project	Plate Girders	Job ref	
Part of structure	Design example - fully restrained	Calc sheet no	C3 / 10 / rev
Drawing ref	Calc by R.A.S.	Check by	Date

Ref BS5950	Calculations	Output
4.4.6.6	<u>Buckling check on inter. stiff at 1.</u>	
	Max. shear adjacent to stiffener 1 is $V = 1228$ kN	
4.4.5.3	Shear buckling resistance of web without tension field action $V_{cr} = q_{cr}.d.t.$	
	For $\frac{a}{d} = \frac{1500}{1320} = 1.14$ $\frac{d}{t} = \frac{1320}{10} = 132$	
Table 21b	For $p_y = 275$ $q_{cr} = 91.96\ N/mm^2$	
	$V_{cr} = \frac{91.96\times1320\times10}{10^3} = 1213$ kN	
	Force on stiffener $F_q = V - V_s \leq P_q$	
	V_s = smaller value of V_{cr} for panels A1 or A2	
	$V_s = V_{cr} = 1213$ kN	
	$\therefore F_q = 1228 - 1213 = 15$ kN.	
4.5.1.5	Buckling resistance of stiffener	
	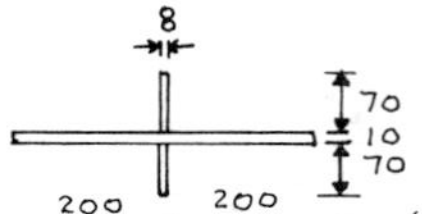 Effective section of strut	
	Area $A = (2\times70\times8) + (400\times10) = 1120 + 4000 = 5120\ mm^2$	
	$I = \frac{8\times150^3}{12} + \frac{400\times10^3}{12} = 2250000 + 33333 = 2283333\ mm^4$	
	$r = \sqrt{\frac{I}{A}} = \sqrt{\frac{2283333}{5120}} = 21.1$	
	$\lambda = \frac{L_E}{r} = \frac{0.7\times1320}{21.1} = 43$	
Table 27c	For $p_{ys} = 255$ $\lambda = 43$ $p_c = 217.5\ N/mm^2$	
	$P_q = p_c.A = \frac{217.5\times5120}{10^3} = 1113.6$ kN	
	Since $F_q < P_q$ $15 < 1113.6$ OK	Inter. stiffs Buckling O.K Use 2 flats 70×8.

CURPE CONSULTANTS
46 Orburn Road
Dunfield

Project	Plate Girders	Job ref
Part of structure	Design example – fully restrained	Calc sheet no C3 / 11 / rev
Drawing ref	Calc by R.A.S. Date	Check by Date

Ref BS5950	Calculations	Output
	Stiffener at B	
4.4.6.6.	Inter. stiffener subject to external load.	
	$\frac{F_q - F_x}{P_q} + \frac{F_x}{P_x} + \frac{M_s}{M_{ys}} \not> 1 \cdot 0$	
	$F_q = V - V_s$ V at B = 1155 kN	
	$V_s = V_{cr}$ for panel 2-B	
	$\frac{a}{d} = \frac{1500}{1320} = 1 \cdot 14$ $\frac{d}{t} = \frac{1320}{10} = 132$	
Table 21b	For $p_y = 275$ $q_{cr} = 91 \cdot 96\ N/mm^2$	
	$V_{cr} = q_{cr}.d.t = \frac{91 \cdot 96 \times 1320 \times 10}{10^3} = 1213\ kN$	
	$F_q = V - V_s = 1155 - 1213$ negative	
	i.e. no tension field action	
4.4.6.6	$F_q - F_x = 0$	
	$M_s = 0$	
	$F_x = 1100\ kN.$	
4.5.1.5	Buckling resistance	
	$P_x = P_q$ as calculated for stiff at 1	
	$P_x = 1113 \cdot 6\ kN$	
	Since $F_x < P_x$ $1100 < 1113 \cdot 6\ kN$ OK	Buckling O.K. All inter. stiffs use 2 flats 70×8
4.5.2.2	**Web check between stiffeners**	
	U.D.L. applied between stiffeners	
	Comp. stress on comp. edge f_{ed} of the web $\not>$ comp. strength for edge loading P_{ed}	
	Check panel	
	$f_{ed} = \frac{24 \cdot 44 \times 10^3}{10 \times 10^3} = 2 \cdot 444\ N/mm^2$	
	$P_{ed} = \left[2 \cdot 75 + \frac{2}{(a/d)^2}\right]\frac{E}{(d/t)^2} = \left[275 + \frac{2}{1 \cdot 14^2}\right]\frac{20 \cdot 5 \times 10^4}{132^2} = 50 \cdot 43$	
	Since $f_{ed} < P_{ed}$ O.K.	Comp. stress on web edge O.K.

CURPE CONSULTANTS
46 Orburn Road
Dunfield

Project	Plate Girders	Job ref
Part of structure	Design example – fully restrained	Calc sheet no C3 / 12 / rev
Drawing ref	Calc by R.A.S. Date	Check by Date

Ref BS5950	Calculations	Output

Arrangement of stiffeners

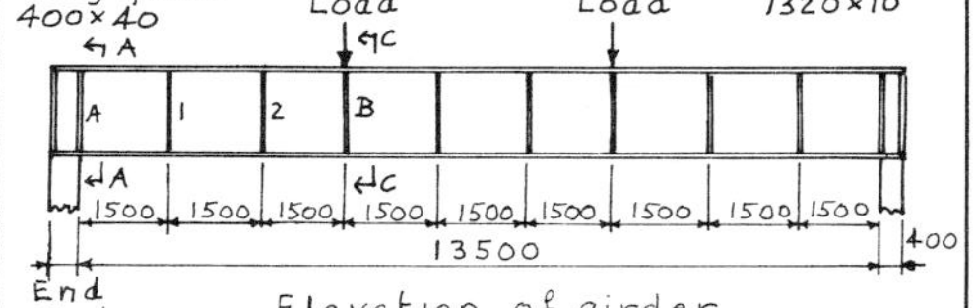

Elevation of girder

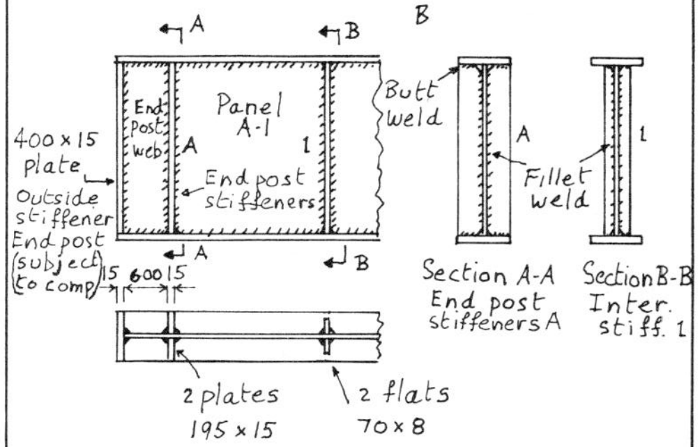

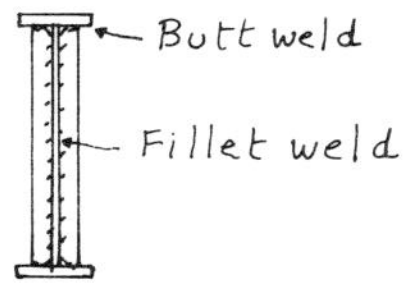

Section C-C
Load carrying inter. stiff. B

Allow 12 mm×12 mm chamfer ends of stiffeners to clear fillet welds connecting flange to web

CURPE CONSULTANTS
46 Orburn Road
Dunfield

Project: Plate Girders			Job ref	
Part of structure: Design example — If restrained at A, B, D, E only			Calc sheet no: C3/13/	rev
Drawing ref	Calc by: R.A.S.	Date	Check by	Date

Ref BS5950	Calculations	Output

Case (2)

Girder restrained at ends and at point loads (at A, B, D, E) only

Girder checked for lateral torsional buckling

$M_b = S_x . p_b$

$\bar{M} \not> M_b$

$\bar{M} = m . M_A$

Plan on buckled profile-comp. flange

Lengths AB and BD to be checked

Section properties: $S_x = 26116\ cm^3$, $r_y = 9.7\ cm$, $u = 0.9$, $x = 39.1$

Length AB — Length loaded $\therefore m = 1.0$

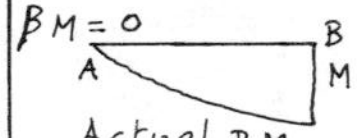

Actual B.M.

$\bar{M} = m . M_A = 1.0 \times 5445$ kNm

$\beta = 0.00$

$M_o = \frac{330}{3} \times \frac{4.5}{8} = 61.9$ kNm

$\gamma = \frac{M}{M_o} = \frac{5445}{61.9} = 87.96$

Table 16 — $\therefore n = 0.77$

$\lambda = \frac{L_E}{r_y} = \frac{4.5 \times 10^3}{9.7 \times 10} = 46.3$

Table 14 — $\frac{\lambda}{x} = \frac{46.3}{39.1} = 1.2 \quad \therefore v = 0.98$

$\lambda_{LT} = n u v \lambda = 0.77 \times 0.9 \times 0.98 \times 46.3 = 35$

CURPE CONSULTANTS
46 Orburn Road
Dunfield

Project: Plate Girders			Job ref	
Part of structure: Design example — If restrained at A, B, D, E only			Calc sheet no: C3/14/	rev
Drawing ref	Calc by: RAS	Date	Check by	Date

BS 5950 Ref	Calculations	Output

Table 12 — $\therefore p_b = 265\ N/mm^2$

$M_b = S_x p_b = \frac{26116 \times 10^3 \times 265}{10^6} = 6920$ kNm

$\bar{M} < M_b \quad 5445 < 6920$ — **Output:** Length AB O.K.

Length BD — Length loaded $\therefore m = 1.0$

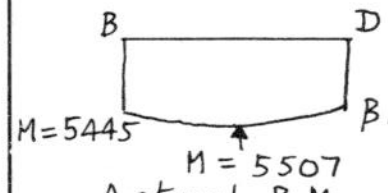

Actual B.M.

$\bar{M} = m M_A = 1.0 \times 5507$ kNm

$\beta = 1.00$

$M_o = \frac{330}{3} \times \frac{4.5}{8} = 61.9$ kNm

$\gamma = \frac{M}{M_o} = \frac{5445}{61.9} = 87.96$

Table 16 — $\therefore n = 1.00$

$\lambda = \frac{L_E}{r_y} = 46.3$ as before

Table 14 — $\frac{\lambda}{x} = 1.2$ as before $\therefore v = 0.98$

$\lambda_{LT} = n u v \lambda = 1.0 \times 0.9 \times 0.98 \times 46.3 = 41.7$

Table 12 — $P_b = 236.9\ N/mm^2$

$M_b = \frac{26116 \times 10^3 \times 236.9}{10^6} = 6186.8$ kNm

$\bar{M} < M_b \quad 5507 < 6186.8$ — **Output:** Length BD O.K.

Length DE — Same as AB

Plate girder is adequate for lateral torsional buckling if restraints are provided at A, B, D, E only

4

Compression members – stanchions

General requirements

Strut, stanchion and column are the terms applied most commonly to compression members. Strut is a very general term, but column and stanchion are reserved for the main vertical members carrying loads from the roof, floors and walls in a building. These loads are delivered to the stanchions by systems of simply supported beams, girders and trusses, where each load is interpreted, in accordance with BS 5950, as being applied axially or eccentrically for the purpose of design.

At one level of an arrangement of beams, stanchions are axially or concentrically loaded if the applied loads are symmetrical or 'balanced' (i.e. of equal magnitude) about both axes, and eccentrically loaded if the loads are not balanced about one or both axes. Eccentric loading produces bending moments in the stanchions as well as compression load. A short, stocky stanchion fails by squashing or crushing of the material, and for long slender stanchions failure occurs by overall flexural buckling. Other forms of failure are torsional buckling and local buckling.

Shear walls, core structures and bracing acting in conjunction with floors and roof, may be used to resist wind loads and consequently, in this case, stanchions carry only vertical loads and any associated moments. Where no resistance is available from these sources, then the whole of the wind forces may be resisted entirely by the structural steel framework (i.e. beams and stanchions) acting in conjunction with the floors and roofs. This chapter is confined to the design of stanchions; compression members acting as struts in latticed frameworks such as roof trusses, lattice girders and bracing, are dealt with in a subsequent chapter.

Classification of sections

To prevent local buckling, limiting width to thickness ratios for flanges and webs in compression are given in Table 7 of BS 5950. Classification depends not only on the dimensions of the section but also on the distribution of stress across the section and the position of the neutral axis.

Radius of gyration

This is a geometrical property depending upon the size and shape of the section, and is one measure of effectiveness in resisting buckling. It is used in stanchion design and is expressed by the formula:

$$r = (I/A)^{\frac{1}{2}}$$

where I = the second moment of area (normally referred to in structural steel as the moment of inertia),

A = area of the section.

Effective length

There are two other factors which affect the buckling tendency of stanchions:

1) Actual length L, which is taken as the distance between the points of effective restraints forming into the member.
2) End conditions.

These combine to produce what is termed the effective length L_E. In design, end conditions are accounted for through the use of an effective length factor which, when multiplied by the actual length L, gives the effective length L_E for buckling.

Guidance in the choice of factors is given in Cl. 4.7.2 and Table 24 of BS 5950. Figure 4.1 shows the Code values as either equal to or higher than the equivalent theoretical values. BS 5950 states that for members other than angles, channels and T-sections, L_E should be determined from the actual length L and the conditions of restraint in the relevant plane, as follows:

1) Restraining members carrying more than 90% of their reduced moment capacity after reduction for axial load should be regarded as incapable of providing directional restraint.
2) Table 24 of BS 5950 gives standard conditions of restraint.
3) Appendix D1 relates to single-storey stanchions of simple construction. It can be seen that a stanchion may have different effective lengths about the X–X and Y–Y axes.
4) Appendix E is used for members forming part of a rigid jointed frame.

Nominal effective lengths L_E are given in Table 24 of BS 5950.

Slenderness

This is the absolute measurement of the stanchion's tendency to buckle, and is given by:

$$\lambda = \frac{L_E}{r}$$

In practice it is necessary to consider slenderness about both axes X–X and Y–Y:

$$\lambda_x = \frac{L_{EX}}{r_x} \text{ and } \lambda_y = \frac{L_{EY}}{r_y}$$

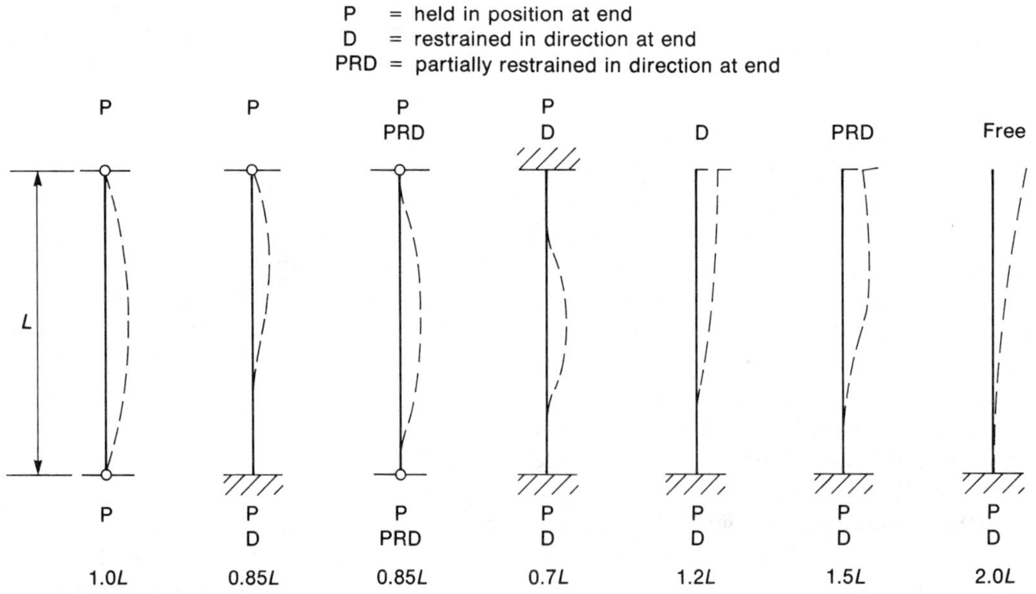

Fig. 4.1 *Effective lengths of stanchions*

At high values of slenderness, struts become very flexible and deflect more readily, and for this reason upper limits are imposed. The value of λ should not exceed 180 for members resisting loads other than wind load.

Compression resistance

The compressive resistance P_c should be taken as:

$$P_c = A_g p_c$$

where A_g = gross sectional area,
p_c = compressive strength from Tables 27(a)–(d) of BS 5950.

Compressive strength

The compressive strength depends on:

1) the slenderness λ,
2) the design strength P_y (reduced for slender sections),
3) the relevant strut curve.

Tables 25 and 26 of BS 5950 indicate for any section shape, thickness of material and axis of buckling (X–X or Y–Y) which of the four strut Tables 27(a)–(d) is relevant.

Eccentric connections

In practice most stanchions are eccentrically loaded. For stanchions in simple construction, the magnitude of the eccentricity of the beam and truss end reactions to be used for the purpose of calculating nominal applied bending moments to the stanchion, are shown in Fig. 4.2.

Stanchions in simple multi-storey construction

In structures of simple multi-storey construction it is not necessary to consider the effect on stanchions of pattern loading. In designing a stanchion, all supported beams at one level may be assumed as fully loaded. Nominal moments applied by simple beams should be calculated using the appropriate eccentricities.

In multi-storey stanchions which are effectively continuous at splices, the net moment applied at any one level should be divided between the stanchion lengths above and below that level in proportion to the stiffness I/L of each length, except that when the ratio of the stiffness does not exceed 1.5, the moment may be divided equally. I is the moment of inertia of the stanchion about its relevant axis, and L is the length of the stanchion.

The nominal moments applied to the stanchion are assumed to have no effect at the levels above and below their applied level. When nominal moments are applied, the stanchion should satisfy the relationship given in 4.8.3.3.1 for the 'simplified' approach, with m and n both taken as 1.0. A 'more exact' approach should not be used in conjunction with nominal moments. Cased stanchions should satisfy the relationship given in 4.14.4 with m and n both taken as 1.0 in 4.14.4(b), BS 5950.

In calculating M_b only, with only applied nominal moments, the equivalent slenderness of the stanchion is taken as:

$$\lambda_{LT} = \frac{0.5L}{r_y}$$

where L = distance between levels at which both axes are restrained,
r_y = radius of gyration about the minor axis.

Compression members with moments

These should be checked for local capacity at the points of greatest bending moment and axial load (usually at the ends). This capacity may be limited either by yielding or local buckling, depending on the properties of the section. The member should also be checked for overall buckling.

Local capacity check

For semi-compact and slender cross-sections and for the simplified approach for compact cross-sections, the following relationship should be satisfied:

$$\frac{F}{A_g p_y} + \frac{M_x}{M_{cx}} + \frac{M_y}{M_{cy}} \ngtr 1$$

where F = applied axial load,
A_g = gross cross-sectional area,
M_x = applied moment about major axis X–X,

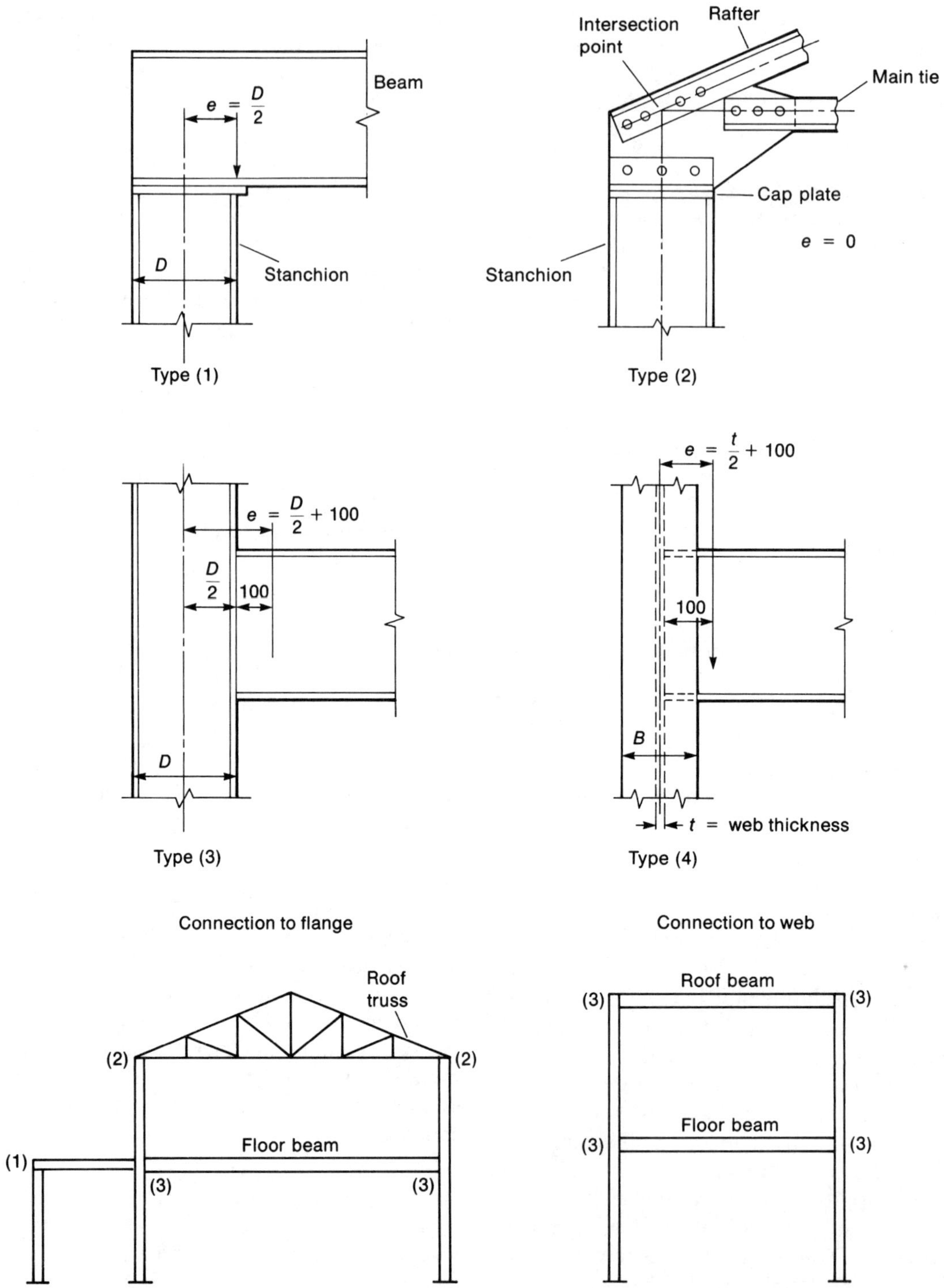

Fig. 4.2 *Key elevations illustrating eccentricity types for stanchions in simple construction only, i.e. wind forces resisted by bracing, shear walls or cores*

M_{cx} = moment capacity about major axis X–X in absence of axial load,
M_y = applied moment about minor axis Y–Y,
M_{cy} = moment capacity about minor axis Y–Y in absence of axial load.

Overall buckling check
The simplified relationship to be satisfied is:

$$\frac{F}{A_g p_c} + \frac{mM_x}{M_b} + \frac{mM_y}{p_y Z_y}$$

where p_c = compressive strength,
m = equivalent uniform moment factor from Table 18 of BS 5950,
M_b = buckling resistance moment capacity about major axis X–X,
Z_y = elastic modulus of section about minor axis Y–Y.

Reduction in imposed load

In the case of multi-storey stanchions, the imposed load on storey-height stanchions may be reduced in accordance with BS 6399: Part 1, as follows:

No reduction for one floor carried by a member
10% reduction for two floors carried by a member
20% reduction for three floors carried by a member
30% reduction for four floors carried by a member

For this purpose, the roof is counted as a floor.

This reduction applies only to the axial load, but the full imposed load is taken in determining the moments due to the assumed eccentricity taken for the beam reactions. It is therefore convenient to treat dead and imposed loads separately.

Note that this reduction is not always applicable, and reference should be made to BS 6399: Part 1.

Stanchion design – axially loaded

Design is an indirect process, and is accomplished by trial and error.

1) Determine beam reactions, separating dead and imposed loading. Accumulate these separate loads starting at the roof and working down to the base.
2) Apply reductions in imposed loads.
3) Apply load factors to obtain design loads.
4) Select a trial section and steel grade.
5) Design strength p_y is taken from Table 6.
6) Classify section, Table 7.
7) Effective lengths L_{EX} and L_{EY} are estimated using Table 24.
8) Slenderness λ_x and λ_y are calculated.
9) Select strut curves from Table 25.
10) Compressive strengths p_{cx} and p_{cy} are read from appropriate Tables 27(a)–(d).
11) Compression resistances P_{cx} and P_{cy} are calculated from $P_c = A_g p_c$.
The compression resistance P_c is the lesser of P_{cx} and P_{cy}.

(The tables referred to above are from BS 5950.)

In the multi-storey axially loaded stanchion design example, Ex. 1, the beam to stanchions connections are:

1) at floor levels, bolted top and bottom flange cleats;
2) at roof, bolted to cap plate, with beams bolted together by web cleats.

At base level, the machined ends of the slab base and stanchion are connected together by bolted flange cleats. H.D. bolts secure the stanchion to the foundation.

The effective lengths are consequently interpreted and taken as:

1) Top and bottom lengths = 1.0L, since they are effectively held in position at both ends and not restrained in direction at either end.
2) Intermediate lengths = 0.85L, since they are effectively held in position at both ends and partially restrained in direction at both ends.

Stanchion design – eccentrically loaded

Design again is an indirect process and is accomplished by trial and error.

1) Determine beam reactions, separating dead and imposed loading. Accumulate these

separate loads starting at the roof and working down to the base.

2) Apply reductions in imposed loads.
3) Select a trial section and steel grade.
4) Determine the moments at each level, using appropriate eccentricity e values:

$$e_x = \frac{D}{2} + 100 \text{ mm for X–X axis.}$$

$$e_y = \frac{t}{2} + 100 \text{ mm for Y–Y axis.}$$

5) Determine stiffness $\frac{I}{L}$ for both axes for each storey length.
6) Distribute moments.
7) Apply load factors to obtain design loads and moments.
8) Now check trial section.
9) Design strength is taken from Table 6.
10) Classify section.
11) Determine moment capacities of section.
12) Check local capacity.
13) Effective lengths L_{EX} and L_{EY} are estimated using Table 24.
14) Slenderness λ_x and λ_y are calculated.
15) Select strut curves from Table 25.
16) Compressive strengths p_{cx} and p_{cy} are read from appropriate Tables 27(a)–(d). Use lesser value for p_c.
17) Determine equivalent slenderness λ_{LT} and read the bending strength p_b from Table 11.
18) Determine the buckling resistance moment M_b.
19) Check for overall buckling.

(The tables referred to above are from BS 5950.)

In the multi-storey eccentrically loaded stanchion design example (Example 2, page 37) the floor and roof beam to stanchion connections are provided by flexible end plates welded to the beam and bolted to the stanchion. The depth of the end plates is suggested as a maximum of 0.6D, where D is the overall depth of the beam, to restrict the development of beam end fixity. At the base level, the machined ends of the slab base and stanchion are connected together by bolted flange cleats. H.D. bolts secure the stanchion to the foundation.

The effective lengths are consequently interpreted and taken as:

1) Bottom length = 1.0L, since it is effectively held in position at both ends and not restrained in direction at either end.
2) Remainder of lengths = 0.85L, since they are effectively held in position at both ends and partially restrained in direction at both ends.

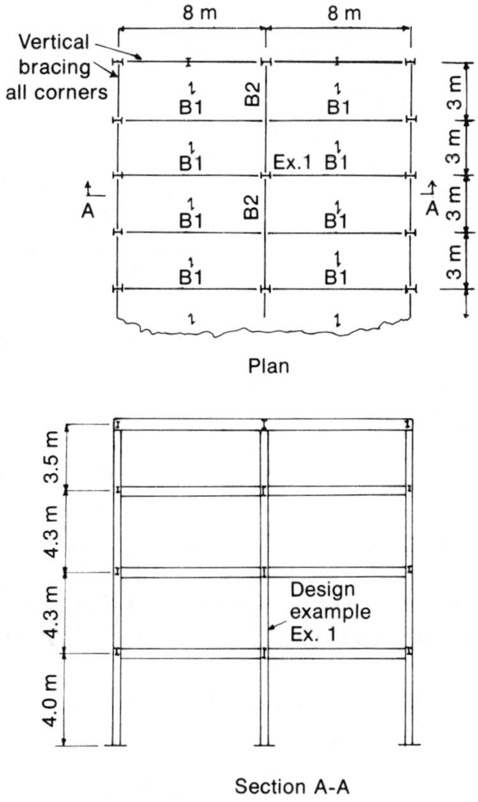

Fig. 4.3 *Sketch illustrating the stanchion designed in Example 1*

CURPE CONSULTANTS 46 Orburn Road Dunfield	Project: Multi-Storey Structure		Job ref
	Part of structure: Multi-storey stanchion-axial load		Calc sheet no: C4 / 1 / 1; rev
	Drawing ref	Calc by: R A S. ; Date	Check by ; Date

Ref	Calculations	Output
BS5950		

Ex. 1. **Multi-storey stanchion-axial load**

Roof loading

Asphalt	0·5
Screed	0·5
Precast units (100)	2·1
Steel & lightweight casing	0·4
Ceiling	0·5
Dead =	4·0 kN/m²

Imposed = 1·5 kN/m²

Floor loading

Tiles	0·05
Screed	0·60
Precast units (150)	3·80
Steel & lightweight casing	0·45
Ceiling	0·60
	5·50
Allow for partitions	1·00
Dead =	6·50 kN/m²

Imposed = 3·0 kN/m²

Roof beam B1. U.D.L. loading.

Dead load = 4·0 × 8 × 3 = 96·0 kN, R = 48 kN

Imposed " = 1·5 × 8 × 3 = 36·0 kN, R = 18 kN

Roof beam B2

96 (D) 36 (I)

R = 48 (D) 18 (I) R = 48 (D) 18 (I).

Floor beam B1 U.D.L. loading.

Dead load = 6·5 × 8 × 3 = 156 kN, R = 78 kN

Imposed " = 3 × 8 × 3 = 72 kN, R = 36 kN

Floor beam B2

156 (D) 72 (I)

R = 78 (D) 36 (I) R = 78 (D) 36 (I)

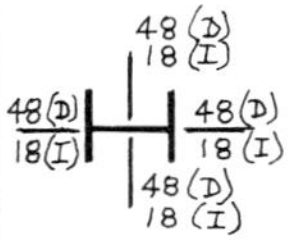

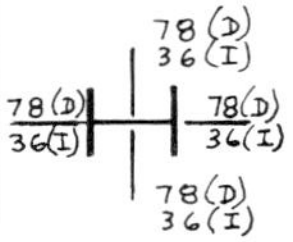

C.4/2

Ex. 1. Axial load

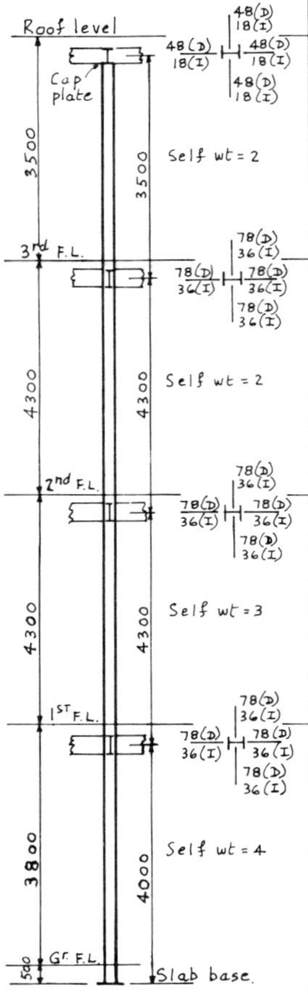

Level	Dead load (D) kN	Imposed load (I) kN	Reduction Imposed load kN	Factored loads (D)+(I) kN.
Roof level – 3rd F.L.	48 48 48 48 2 194	18 18 18 18 72	0	1·4 (194) = 271·6 1·6 (72) = 115·2 386·8
3rd F.L. – 2nd F.L.	78 78 78 78 2 508	36 36 36 36 216	10% of 216 = 22	1·4 (508) = 711·2 1·6 (216−22) = 310·4 1021·6
2nd F.L. – 1st F.L.	78 78 78 78 3 823	36 36 36 36 360	20% of 360 = 72	1·4 (823) = 1152·2 1·6 (360−72) = 460·8 1613·0
1st F.L. – Gr. F.L.	78 78 78 78 4 1139	36 36 36 36 504	30% of 504 = 151	1·4 (1139) = 1594·6 1·6 (504−151) = 564·8 2159·4

CURPE CONSULTANTS 46 Orburn Road Dunfield	Project: Multi-Storey Structure	Job ref
	Part of structure: Multi-storey stanchion-axial load	Calc sheet no: C4 / 3 rev: 1
	Drawing ref: — Calc by: R.A.S. Date: —	Check by: — Date: —

Ref BS5950	Calculations	Output
	Ex. 1	
	Stanchion design – axial loading	
	Top length – Roof to third floor	
	Design load = 386·8 kN $L_x = L_y = 3{\cdot}5$ m.	
	Try 152 × 152 × 23 U.C.	
	Section classification: $\varepsilon = 1{\cdot}0$	
Table 6	For $T = 6{\cdot}8 < 16$ $p_y = 275\ \text{N/mm}^2$	
	$\frac{b}{T} = \frac{76{\cdot}2}{6{\cdot}8} = 11{\cdot}2 < 15\varepsilon$	
	$\frac{d}{t} = \frac{123{\cdot}5}{6{\cdot}1} = 20{\cdot}2 < 39\varepsilon$	
	Since $\frac{b}{T} < 15\varepsilon$ and $\frac{d}{t} < 39\varepsilon$	
Table 7	Section is not slender	Semi-compact
Table 24	Effective lengths $L_{EX} = L_{EY} = 1{\cdot}0L$	
	Slenderness $\lambda x = \frac{1{\cdot}0 \times 3{\cdot}5 \times 10^3}{6{\cdot}51 \times 10} = 53{\cdot}75$	
	$\lambda y = \frac{1{\cdot}0 \times 3{\cdot}5 \times 10^3}{3{\cdot}68 \times 10} = 95$	
Table 25	For a rolled H section of thickness < 40 use Table 27 (b) for x-x axis 27 (c) for y-y axis.	
Table 27(b)	$p_{cx} = 230\ \text{N/mm}^2$	
Table 27(c)	$p_{cy} = 133{\cdot}5\ \text{N/mm}^2$	
	∴ Comp. Strength $p_c = 133{\cdot}5\ \text{N/mm}^2$ (lesser)	$p_c = 133{\cdot}5\ \text{N/mm}^2$
4.7.4.	Compressive resistance $P_c = p_c . A_g$.	
	$P_c = \frac{133{\cdot}5 \times 29{\cdot}8 \times 10^2}{10^3} = 397{\cdot}8\ \text{kN} > 386{\cdot}8$	$P_c = 397{\cdot}8$ kN satisfactory 152 × 152 × 23 UC

CURPE CONSULTANTS 46 Orburn Road Dunfield	Project: Multi-Storey Structure	Job ref
	Part of structure: Multi-storey stanchion-axial load	Calc sheet no: C4 / 4 rev: 1
	Drawing ref: — Calc by: R.A.S. Date: —	Check by: — Date: —

Ref BS5950	Calculations	Output
	Ex. 1.	
	Third floor to second floor	
	Design load = 1021·6 kN $L_x = L_y = 4{\cdot}3$ m	
	Try 203 × 203 × 46 UC	
	Section classification: $\varepsilon = 1{\cdot}0$, $p_y = 275$	
Table 6	$\frac{b}{T} = \frac{101{\cdot}6}{11} = 9{\cdot}24 < 15\varepsilon$	
	$\frac{d}{t} = \frac{160{\cdot}9}{7{\cdot}3} = 22{\cdot}0 < 39\varepsilon$	
	Since $\frac{b}{T} < 15\varepsilon$ and $\frac{d}{t} < 39\varepsilon$	
Table 7	Section is not slender	Compact
Table 24	Effective lengths $L_{EX} = L_{EY} = 0{\cdot}85L$	
Table 27(b)	$\lambda x = \frac{0{\cdot}85 \times 4{\cdot}3 \times 10^3}{8{\cdot}81 \times 10} = 41{\cdot}5$ ∴ $p_{cx} = 248{\cdot}5\ \text{N/mm}^2$	
Table 27(c)	$\lambda y = \frac{0{\cdot}85 \times 4{\cdot}3 \times 10^3}{5{\cdot}11 \times 10} = 71{\cdot}5$ ∴ $p_{cy} = 178\ \text{N/mm}^2$	$p_c = 178\ \text{N/mm}^2$
4.7.4.	$P_c = \frac{178 \times 58{\cdot}8 \times 10^2}{10^3} = 1046{\cdot}6 > 1021{\cdot}6$	$P_c = 1046{\cdot}6$ kN 203 × 203 × 46 UC satisfactory
	Second floor to first floor	
	Design load = 1613 kN $L_x = L_y = 4{\cdot}3$ m	
	Try 203 × 203 × 71 UC	
	Section classification:	
Table 6	For T = 17·3, $p_y = 265\ \text{N/mm}^2$, $\varepsilon = \sqrt{\frac{275}{p_y}} = 1{\cdot}02$	
	$\frac{b}{T} = \frac{103{\cdot}1}{17{\cdot}3} = 5{\cdot}96 < 15\varepsilon$	
	$\frac{d}{t} = \frac{160{\cdot}9}{10{\cdot}3} = 15{\cdot}6 < 39\varepsilon$	
	Since $\frac{b}{T} < 15\varepsilon$ and $\frac{d}{t} < 39\varepsilon$	
Table 7	Section is not slender	Plastic
Table 24	$L_{EX} = L_{EY} = 0{\cdot}85\ L$	

CURPE CONSULTANTS 46 Orburn Road Dunfield	Project: Multi-Storey Structure		Job ref
	Part of structure: Multi-storey stanchion - axial load		Calc sheet no: C4 / 5 / rev
	Drawing ref	Calc by: R.A.S. Date	Check by Date

Ref	Calculations	Output
	Ex. 1.	
Table 27(b)	$\lambda x = \frac{0.85 \times 4.3 \times 10^3}{9.16 \times 10} = 39.9 \quad \therefore pcx = 241\ N/mm^2$	
Table 27(c)	$\lambda y = \frac{0.85 \times 4.3 \times 10^3}{5.28 \times 10} = 69.2 \quad \therefore pcy = 178\ N/mm^2$	
	$\therefore pc = 178\ N/mm^2$	
	$Pc = \frac{178 \times 91.1 \times 10^2}{10^3} = 1621 > 1613$	Pc = 1621 kN 203×203×71 UC satisfactory
	First floor to base	
	Design load = 2159.4 kN. $\quad Lx = Ly = 4\ m$	
	Try 254×254×89 UC.	
	Section classification:	
Table 6	For T = 17.3, $py = 265\ N/mm^2$, $\varepsilon = 1.02$	
	$\frac{b}{T} = \frac{127.95}{17.3} = 7.4 < 15\varepsilon$	
	$\frac{d}{t} = \frac{200.3}{10.5} = 19.1 < 39\varepsilon$	
Table 7	section is not slender	Plastic
Table 24	$L_{EX} = L_{EY} = 1.0\ L$	
Table 27(b)	$\lambda x = \frac{1.0 \times 4.0 \times 10^3}{11.2 \times 10} = 35.7 \quad \therefore pcx = 246\ N/mm^2$	
Table 27(c)	$\lambda y = \frac{1.0 \times 4.0 \times 10^3}{6.52 \times 10} = 61.3 \quad \therefore pcy = 193\ N/mm^2$	$pc = 193\ N/mm^2$
4.7.4.	$\therefore Pc = \frac{193 \times 114 \times 10^2}{10^3} = 2200\ kN$	Pc = 2200 kN 254×254×89 UC Satisfactory
	Since 2200 > 2159.4 sat.	
	For practical purposes splice stanchion above second floor:	
	Top section use 203×203×46 UC	
	Bottom section use 254×254×89 UC.	

C4/6

Ex.2. Eccentric loads.

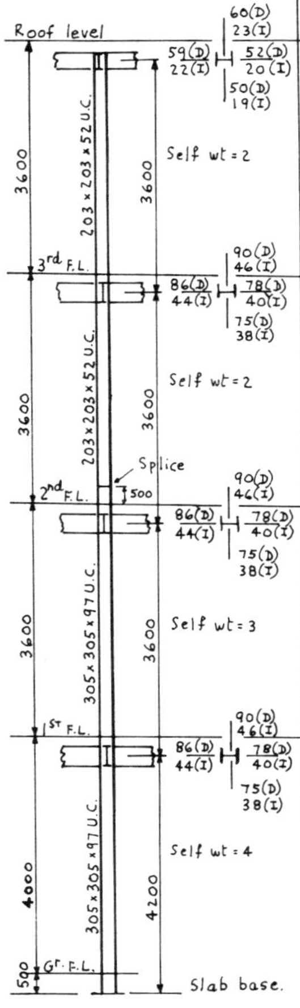

Dead load (D) kN	Imposed load (I) kN	Reduction Imposed load kN	Moments Before distribution Mx kN.mm (Dead)	Mx kN.mm (Imposed)
59 52 60 50 2 223 ↓	22 20 23 19 84 ↓	0	(59 − 52) 203 = 1421	(22 − 20) 203 = 406
86 78 90 75 2 554 ↓	44 40 46 38 252 ↓	10% of 252 = 25	(86 − 78) 203 = 1624	(44 − 40) 203 = 812
86 78 90 75 3 886 ↓	44 40 46 38 420 ↓	20% of 420 = 84	(86 − 78) 254 = 2032	(44 − 40) 254 = 1016
86 78 90 75 4 1219	44 40 46 38 588	30% of 588 = 176	(86 − 78) 254 = 2032	(44 − 40) 254 = 1016

Ex. 2. C4/7

Moments Before distribution.		Stanchion Stiffness		Moments After distribution			
M_Y kN.mm (Dead)	M_Y kN.mm. (Imposed)	$\frac{I_X}{L_X}$ X-X axis	$\frac{I_Y}{L_Y}$ Y-Y axis	M_X kN.mm (Dead)	M_X kN.mm (Imposed)	M_Y kN.mm. (Dead)	M_Y kN.mm. (Imposed)
(60−50) 104 = 1040	(23−19) 104 = 416			1421	406	1040	416
		$\frac{5263}{3600}$ = 1.46	$\frac{1770}{3600}$ = 0.49				
				½(1624) = 812	½(812) = 406	½(1560) = 780	½(832) = 416
(90−75) 104 = 1560	(46−38) 104 = 832			812	406	780	416
		$\frac{5263}{3600}$ = 1.46	$\frac{1770}{3600}$ = 0.49				
				$\frac{2032 \times 1.46}{1.46 + 6.17}$ = 389	$\frac{1016 \times 1.46}{1.46 + 6.17}$ = 194	$\frac{1575 \times 0.49}{0.49 + 2.02}$ = 307	$\frac{840 \times 0.49}{0.49 + 2.02}$ = 164
(90−75) 105 = 1575	(46−38) 105 = 840			1643	822	1268	676
		$\frac{22202}{3600}$ = 6.17	$\frac{7268}{3600}$ = 2.02				
				½(2032) = 1016	½(1016) = 508	½(1575) = 788	½(840) = 420
(90−75) 105 = 1575	(46−38) 105 = 840			1016	508	788	420
		$\frac{22202}{4200}$ = 5.29	$\frac{7268}{4200}$ = 1.73				

Ex. 2. C4/8

Factored Loads kN (D) + (I)	Factored Moments	
	M_X kN.mm. (D) + (I)	M_Y kN.mm. (D) + (I)
Roof level	1.4(1421) = 1989 1.6(406) = 649 2638	1.4(1040) = 1456 1.6(416) = 666 2122
1.4 (223) = 312 1.6 (84) = 134 446		
3rd F.L.	1.4(812) = 1137 1.6(406) = 650 1787	1.4(780) = 1092 1.6(416) = 666 1758
	1787	1758
1.4 (554) = 775 1.6 (252−25) = 363 1138		
2nd F.L.	1.4(389) = 544 1.6(194) = 310 854	1.4(307) = 430 1.6(164) = 262 692
	1.4(1643) = 2300 1.6(822) = 1315 3615	1.4(1268) = 1775 1.6(676) = 1082 2857
1.4 (886) = 1240 1.6 (420−84) = 537 1777		
1st F.L.	1.4(1016) = 1422 1.6(508) = 813 2235	1.4(788) = 1103 1.6(420) = 672 1775
	2235	1775
1.4(1219) = 1706 1.6(588−176) = 659 2365		
Base level.		

Bending Moment Diagrams

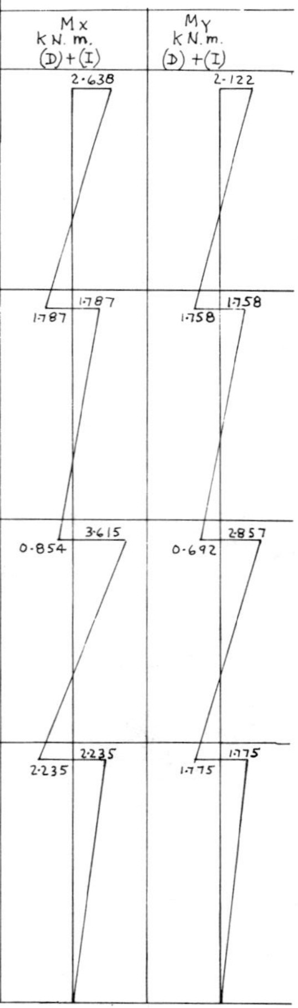

CURPE CONSULTANTS
46 Orburn Road
Dunfield

Project: Multi-Storey Stanchion | Job ref:
Part of structure: Eccentric loads. | Calc sheet no: C4 / 9 | rev: /
Drawing ref: | Calc by: R.A.S. | Date: | Check by: | Date:

Ref BS5950	Calculations	Output
	Ex. 2.	
	Design Loads and Moments	

Storey Height	Load F	M_x (kNm)	M_Y (kNm).
Roof – 3rd.	446 kN	2·638	2·122
3rd – 2nd	1138 kN	1·787	1·758
2nd – 1st	1777 kN	3·615	2·857
1st – base	2365 kN	2·235	1·775

Stanchion Design.

Roof – 3rd: $F = 446\,kN$, $M_x = 2{\cdot}638\ kN.m.$, $M_Y = 2{\cdot}122\ kN.m.$

Try 203 × 203 × 52 U.C., grade 43.

Properties: $A_g = 66{\cdot}4\,cm^2$, $r_x = 8{\cdot}9\,cm$, $r_Y = 5{\cdot}16\,cm$, $Z_x = 510\,cm^3$, $Z_Y = 174\,cm^3$, $S_x = 568\,cm^3$, $S_Y = 264\,cm^3$.

Table 6 — Design strength $p_Y = 275\ N/mm^2$

Table 7 — Flange $\frac{b}{T} = \frac{101{\cdot}95}{12{\cdot}5} = 8{\cdot}16 < 8{\cdot}5\varepsilon$ Plastic

Web $\frac{d}{t} = \frac{160{\cdot}9}{8} = 20{\cdot}1 < 39\varepsilon$ Semi-compact

Output: Plastic, since $\alpha > 2$, $\frac{b}{T} < 8{\cdot}5\varepsilon$, $\frac{d}{t} < 39\varepsilon$

4.8.3.2 — Local capacity check:

4.2.5 — Moment capacities are:

$$M_{cx} = p_Y S_x = \frac{275 \times 568 \times 10^3}{10^6} = 156{\cdot}2\ kN\,m$$

$$\not> 1{\cdot}2\,p_Y Z_x = \frac{1{\cdot}2 \times 275 \times 10^3 \times 510}{10^6} = 168{\cdot}3\ kN\,m$$

$$M_{cY} = p_Y S_Y = \frac{275 \times 264 \times 10^3}{10^6} = 72{\cdot}6\ kN\,m$$

$$\not> 1{\cdot}2\,p_Y Z_Y = \frac{1{\cdot}2 \times 275 \times 174 \times 10^3}{10^6} = 57{\cdot}4\ kN\,m$$

CURPE CONSULTANTS
46 Orburn Road
Dunfield

Project: Multi-Storey Stanchion | Job ref:
Part of structure: Eccentric loads. | Calc sheet no: C4 / 10 | rev: /
Drawing ref: | Calc by: R.A.S. | Date: | Check by: | Date:

Ref BS 5950	Calculations	Output
	Ex. 2.	

4.8.3.2

$$\frac{F}{A_g\,p_Y} + \frac{M_x}{M_{cx}} + \frac{M_Y}{M_{cY}} \not> 1{\cdot}0$$

$$\frac{446 \times 10^3}{66{\cdot}4 \times 10^2 \times 275} + \frac{2{\cdot}638}{156{\cdot}2} + \frac{2{\cdot}122}{57{\cdot}4}$$

$$= 0{\cdot}244 + 0{\cdot}017 + 0{\cdot}037 = 0{\cdot}298 < 1{\cdot}0 \text{ satisfactory}$$

Output: Local capacity sat.

4.8.3.3.1 — Overall buckling check:

Table 24 — The stanchion is effectively held in position and partially restrained in direction at both ends and effective length L_E is taken as 0·85 L for both axes.

Table 27(b) — $\lambda_x = \frac{L_{Ex}}{r_x} = \frac{0{\cdot}85 \times 3600}{8{\cdot}9 \times 10} = 34{\cdot}4 \ \therefore p_c = 257\ N/mm^2$

Table 27(c) — $\lambda_Y = \frac{L_{EY}}{r_Y} = \frac{0{\cdot}85 \times 3600}{5{\cdot}16 \times 10} = 59{\cdot}3 \ \therefore p_c = 202\ N/mm^2$

$\therefore p_c = 202\ N/mm^2$

Table 11, 4.7.7. — $\lambda_{LT} = \frac{0{\cdot}5\,l}{r_Y} = \frac{0{\cdot}5 \times 3600}{5{\cdot}16 \times 10} = 34{\cdot}88 \ \therefore p_b = 273\ N/mm^2$

4.3.7.3. — $M_b = S_x\,p_b = \frac{568 \times 10^3 \times 273}{10^6} = 155{\cdot}1\ kN\,m$

4.8.3.3.1

$$\frac{F}{A_g\,p_c} + \frac{m\,M_x}{M_b} + \frac{m\,M_Y}{p_Y\,Z_Y} \not> 1{\cdot}0$$

$$\frac{446 \times 10^3}{66{\cdot}4 \times 10^2 \times 202} + \frac{1{\cdot}0 \times 2{\cdot}638}{155{\cdot}1} + \frac{1{\cdot}0 \times 2{\cdot}122 \times 10^6}{275 \times 174 \times 10^3}$$

$$= 0{\cdot}332 + 0{\cdot}017 + 0{\cdot}044 = 0{\cdot}393 < 1{\cdot}0 \text{ Satisfactory}$$

Output: Overall buckling sat.

Third to second floor: $F = 1138\ kN$, $M_x = 1{\cdot}787\ kN\,m$, $M_Y = 1{\cdot}758\ kN\,m$

Try 203 × 203 × 52 U.C.

CURPE CONSULTANTS
46 Orburn Road
Dunfield

Project: Multi-Storey Stanchion | Job ref:
Part of structure: Eccentric loads | Calc sheet no: C4 / 11 / rev
Drawing ref: | Calc by: R.A.S. | Date: | Check by: | Date:

Ref BS 5950	Calculations	Output
	Ex. 2.	
4.8.3.2	Local capacity check:	
	$\frac{1138 \times 10^3}{6.4 \times 10^2 \times 275} + \frac{1.787}{156.2} + \frac{1.758}{57.4}$	
	$= 0.623 + 0.011 + 0.03 = 0.664 < 1.0$ satisfactory	Local cap. sat.
4.8.3.3.1	Overall buckling check:	
	$L_{EX} = L_{EY} = 0.85L$	
	$\frac{1138 \times 10^3}{66.4 \times 10^2 \times 202} + \frac{1.0 \times 1.787}{155.1} + \frac{1.0 \times 1.758 \times 10^6}{275 \times 174 \times 10^3}$	
	$= 0.848 + 0.0115 + 0.0367 = 0.8962 < 1.0$ satisfactory	Overall buckling sat.
	Second to first floor	
	This is the same section as bottom length which is the criterion for design. because of its greater length and heavier loading.	
	First floor to base $F = 2365$ kN, $M_x = 2.235$ kNm, $M_y = 1.775$ KNm	
	Try 305 × 305 × 97 U.C, grade 43	
	Properties: $A_g = 123\,cm^2$, $r_x = 13.4\,cm$, $r_y = 7.68\,cm$, $Z_x = 1440\,cm^3$, $Z_y = 477\,cm^3$, $S_x = 1590\,cm^3$, $S_y = 723\,cm^3$	
Table 6	$p_y = 275\ N/mm^2$	
	Flange $\frac{b}{T} = \frac{152.4}{15.4} = 9.9 < 15\varepsilon$ Semi-compact	
	Web $\frac{d}{t} = \frac{246.6}{9.9} = 24.9 < 39\varepsilon$ Semi-compact	Semi-compact
4.8.3.2	Local capacity check:	
4.2.5	$M_{cx} = p_y Z_x = \frac{275 \times 1440 \times 10^3}{10^6} = 396$ kN m	
	$M_{cy} = p_y Z_y = \frac{275 \times 477 \times 10^3}{10^6} = 131$ KN m	
	$\frac{F}{A_g p_y} + \frac{M_x}{M_{cx}} + \frac{M_y}{M_{cy}} \ngtr 1.0$	
	$\frac{2365 \times 10^3}{123 \times 10^2 \times 275} + \frac{2.235}{396} + \frac{1.775}{131}$	

CURPE CONSULTANTS
46 Orburn Road
Dunfield

Project: Multi-Storey Stanchion | Job ref:
Part of structure: Eccentric loads. Splice connection | Calc sheet no: C4 / 12 / rev
Drawing ref: | Calc by: R.A.S. | Date: | Check by: | Date:

Ref	Calculations	Output
	Ex. 2.	
	$= 0.699 + 0.0056 + 0.0135 = 0.7181 < 1.0$ satisfactory	Local capacity sat.
4.8.3.3.1	Overall buckling check:	
	$L_{EX} = L_{EY} = 1.0L$	
Table 27(b)	$\lambda_x = \frac{1.0 \times 4200}{13.4 \times 10} = 31.34 \therefore p_c = 261\ N/mm^2$	
Table 27(c)	$\lambda_y = \frac{1.0 \times 4200}{7.68 \times 10} = 54.68 \therefore p_c = 212\ N/mm^2$	
	$\therefore p_c = 212\ N/mm^2$	
Table 11	$\lambda_{LT} = \frac{0.5L}{r_y} = \frac{0.5 \times 4200}{7.68 \times 10} = 27.34 \therefore p_b = 275\ N/mm^2$	
4.3.7.3.	$M_b = S_x p_b$	
	$= \frac{1590 \times 10^3 \times 275}{10^6} = 437$ kN m $\ngtr M_{cx}$	
4.8.3.3.1.	$\frac{F}{A_g p_c} + \frac{m M_x}{M_b} + \frac{m M_y}{p_y Z_y} \ngtr 1.0$	
	$\frac{2365 \times 10^3}{123 \times 10^2 \times 212} + \frac{1.0 \times 2.235}{396} + \frac{1.0 \times 1.775 \times 10^6}{275 \times 477 \times 10^3}$	
	$0.907 + 0.0056 + 0.0135 = 0.9261 < 1.0$ satisfactory	Overall buckling sat.
	Stanchion Splice	
	This is located 500 mm above second floor level.	
	Due to factored loads:	
	Axial compression load $F_c = 1138$ kN	
	Bending moments $M_x = 0.267$ kNm, $M_y = 0.156$ kNm	
	obtained by proportion from bending mom diagram	
	Shear force $F_v = 1.4$ kN	
	Assumptions:	
	(a) The ends of the stanchions and the	

CURPE CONSULTANTS 46 Orburn Road Dunfield	Project: Multi-Storey Stanchion	Job ref
	Part of structure: Splice connection	Calc sheet no: C4 / 13 / rev
	Drawing ref: / Calc by: R.A.S. / Date:	Check by: / Date:

Ref BS 5950	Calculations	Output
	Ex. 2.	

surfaces of the division plate are machined for full contact in bearing.

(b) The splice is located 500 mm above F.L. and hence the moment due to strut action is considered insignificant.

(c) M_x, M_y, F_v are insignificant and are ignored.

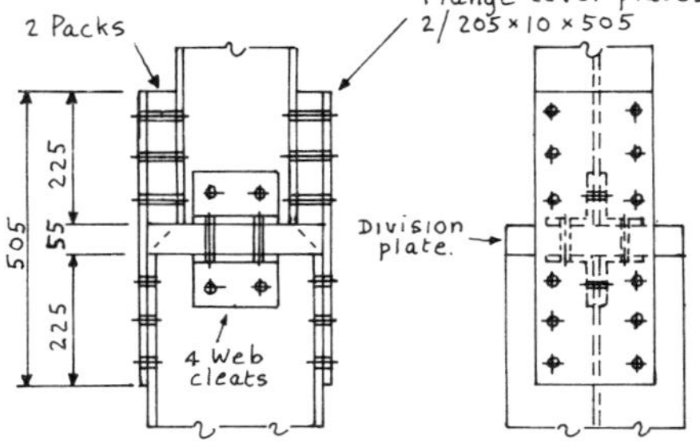

All bolts are M20 grade 4.6 bolts.

Output: Splice detail as shown

Proportioning of splice.

(1) Projection of flange cover plates beyond the ends of the members to be equal to the upper flange width or 225 mm, whichever is the greater.

(2) Thickness of flange cover plates to be equal to half the thickness of the upper flange or 10 mm, whichever is the greater.

(3) A division plate and web cleats (to give a load dispersion of 45° through the plate), the section being of different serial size.

CURPE CONSULTANTS 46 Orburn Road Dunfield	Project: Multi-Storey Structure	Job ref
	Part of structure: Multi-Storey Stanchion - Slab Base	Calc sheet no: C4 / 14 / rev
	Drawing ref: / Calc by: R.A.S. / Date:	Check by: / Date:

Ref BS 5950	Calculations	Output
	Ex. 2.	

4.13.2.2. Design of Slab Base · Axial Load

F = 2365 kN (Factored)

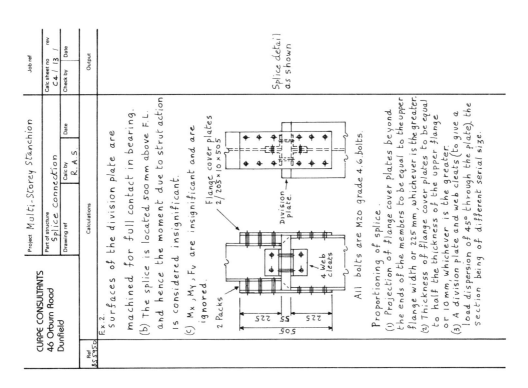

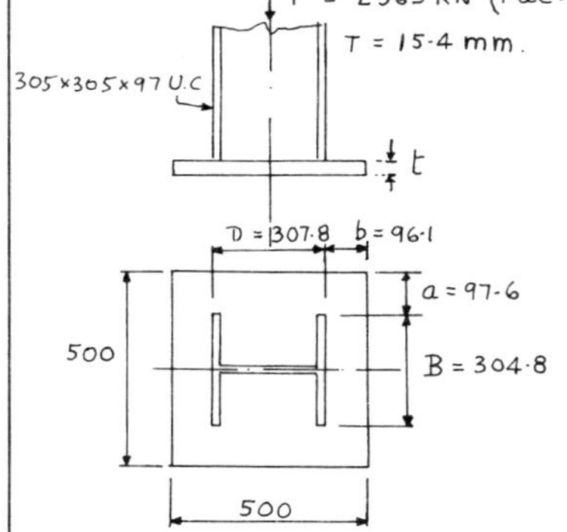

4.13.1. Bearing strength of concrete = $0{\cdot}4 f_{cu}$

Assuming $f_{cu} = 25\ \text{N/mm}^2$

Minimum area of baseplate $= \dfrac{F}{0{\cdot}4 f_{cu}}$

$= \dfrac{2365 \times 10^3}{0{\cdot}4 \times 25} = 236500\ \text{mm}^2$

For a square plate side $= \sqrt{236500} = 486{\cdot}3$ mm

For practical purposes say 500 × 500 — **Output:** Area 500×500

Then $w = \dfrac{\text{Load}}{\text{Area}} = \dfrac{2365 \times 10^3}{500 \times 500} = 9{\cdot}46\ \text{N/mm}^2$

Assuming thickness of baseplate < 40 mm

Table 6: Then $p_{yp} = 265\ \text{N/mm}^2$

$t = \left[\dfrac{2{\cdot}5}{p_{yp}}\, w\,(a^2 - 0{\cdot}3b^2)\right]^{\frac{1}{2}}$

CURPE CONSULTANTS 46 Orburn Road Dunfield	Project Multi. Storey Structure			Job ref	
	Part of structure Multi-Storey Stanchion - Slab Base			Calc sheet no rev C4 / 15 1	
	Drawing ref	Calc by R.A.S.	Date	Check by	Date

Ref BS5950	Calculations	Output
	Ex.2. $$t = \left[\frac{2.5}{265} \times 9.46 \left(97.6^2 - 0.3 \times 96.1^2\right)\right]^{\frac{1}{2}}$$ $$= \left[0.0892\left(9525.76 - 2770.56\right)\right]^{\frac{1}{2}}$$ $$= \sqrt{602.56}$$ $$= 24.54 \text{ mm}.$$ Make baseplate $500 \times 500 \times 25$ mm	Baseplate $500 \times 500 \times 25$
4, 13.3.	The bottom of the stanchion and the upper surface of the plate are machined to achieve full contact bearing. Flange cleats are used to connect the baseplate to the stanchion. H.D (holding down) bolts 4.6 grade are used to connect stanchion to foundation.	
	 M20 bolts 4.6 grade 25 ⊥ Flange cleats $150 \times 100 \times 12$ L Base plate $500 \times 500 \times 25$ FOR H.D bolts Bolts countersunk U.S. base plate Base Detail.	Base detail as shown.

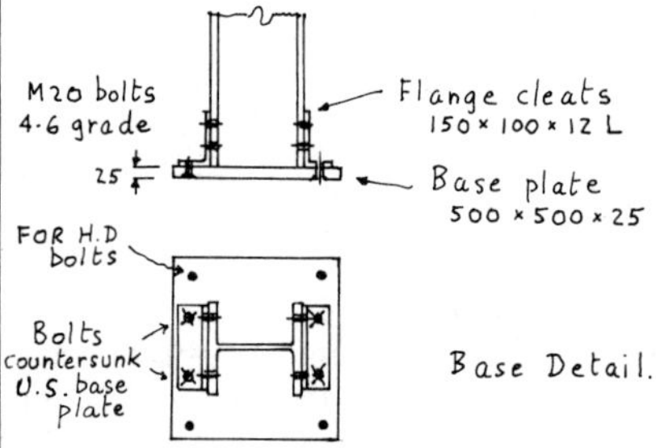

5

Cased sections

General requirements

Solid concrete casing contributes to the strength and stiffness of a steel section and also acts as fire protection. A rolled steel I or H section encased in concrete may be designed by empirical methods presented in Cl. 4.14.2, 4.14.3, and 4.14.4 of BS 5950 for each of the following:

1) Cased member subject to bending.
2) Cased struts.
3) Cased members subject to axial load and moment.

For these methods to be used, however, it is necessary that certain requirements of BS 5950 be met. Some of these requirements are shown in Figure 5.1 but reference should be made to other associated details as outlined in BS 5950.

The effective length L_E of the cased section is limited to the least of:

$$40b_c; \quad \frac{100b_c^2}{d_c}; \quad 250r$$

where b_c and d_c are as shown on Fig. 5.1.
r is the minimum radius of gyration of the steel section alone.

Cased members subject to bending

Cased beams which meet the stated requirements should be designed as for an uncased section, except that r_y may be taken as the greater of:

$0.2\,(B + 100)$ mm; r_y of the uncased section

All other properties should be taken as for the uncased section.

The buckling resistance moment $M_b \not> 1\frac{1}{2}$ times permitted M_b for the uncased section, where B is the width of the flange.

For deflection calculation, the effective moment of inertia of the cased section may be taken as that of the steel section plus the transformed net area of the concrete, i.e.

$$I_{cs} = I_s + \frac{I_c - I_s}{\alpha_e}$$

where I_{cs} = second moment of area of cased section,
I_s = second moment of area of steel section,
I_c = second moment of area of gross concrete section,
α_e = modular ratio.

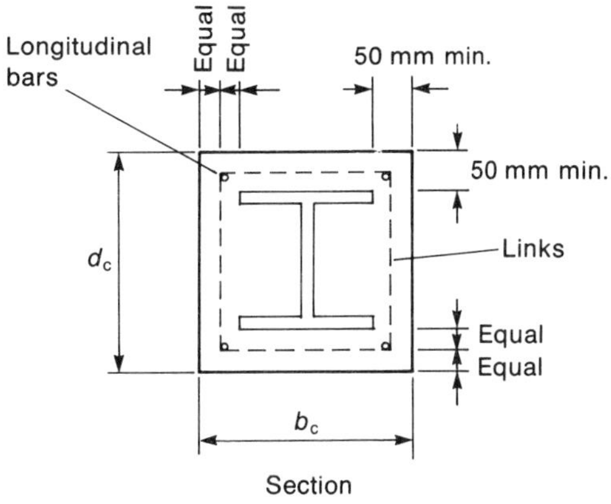

Links: Steel fabric or 5 mm dia. steel reinf. 200 mm max. spacing.

Fig. 5.1

Cased struts

The design basis set out in Cl. 4.14.3 of BS 5950 is as follows:

1) r_y should be taken as $0.2\,b_c \not> 0.2(B + 150)$ mm
   r_x should be taken as that of the steel section alone.

2) The compression resistance of the cased section is

$$P_c = \left(A_g + 0.45\frac{f_{cu}}{p_y}A_c\right)p_c$$

$$\not> \text{ the short strut capacity } P_{cs}$$

where A_c = gross sectional area of the concrete. Casing in excess of 75 mm from the overall dimensions of the steel section is neglected as is any applied finish.
A_g = gross area of steel section.
f_{cu} = characteristic strength of the concrete at 28 days, $\not>$ 40 N/mm^2.
p_c = compressive strength of the steel section determined using r_x and r_y, as defined in (a), and taking $p_y <$ 355 N/mm^2.
p_y = design strength of the steel.

$$P_{cs} = \left(A_g + 0.25\frac{f_{cu}}{p_y}A_c\right)p_y$$

Cased members subject to axial load and moment

The design of these members is set out in Cl. 4.14.4 of BS 5950. The members must satisfy two relationships:

1) For capacity:

$$\frac{F_c}{P_{cs}} + \frac{M_x}{M_{cx}} + \frac{M_y}{M_{cy}} \not> 1.0$$

2) For buckling resistance:

$$\frac{F_c}{P_c} + \frac{mM_x}{M_b} + \frac{mM_y}{M_{cy}} \not> 1.0$$

where F_c = compressive force due to axial load,
P_{cs} = short strut capacity, i.e. the compression resistance of a cased strut of zero slenderness,
M_x = applied moment about the major axis,
M_y = applied moment about the minor axis,
M_{cx} = moment capacity of the steel section about X–X axis,
M_{cy} = moment capacity of the steel section about Y–Y axis,
P_c = compression resistance,
m = equivalent uniform moment factor from Table 18, BS 5950.
M_b = buckling resistance moment obtained using r_y for a cased section.

CURPE CONSULTANTS 46 Orburn Road Dunfield	Project: Cased Steel Stanchion	Job ref
	Part of structure: Cased stanchion-axial load	Calc sheet no: C5 / 1 rev: 1
	Drawing ref: — Calc by: RAS Date: —	Check by: — Date: —

Ref BS5950	Calculations	Output

4.14.3. Ex. 1. Design of cased stanchion

Bottom length of a stanchion in a multi-storey building. Axially loaded.

Actual length 4·8 m – Base to centre of floor beams.

Table 24 — Effective lengths $L_{EX} = L_{EY} = 0{\cdot}85L$

$L_{EX} = L_{EY} = 0{\cdot}85 \times 4800 = 4080$ mm

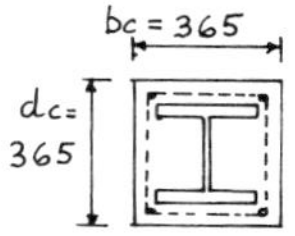

254 × 254 × 89 U.C, grade 43

Concrete grade 20.

Table 7 — Section is not slender.

Properties of steel section:

$A_g = 114\ cm^2$; $r_x = 11{\cdot}2$ cm ; $r_y = 6{\cdot}52$ cm.

4.14.3(a) Cased section:

$r_x = 11{\cdot}2$ cm

$r_y = 0{\cdot}2\,b_c = 0{\cdot}2 \times 365 = 73$ mm.

$\not> 0{\cdot}2(B + 150)$

$0{\cdot}2(255{\cdot}9 + 150) = 81{\cdot}18$ mm.

$\therefore r_y = 73$ mm.

4.14.1(i) Effective length L_E is limited to least of:

$40\,b_c = 40 \times 365 = 14600$ mm

$\dfrac{100\,b_c^2}{d_c} = \dfrac{100 \times 365^2}{365} = 36500$ mm

$250\,r = 250 \times 6{\cdot}52 \times 10 = 16300$ mm

Table 6 — Since T = 17·3 mm , $p_y = 265\ N/mm^2$

Cased section:

Slenderness $\lambda_x = \dfrac{L_{EX}}{r_x} = \dfrac{4080}{11{\cdot}2 \times 10} = 36{\cdot}4$

Table 27(b) — $\therefore p_{cx} = 245{\cdot}5\ N/mm^2$

Slenderness $\lambda_y = \dfrac{L_{EY}}{r_y} = \dfrac{4080}{73} = 55{\cdot}9$.

CURPE CONSULTANTS 46 Orburn Road Dunfield	Project: Cased Steel Stanchion	Job ref
	Part of structure: Cased stanchion-axial load	Calc sheet no: C5 / 2 rev: 1
	Drawing ref: — Calc by: R.A.S. Date: —	Check by: — Date: —

Ref	Calculations	Output

Table 27(c) — $\therefore p_{cy} = 202\ N/mm^2$

$\therefore p_c = 202\ N/mm^2$, the lesser value.

Gross sectional area of the concrete is:

$A_c = 365 \times 365 = 133225\ mm^2$

4.14.3(b) Compression resistance:

$P_c = \left(A_g + 0{\cdot}45\,\dfrac{f_{cu}}{p_y}.\,A_c\right)p_c.$

$= \left(114 \times 10^2 + \dfrac{0{\cdot}45 \times 20 \times 133225}{265}\right)\dfrac{202}{10^3}$

$= 3216{\cdot}8$ kN

$\not>$ Short strut capacity:

$P_{cs} = \left(A_g + 0{\cdot}25\,\dfrac{f_{cu}}{p_y}.\,A_c\right)p_y.$

$= \left(114 \times 10^2 + \dfrac{0{\cdot}25 \times 20 \times 133225}{265}\right)\dfrac{265}{10^3}$

$= 3687{\cdot}1$ kN.

$\therefore$ Compression resistance = 3216·8 kN — Output: Comp. resistance 3216·8 kN

Note

The effect of encasing the stanchion in concrete in accordance with the code is that the compression resistance has increased from 2166 kN to 3216·8 kN. See a previous example for reference.

CURPE CONSULTANTS 46 Orburn Road Dunfield	Project: Cased Steel Stanchion	Job ref
	Part of structure: Cased stanchion - eccentric load	Calc sheet no: C5 / 3 rev: 1
	Drawing ref: — Calc by: RAS Date: —	Check by: — Date: —

Ref BS5950	Calculations	Output
4.14.4.	Ex. 2. Design of cased stanchion	
	Bottom length of a stanchion in a multi-storey building. Eccentric loading	
	$F_c = 3000kN$, $M_x = 20kNm$, $M_y = 15kNm$ } Design load and Moments.	
	Length $L = 4.9$ m	
	Effective length $L_{EX} = L_{EY} = 0.85L = 4.2$ m.	
	Section 305×305×97 UC. grade 43 steel. Semi-compact section.	
	$b_c = 410$; $d_c = 410$ 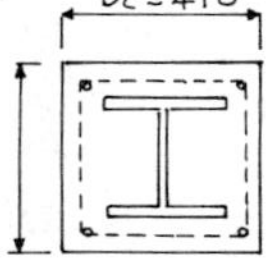	
	Concrete grade 20	
	Properties of steel section:	
	$A_g = 123\ cm^2$; $r_x = 13.4cm$; $r_y = 7.68cm$.	
	$Z_x = 1440\ cm^3$; $Z_y = 477\ cm^3$; $S_x = 1590\ cm^3$	
Table 6	For T = 15.4 mm, $p_y = 275\ N/mm^2$	
4.14.4(a)	(a) Capacity check:	
	$P_{cs} = \left(A_g + 0.25\frac{f_{cu}}{p_y}.A_c\right)p_y$	
	$= \left(123\times10^2 + \frac{0.25\times20\times410\times410}{275}\right)\frac{275}{10^3} = 4223\ kN$	
	$M_{cx} = p_y.Z_x = \frac{275\times1440\times10^3}{10^6} = 396\ kNm$	
	$M_{cy} = p_y.Z_y = \frac{275\times477\times10^3}{10^6} = 131\ kNm$.	
	$\frac{F_c}{P_{cs}} + \frac{M_x}{M_{cx}} + \frac{M_y}{M_{cy}}$	
	$= \frac{3000}{4223} + \frac{20}{396} + \frac{15}{131} = 0.875 < 1$ Satisfactory	Capacity sat.

CURPE CONSULTANTS 46 Orburn Road Dunfield	Project: Cased Steel Stanchion	Job ref
	Part of structure: Cased Stanchion - eccentric load	Calc sheet no: C5 / 4 rev: 1
	Drawing ref: — Calc by: R.A.S. Date: —	Check by: — Date: —

Ref BS5950	Calculations	Output
4.14.4(b)	(b) Buckling resistance	
	Cased section $r_y = 0.2b_c = 0.2\times410 = 82mm$	
	Slenderness:	
Table 27(b)	$\lambda_x = \frac{4200}{13.4\times10} = 31.34$ $\therefore p_{cx} = 261\ N/mm^2$	
Table 27(c)	$\lambda_y = \frac{4200}{82} = 51.2$ $\therefore p_{cy} = 218.5\ N/mm^2$	
	$\therefore p_c = 218.5\ N/mm^2$	
4.14.3(b)	Compression resistance:	
	$P_c = \left(A_g + \frac{0.45f_{cu}}{p_y}A_c\right)p_c$	
	$= \left(123\times10^2 + \frac{0.45\times20\times410\times410}{275}\right)\frac{218.5}{10^3}$	
	$= 3889.6\ kN \not> P_{cs}$	
4.14.4(b)	$\frac{F_c}{P_c} + \frac{m\,M_x}{M_b} + \frac{m\,M_y}{M_{cy}} \not> 1.0$	
	To find M_b	
4.7.7	Uncased $\lambda_{LT} = \frac{0.5L}{r_y} = \frac{0.5\times4.9\times10^3}{7.68\times10} = 31.9$	
	$\therefore p_b = 274\ N/mm^2$	
	Cased $\lambda_{LT} = \frac{0.5L}{r_y} = \frac{0.5\times4.9\times10^3}{82} = 29.8$	
	M_b for cased section $\not> 1\frac{1}{2} M_b$ uncased section	
4.3.7.3	$M_b = S_x\,p_b$	
	$= \frac{1590\times10^3\times275}{10^6} = 437\ kNm \not> M_{cx}$	
4.14.4(b)	$\frac{3000}{3889.6} + \frac{20}{396} + \frac{15}{131} = 0.9362 < 1.0$ Satisfactory	USE 305×305×97 UC grade 43 struct. cased

CURPE CONSULTANTS
46 Orburn Road
Dunfield

Project: Cased Steel Beam | Job ref:
Part of structure: Unrestrained cased beam | Calc sheet no: C5 / 5 | rev: /
Drawing ref: | Calc by: R.A.S. | Date: | Check by: | Date:

Ref BS5950	Calculations	Output
	Ex. 3. **Cased beam restrained at ends only by top & bottom cleats**	
	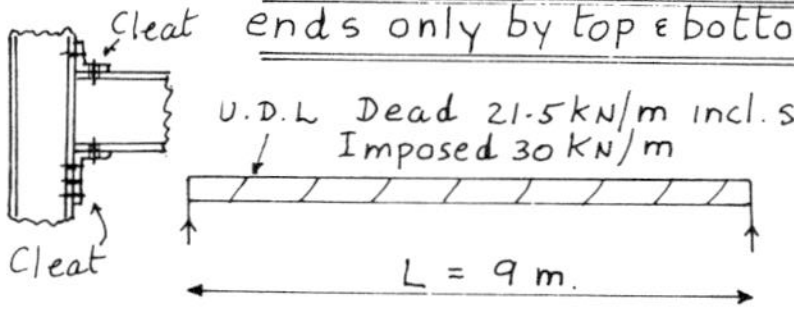	
Table 2	Factored loads: $1{\cdot}4 \times 21{\cdot}5 = 30{\cdot}1$ kN/m	
	$1{\cdot}6 \times 30 = 48{\cdot}0$ kN/m	
	$78{\cdot}1$ kN/m	
	Max. bending moment $= \dfrac{78{\cdot}1 \times 9^2}{8} = 790{\cdot}76$ kNm	
	Try uncased section 610×305×149 UB, grade 43 steel for lateral torsional buckling.	
Table 6	Design strength $p_y = 265$ N/mm²	
Table 7	Section is compact	
4.3.7.1.	$\bar{M} \not> M_b$	
4.3.7.2.	$\bar{M} = m.M_A$ where M_A is max moment	
4.3.7.6	Beam is loaded between lateral restraints	
Table 13	$\therefore m = 1{\cdot}0$	
	Then $\bar{M} = 1{\cdot}0 \times 790{\cdot}76 = 790{\cdot}76$ kN m	
4.3.7.3	$M_b = S_x.p_b$.	
4.3.7.5	$\lambda_{LT} = n u v \lambda$	
4.3.5(a)	$\lambda = \dfrac{L_E}{r_y} = \dfrac{0{\cdot}85 \times 9 \times 10^3}{6{\cdot}99 \times 10} = 109{\cdot}4$	
	$X = 32{\cdot}5$	
Table 14	$\dfrac{\lambda}{X} = \dfrac{109{\cdot}4}{32{\cdot}5} = 3{\cdot}36 \quad \therefore v = 0{\cdot}9$	
	$u = 0{\cdot}886$	

CURPE CONSULTANTS
46 Orburn Road
Dunfield

Project: Cased Steel Beam | Job ref:
Part of structure: Unrestrained cased beam | Calc sheet no: C5 / 6 | rev: /
Drawing ref: | Calc by: R.A.S. | Date: | Check by: | Date:

Ref BS5950	Calculations	Output
4.3.7.6	To find n the slenderness correction factor	
	$\gamma = \dfrac{M}{M_o} = \dfrac{0}{790{\cdot}76} = 0$	
	β = ratio of end moments = 0	
Table 16	$\therefore n = 0{\cdot}94$	
	$\lambda_{LT} = 0{\cdot}94 \times 0{\cdot}886 \times 0{\cdot}9 \times 109{\cdot}4 = 82$	
Table 11	$p_b = 157$ N/mm²	
	$M_b = S_x.p_b = \dfrac{4570 \times 10^3 \times 157}{10^6} = 717$ kNm	Uncased section inadequate for lateral torsional buckling
	Since $790{\cdot}76 > 717$ Inadequate	
4.14.1. 4.14.2.	Try same section cased	
4.3.5.(a)	$L_E = 0{\cdot}85 \times 9 \times 10^3 = 7650$ mm	
4.14.1.(i)	Check L_E for:	
	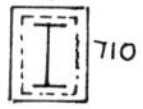 $40\,b_c = 40 \times 405 = 16200$ mm	
	$\dfrac{100\,b_c^2}{d_c} = \dfrac{100 \times 405^2}{710} = 23102$ mm	
	$250\,r = 250 \times 6{\cdot}99 \times 10 = 17475$ mm	
4.14.2	r_y cased section $= 0{\cdot}2(B + 100)$ mm	
	$= 0{\cdot}2(304{\cdot}8 + 100) = 80{\cdot}96$ mm	
4.3.7.3	$M_b = S_x.p_b$.	
4.3.7.5	$\lambda_{LT} = n u v \lambda \qquad \lambda = \dfrac{L_E}{r_y} = \dfrac{7650}{80{\cdot}96} = 94{\cdot}49$	
Table 14	$\dfrac{\lambda}{X} = \dfrac{94{\cdot}49}{32{\cdot}5} = 2{\cdot}91 \qquad \therefore v = 0{\cdot}91$	
4.3.7.6	To find n	
	$\gamma = \dfrac{M}{M_o} = \dfrac{0}{790{\cdot}96} = 0$	
Table 16	$\beta = 0 \qquad \therefore n = 0{\cdot}94$	
	$\lambda_{LT} = 0{\cdot}94 \times 0{\cdot}886 \times 0{\cdot}91 \times 94{\cdot}49 = 71{\cdot}6$	
Table 11	$p_b = 180{\cdot}4$ N/mm²	
	$M_b = S_x.p_b = \dfrac{4570 \times 10^3 \times 180{\cdot}4}{10^6} = 824{\cdot}4$ kNm	

CURPE CONSULTANTS 46 Orburn Road Dunfield	Project: Cased Steel Beam			Job ref	
	Part of structure: Unrestrained cased beam			Calc sheet no: CS / 7	rev: 1
	Drawing ref	Calc by: RAS	Date	Check by	Date

Ref	Calculations	Output
4.14.2	M_b for cased section $\not> 1\frac{1}{2}$ M_b uncased section $824.4 \not> 1\frac{1}{2} \times 717$ $\therefore M_b = 824.4$ kNm $824.4 > 790.76$ satisfactory	Cased section adequate for lateral torsional buckling.
4.14.2.	<u>Check cased section for deflection</u> $I_s = 125000\ \text{cm}^4$ $I_c = \frac{40.5 \times 71^3}{12} = 1207949.6\ \text{cm}^4$ $I_{cs} = I_s + \frac{I_c - I_s}{\alpha_e}$ $= 125000 + \left(\frac{1207949.6 - 125000}{15}\right)$ $= 125000 + 72196.6 = 197196.6\ \text{cm}^4$	
Table 5	Allowable deflection $= \frac{\text{span}}{360}$ brittle finishes $= \frac{9 \times 10^3}{360} = 25$ mm Actual deflection due to unfactored imposed load $= \frac{5}{384} \times \frac{270 \times 10^3 \times 9^3 \times 10^9}{20.5 \times 10^4 \times 197196.6 \times 10^4}$ $= 6.33$ mm Since 6.33 mm < 25 mm satisfactory	Cased section deflection satisfactory
	The beam would also be checked for: shear max. moment & co-existent shear max. shear & co-existent moment web buckling at end supports web bearing " " "	All sat.

6

Steelwork connections

General requirements

In structural steelwork, connections join individual members together to form a structural framework. Bolts and welds in conjunction with fabricated angle cleats, plates, T-sections and other section cuttings are used to make a variety of connections.

Whenever possible, members meeting at a joint should be arranged with their centroidal axes meeting at a point. Where eccentricity occurs at intersections the members and connections should be designed to accommodate resulting moments. In the case of bolted frames of angles and tees the setting-out lines of the bolts may be used instead of the centroidal axis.

Joints

1) Joints in simple construction should not develop significant moments adversely affecting members.
2) Joints between members in rigid construction should be designed to transmit the calculated forces and moments.
3) Where a connection is subject to impact, vibration, or load reversal, welding or HSFG bolts should be used.

Splices

1) Splices should be designed to hold the connected members in place and where possible the members, arranged so that the centroidal axis of the splice coincides with the centroidal axis of the spliced members. Any eccentricity should be catered for.
2) Splices in compression members may be prepared for full contact in bearing by machining butt ends. The splice should be designed to provide continuity of stiffness and resist any tension where bending is present.
3) Where members are not prepared for full contact in bearing, the splice should be capable of transmitting all the moments and forces at that position.
4) Splices in tension members should be capable of transmitting all moments and forces present at that position.
5) Beam splices should be capable of transmitting all moments and forces present at that position and have adequate stiffness.

Bolts

Bolts are classified as Ordinary bolts or Friction Grip bolts.

Ordinary bolts

Available in 4–6 and 8–8 grades and used in 2 mm clearance holes, they are employed normally for general-purpose connections in simple design.

Effective areas

If the threaded portion of the bolt occurs in a shear plane, then the area A_s for resisting shear should be taken as A_t the tensile stress area taken normally as the area at the bottom of the threads.

Where threads do not occur in a shear plane, A_s may be taken as the shank area A. The area A_t is used in all examples for calculating bolt capacities in shear and tension.

Shear capacity
Where no reductions are required for long joints on large grip lengths, the shear capacity P_s is taken as:

$$P_s = p_s A_s$$

where p_s = shear strength, Table 32 of BS 5950,
A_s = shear area A or A_t.

Bearing capacity
The capacity should be taken as the lesser of the bearing capacity of the bolt and the bearing capacity of the connected ply. The bearing capacity of the bolt is taken as:

$$P_{bb} = dtp_{bb}$$

where d = nominal diameter of bolt,
t = thickness of the connected ply,
p_{bb} = bearing strength of bolt, Table 32 of BS 5950.

The bearing capacity P_{bs} of the connected ply is taken as:

$$P_{bs} = dtp_{bs} \leqslant \tfrac{1}{2}etp_{bs}$$

where p_{bs} = bearing strength of connected parts, Table 33 of BS 5950,
d = nominal diameter of bolt,
e = end distance (i.e. edge distance in the direction in which the bolt bears),
t = thickness of ply.

Tension capacity
The tension capacity is taken as:

$$P_t = p_t A_t$$

where p_t = tension strength, Table 32 of BS 5950,
A_t = tensile strength area, i.e. area at the bottom of the threads.

In connections subject to tension, prying action need not be taken into account provided the strengths given in Table 32 of BS 5950 are used.

Combined shear and tension
The following relationship should be satisfied:

$$\frac{F_s}{P_s} + \frac{F_t}{P_t} \leqslant 1.4$$

where F_s = applied shear,
F_t = applied tension,
P_s = shear capacity,
P_t = tension capacity.

Strengths of bolts in clearance holes
Bolts Grade 4.6 p_s = 160 N/mm^2
P_{bb} = 435 N/mm^2
P_t = 195 N/mm^2

Bearing strengths of connected parts
For Grade 43 Steel p_{bs} = 460 N/mm^2

Friction grip bolts

HSFG bolts are tightened by a predetermined shank tension in order that the clamping force provided will transfer loads in the connected members by friction between the parts and not by shear in, or bearing on, the bolts or plies of connected parts. They are suitable when the connection is subject to impact, vibration, load reversal or where large moments occur in a plane or member. Parallel-shank friction grip bolts only are considered here and are used in clearance holes.

Where no reductions are required for long joints, then:

Slip resistance P_{sL}

$$P_{sL} = 1.1 K_s \mu P_0$$

where K_s = 1.0 for bolts in clearance holes,
μ = slip factor taken as 0.45,
P_0 = minimum shank tension, e.g. 144 kN for 20 mm diameter bolts.

Bearing resistance P_{bg}

$$P_{bg} = dtp_{bg} \leqslant \tfrac{1}{3}etp_{bg}$$

where d = diameter of bolt,
e = end distance (i.e. the edge distance in the direction in which the bolt bears),
t = thickness of connected ply,
p_{bg} = bearing strength of parts connected from Table 34 of BS 5950, e.g. 825 N/mm^2 for grade 43 steel.

Tension capacity P_t

$$P_t = 0.9P_0$$

Combined shear and tension
The following relationship should be satisfied:

$$\frac{F_s}{P_{sL}} + \frac{0.8F_t}{P_t} \leqslant 1.0$$

where F_s = applied shear,
F_t = external applied tension.

Welds

Welding produces fusion by the application of heat from an electric arc struck between an electrode supplying filler metal and the material being joined, the base or parent metal. Welding offers some advantages over bolting:

1) Elimination to a certain extent of connecting angle cleats, gusset plates and splice plates.
2) A high degree of rigidity is achieved in a framework.
3) It enables smooth lines to be obtained with the absence of protruding bolt heads and nuts and is more aesthetically pleasing to the eye.
4) Effective area of tension members is not reduced due to the presence of bolt holes.
5) Full strength joints can readily be achieved.

Welding, however, has limitations, e.g. quality of welds and their inspection, distortion during welding, fatigue failures and brittle fractures. Site welding is costly and site connections are normally made using HSFG bolts.

The two basic types of weld in structural steelwork are butt welds and fillet welds.

Butt welds

Butt welds are classified according to the shape into which the plates or other structural sections are formed when the edges are prepared for welding, e.g. V, U, J, or bevel butt welds. They are treated as parent metal with a thickness equal to the throat thickness or a reduced throat thickness. In order that the weld may develop the full strength of the members joined, it is necessary that complete penetration of weld metal into the joint is achieved. No design calculations are generally necessary for complete penetration butt welds as they usually provide a strength at the joint equivalent at least to that of the part or member itself.

For full penetration butt weld:

Shear capacity for 6 mm thickness

$$= 0.6p_yt = \frac{0.6 \times 275 \times 6}{10^3} = 0.99 \text{ kN/mm}$$

Tension or compressive capacity for 6 mm

$$\text{thickness} = p_yt = \frac{275 \times 6}{10^3} = 1.65 \text{ kN/mm}$$

Fillet welds

These are roughly triangular in cross-section and are used to connect overlapping or intersecting plates or members. The size of the weld is specified by its leg length. The throat thickness is taken as 0.7 times the leg length for a 90° angle between joint faces. The effective length is given as the actual length less twice the leg length for an open-ended fillet weld, allowing for both end craters.

The design strength p_w of a fillet weld made using specified electrodes for grade 43 steel is taken as 215 N/mm^2.

$$\text{Capacity of weld} = \frac{0.7 \text{ leg length} \times P_w}{10^3} \text{ kN/mm}$$

For a 6 mm fillet weld, capacity

$$= \frac{0.7 \times 6 \times 215}{10^3} = 0.903 \text{ kN/mm}$$

Eccentric bracket connections

1) Bolted connections

When a load P from a beam is applied on a line of action which does not pass through the centroid of a bolt group, the result is an eccentric loading effect. Brackets are fabricated from angles, section cuttings, plates, etc. and are extended to support the beam. Connections may be subjected to shear and torsion or shear and bending.

Shear and torsion

Bolt groups may be analysed by using the traditional elastic method which is a generally conservative approach.

The load P applied at an eccentricity e, produces two components of shear stress at any bolt in the group. These are:

1) Stress due to direct shear.
2) Stress due to moment.

These stresses are added vectorially to give the maximum resultant shear stress at a point. The maximum resultant is found at the critical bolt X in the group. An adequate bolt can then be selected.

Considering direct shear, then for one bolt:

$$\text{Stress} = \frac{P}{NA}$$

where N = the number of bolts in the group,
A = the cross-sectional area of the bolt.

Since force = stress × area, then for one bolt the direct shear force F_s is given by

$$F_s = \frac{P}{NA} \times A = \frac{P}{N}$$

Considering the torsional moment, at any one bolt:

$$\text{Stress } f = \frac{Mr}{\varepsilon A r^2}$$

where M = torsional moment = Pe,
r = the radial distance from the centroid to the bolt,
A = tensile stress area of the bolt.

Since force = stress × area, then force $F_M = fA$

$$= \frac{MrA}{\varepsilon A r^2} = \frac{Mr}{\varepsilon r^2}$$

In general terms, $F_M = \dfrac{Mr}{\varepsilon r^2} = \dfrac{Per}{\varepsilon r^2}$.

The resultant force F_R is found by adding vectorially the direct shear force F_s and the shear force F_M due to the moment. The resultant force is given by:

$$F_R = \sqrt{(F_s^2 + F_M^2 + 2F_sF_M \cos\theta)}$$

This formula is derived using Pythagoras' theorem:

$$F_R = \sqrt{[(F_M \sin\theta)^2 + (F_s + F_M \cos\theta)^2]}$$

$$= \sqrt{(F_M^2 \sin^2\theta + F_s^2 + 2F_sF_M \cos\theta + F_M^2 \cos^2\theta)}$$

$$= \sqrt{[F_M^2(\sin^2\theta + \cos^2\theta) + F_s^2 + 2F_sF_M \cos\theta]}$$

Since $\sin^2\theta + \cos^2\theta = 1$, then

$$F_R = \sqrt{(F_s^2 + F_M^2 + 2F_sF_M \cos\theta)}$$

As an alternative, F_R can be solved graphically.

Shear and bending

Two methods of analysis may be used in this type of connection:

Method 1. This method is conservative, with the centre of rotation assumed at the bottom line of bolts with no compression taken by the contact surface area between the bracket and the face of the stanchion. The load P produces two types of stress at any bolt in the group. These are:

1) Stress due to direct shear,
2) Tensile stress due to bending.

Considering direct shear, then for one bolt:

$$\text{Stress} = \frac{P}{NA}$$

For one bolt, the direct shear force is given by

$$F_s = \frac{P}{NA} \times A = \frac{P}{N}$$

where N = number of bolts in the group,
A = tensile stress area of the bolt.

Considering bending, the tensile stress at any bolt is given by:

$$\frac{MY}{I} = \frac{PeY}{I}$$

where Y = vertical distance from the centre of rotation to the bolt,
I = second moment of area of the bolt group about the axis of rotation,
= $2\Sigma AY^2$ for two vertical columns of bolts.

Since force = stress × area, then for one bolt, the tensile force F_t due to bending is:

$$F_t = \frac{PeYA}{2\Sigma AY^2}$$

$$= \frac{PeY}{2\Sigma Y^2}$$

The bolt is designed for the combined forces F_s and F_T with the top bolts being critical.

Method 2. In this method, the load P produces:

1) stress due to direct shear in each bolt,
2) tensile stress in bolts located above the neutral axis,
3) compression on part of the contact surface area between the bracket and face of the stanchion.

This method is not adopted in the given example.

2) Welded connections

Connections may be subjected to shear and torsion or shear and bending.

Shear and torsion

The load P applied at an eccentricity e, produces two components of shear stress at any point in the welds. These are:

1) stress due to direct shear,
2) stress due to torsional moment.

These shear stresses are added vectorially to give the resultant shear stress at a point. The maximum resultant is found and an adequate weld size selected.

Considering direct shear, the shear force F_s is given by:

$$F_s = \frac{P}{\text{length of weld}}$$

Shear force F_M due to torsional moment is given by:

$$F_M = \frac{Per}{I_p}$$

where I_p = polar moment of inertia of the weld group,
r = distance from centroid of weld group (assumed centre of rotation) to the critical point in the weld.

The resultant shear on a unit length of weld at the critical point, is given by:

$$F_R = \sqrt{(F_s^2 + F_M^2 + 2F_sF_M \cos\theta)}$$

Shear and bending

Two methods of analysis may be used in this type of connection:

Method 1. The eccentric load P produces two components of stress at any point in the weld. These are:

1) stress due to direct shear,
2) stress due to bending, considered for design as shear stress.

The stresses are added vectorially to give a resultant stress at a point. The maximum resultant is found and an adequate size of weld selected.

Considering direct shear, then shear force F_s is given by:

$$F_s = \frac{P}{\text{length of weld}}$$

Considering bending, then the force F_M due to bending is given by:

$$F_M = \frac{MY}{I} = \frac{PeY}{I}$$

Forces F_s and F_M act at right angles one to the other and resultant force F_R is found by using Pythagoras' theorem, or graphically.

Calculations in the example given assume that the neutral axis (N.A.) alternatively occurs

a) at the horizontal level of the centroid of the welds,
b) at a distance of one sixth the depth of the welds measured from the bottom edge.

Method 2. In this method it is assumed that the top region flange weld resists the bending moment rotating about the bottom of the bracket and the vertical shear is resisted by the bottom region flange weld. This method is not adopted in the given example.

Single angle welded connection
The angle is connected by welding along two sides or alternatively by continuous welding along two sides and one end. If welding is carried out along two sides then different lengths of welds are determined to ensure balancing of forces about the centroidal axis of the angle.

Beam connections to stanchions and beams

Simply supported beams may be connected to stanchions and other beams by means of:

1) Flexible end plates.
2) Web cleats.
3) Top and bottom (seating) cleats.

Flexible end plates
These are normally 8 mm or 10 mm thick plates depending on the beam section size suggesting restriction of development of beam and fixity. The depth of the plates may vary from a maximum of d, the depth between beam fillets, to a minimum of 0.6 times the overall depth of the beam. The upper edge of the plate is better positioned adjacent to the compression flange and is shop-welded to the supported beam and bolted to the stanchion or supporting beam.

Web cleats
Angle web cleats, shop-bolted to both sides of the beam at the ends are a common beam-to-stanchion or beam-to-beam connection. The bolts connecting the cleats to the web of the beam are subjected to eccentric loading.

Top and bottom (seating) cleats
Bolts in the vertical leg of the seating cleat resist the shear at the end of the beam. The top cleats or alternatively web cleats positioned adjacent to the top flange of the beam hold the beam in place.

Buckling and bearing of the web require investigation at this position.

Beam splices

Steel beams are rolled at the mill to limited lengths depending on their cross-sectional dimensions. Where long-span beams are required outside the range of normal available lengths, two convenient lengths may be joined or spliced together to form one long beam. Splices in beams ease the problems associated with handling and fabrication in the shop, and transportation of the steelwork to the site. They are generally classified either as shop or site splices. Shop splices are made during the fabrication of the beam in the shop. Site splices are necessary where the beam is too long to be transported in one piece from the shop to the site. Splices are located preferably away from positions of high bending moment and shear.

Splices may be made of bolted or welded construction, see Fig. 6.1(a). A butt-welded splice has the advantage of eliminating plates and protruding bolts and therefore has a more pleasing appearance. Site welding however, is more difficult and expensive than shop welding. For bolted construction, plates are connected to flanges and web by HSFG bolts rather than by ordinary bolts, since any slip produces additional deflection in

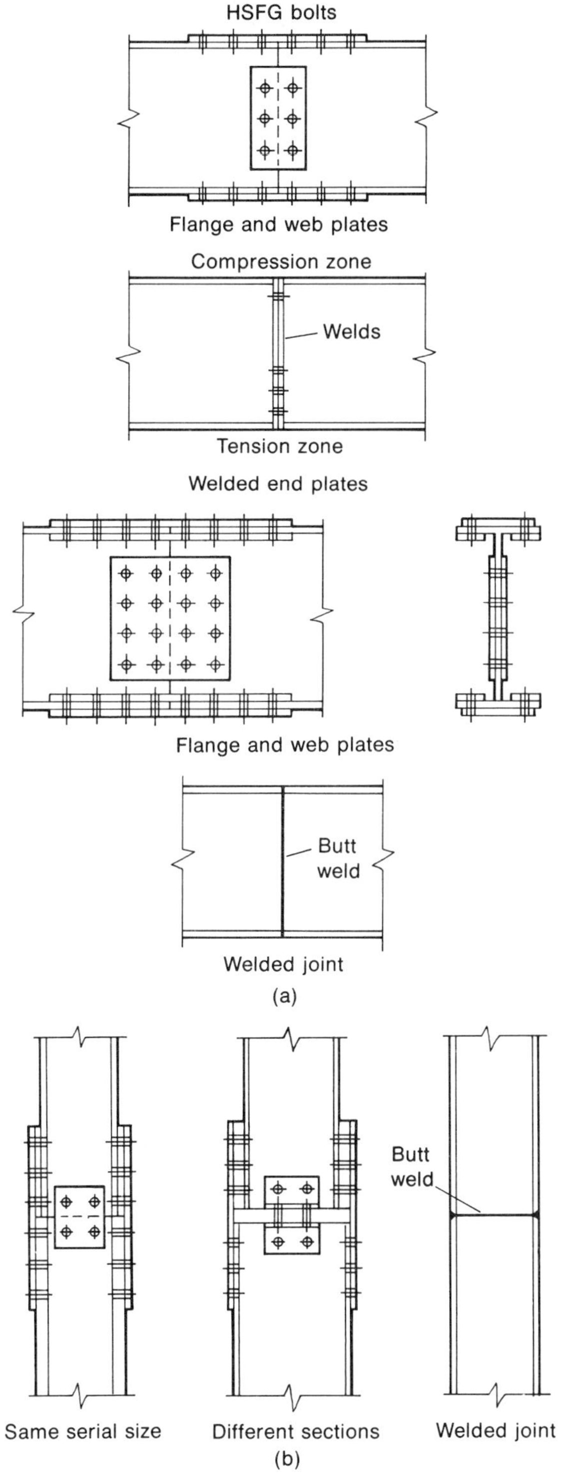

Fig. 6.1 *(a) Beam splices, (b) stanchion splices*

the beam. Tests have proved that despite the handicap of holes in the beam at the splice position, collapse takes place outside the connection area. This is because the formation of a plastic hinge entails buckling of the compression flange and at the joint this flange is stiffened by the splice plate.

Current design of beam splices assumes that the web transmits the shear force and the flange splices resist the moment. These assumptions are reasonable and hence the analysis and design of a beam splice can be divided into:

a) the resistance of the flange splice to the applied moments,
b) the shear resistance of the web splice.

Flange splice
For single flange plates, the bolts in the flanges must resist the force given by:

$$\frac{M}{D}$$

where M = applied moment,
D = overall depth of beam.

A single flange splice plate on each flange is normally sufficient with bolts in single shear. For the deeper section beams, splice plates may be used on both sides of the flanges, thus providing a double shear connection with less bolts and also reducing the thickness of the plates.

Web splice
The bolts should be designed for the eccentric vertical shear force at the spliced section. Normally two plates, one on either side of the web, are recommended. The bolts are therefore in a double shear condition and this reduces the number of bolts necessary and thus the eccentricity.

Stanchion splices

For moderate-height, single-storey or two-storey buildings, stanchions may be made in one length. In multi-storey structures, however, it is impossible to obtain stanchion sections of the full height required. In addition, the difficulty of handling, transporting and erection of these members would prohibit their use.

To minimize these problems and for economy, such stanchions are made in lengths, joined or spliced together on the site. Splices are generally made by bolting or alternatively by welding, this being a more difficult and expensive method. Splices are located normally at 500 mm above floor level ideally adjacent to horizontal constraints provided by the incoming connected floor beams, and where the effects of column buckling are insignificant. This location also leads to easy access for the installation of the connecting site bolts or for the performance of the welding operations, as appropriate (see Fig. 6.1(b)).

The splice must carry the axial load and moment that occurs in the stanchion at the point to be spliced. If the number of bolts or weld dimensions are too great to be acceptable, transfer of load may be made by direct bearing. The ends of the stanchions and any end plates at the splice may be accurately machined square with the axis, so that the load is transferred uniformly by direct bearing over the entire surfaces in contact. In this case splices are usually proportioned from past experience. The splice material and bolts or welds should be sufficient to hold the connected stanchions accurately in place, and to resist any tension resulting from the presence of large bending moments and relatively small axial loads.

If the ends of the stanchions and any end plates are not machined so as to ensure direct bearing over the entire surfaces in contact, then sufficient splicing material and bolts or welds must be provided to transmit all the forces present at the splice.

Stanchion bases

Bases to steel stanchions are designed to transmit all the applied forces and moments to their foundations. The nominal bearing pressure between the baseplate and a concrete foundation determined on the basis of linear distribution should not exceed $0.4 f_{cu}$, where f_{cu} = characteristic concrete cube strength at 28 days.

Baseplates may be designed either by an empirical method given in 4.13.2 of BS 5950 or by other rational means. Two of the most common types are 'slab' and 'gusseted' bases.

Slab bases

With large axial loads, the end of stanchion and the top of the baseplate may be machined to achieve full contact area, and the compression transmitted to the baseplate in direct bearing. Bolts or welds are provided to retain the parts securely in place and to resist any shear or tension developed at the connection. In the case of relatively lighter loads, it may not be necessary to machine the parts, and bolts or welds should be provided to resist all forces and moments to which the base is subjected.

Gusseted plates

These built-up bases consist of a baseplate, gusset plates and stiffeners generally connected together by fillet welds. The contact surfaces between the bottom of the stanchion, gussets and the top of the baseplate may be machined to achieve full contact area, and the compression transmitted to the baseplate in direct bearing. Bolts or welds are provided to transmit any shear or tension developed at the connection. Where the machining is not carried out, the bolts or welds should be provided to transmit all forces and moments to which the base is subjected.

Empirical design of baseplates

Where a rectangular plate is concentrically loaded by I, H, channel, box or RHS, its minimum thickness should be:

$$t = \left[\frac{2.5}{p_{yp}} w(a^2 - 0.3b^2)\right]^{\frac{1}{2}} \nless T$$

where a = greater projection of the plate beyond the column,
b = lesser projection of the plate beyond the column,

w = pressure on the underside of the plate assuming uniform distribution,

p_{yp} = design strength of the plate (from 3.1.1 or Table 6 of BS 5950), but not greater than 270 N/mm^2.

This formula can be derived taking into account that bending occurs in two directions mutually perpendicular to one another, and making allowance for this effect, by taking Poisson's Ratio as 0.3.

Consider point C and two cantilever strips 1 mm wide.

$$\text{Load on horizontal strip} = \text{pressure} \times \text{area}$$

$$= wa1$$

$$M_x \text{ at C} = \text{load} \times \text{lever arm}$$

$$= (wa1)\left(\frac{a}{2}\right)$$

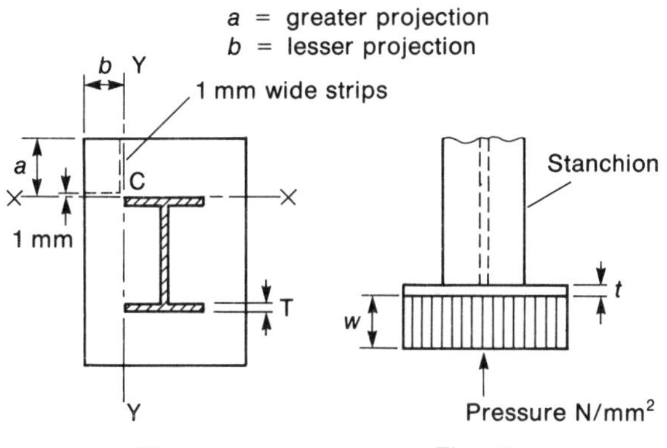

Plan Elevation

$$= \frac{wa^2}{2}$$

$$\text{Similarly } M_y \text{ at C} = \frac{wb^2}{2}$$

With a the greater projection, the net moment is

$$M = \frac{wa^2}{2} - (0.3)\frac{wb^2}{2}$$

$$= \frac{w}{2}(a^2 - 0.3b^2)$$

M_r of strip section $= 1.2\,p_{yp}\,z$, where $z = \dfrac{1 \times t^2}{6}$, the elastic modulus of the baseplate,

$$= 1.2\,p_{yp}\left(\frac{1 \times t^2}{6}\right)$$

$$\text{Equating } \frac{w}{2}(a^2 - 0.3b^2) = 1.2\,p_{yp}\left(\frac{t^2}{6}\right)$$

$$\frac{w}{2}(a^2 - 0.3b^2) = 1.2\,p_{yp}\frac{t^2}{6}$$

$$t^2 = \frac{w \times 6}{2 \times 1.2\,p_{yp}}(a^2 - 0.3b^2)$$

$$= \frac{2.5}{p_{yp}}w(a^2 - 0.3b^2)$$

$$t = \left[\frac{2.5}{p_{yp}}w(a^2 - 0.3b^2)\right]^{\frac{1}{2}}$$

CURPE CONSULTANTS 46 Orburn Road Dunfield	Project: Beam Splices		Job ref
	Part of structure: Beam splice design-bolted		Calc sheet no: C6 / 1, rev: /
	Drawing ref:	Calc by: R.A.S. Date:	Check by: Date:

Ref	Calculations	Output
BS5950		

Ex. 1. **Beam Splice**

Applied moment = 280 kNm

Applied shear = 300 kN.

Beam section 533 x 210 x 82 U.B.

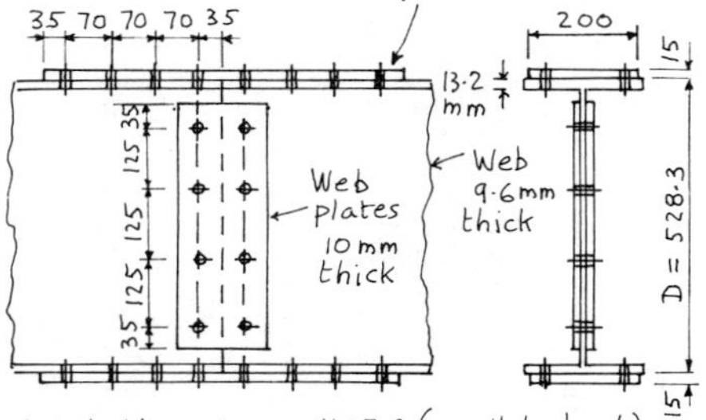

All bolts. M 20. H.S.F.G. (parallel shank)

Basic assumptions:

(i) The flange cover plates resist the bending moment

(ii) The web splice resists the vertical shear plus the torsional moment induced by the eccentricity of this shear on the bolt groups on each side of the joint.

Bolts in flange plates:

Force in bolts each side $= \frac{M}{D} = \frac{280 \times 10^3}{528.3} = 530\,kN$

CURPE CONSULTANTS 46 Orburn Road Dunfield	Project: Beam Splices		Job ref
	Part of structure: Beam splice design-bolted		Calc sheet no: C6 / 2, rev: /
	Drawing ref:	Calc by: R.A.S. Date:	Check by: Date:

Ref	Calculations	Output
BS5950		

Bolts are in single shear

Lever arm = D

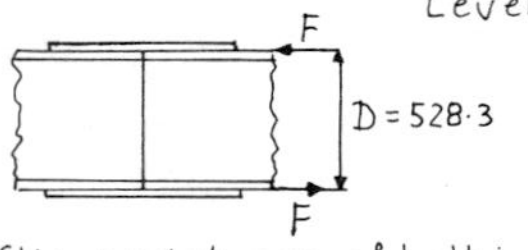

CL. 6.4.2.1 Slip resistance of bolt in single shear

$P_{SL} = 1.1\, K_s\, \mu . P_o$

$= 1.1 \times 1.0 \times 0.45 \times 144 = 71.3$ kN.

CL. 6.4.2.2 Bearing capacity can be found to be greater from $d.t.P_{bg} \not> \frac{1}{3} e.t.p_{bg}$ $e = 35$

∴ Number of bolts reqd $= \frac{530}{71.3} = 7.43$

8 bolts satisfactory — Output: Flange bolts sat.

Flange cover plates

Bottom plate in tension

Gross area $A_g = 200 \times 15 = 3000\,mm^2$

Net area $A_{net} = 3000 - (2 \times 22 \times 15) = 2340\,mm^2$

Tensile force $F_t = \frac{M}{\text{Lever arm}} = \frac{280 \times 10^3}{528.3 + 15} = 515.3\,kN$

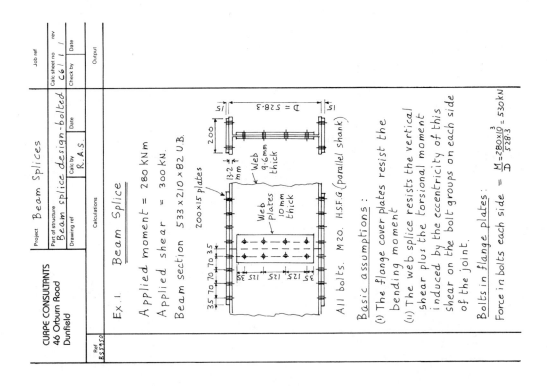

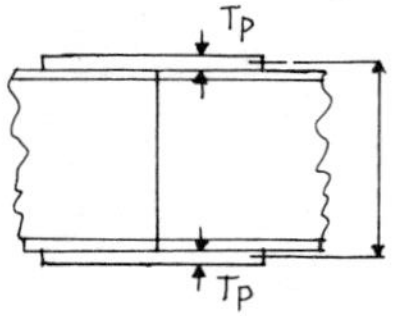

Lever arm $= D + \frac{T_p}{2} + \frac{T_p}{2}$

$= 528.3 + \frac{15}{2} + \frac{15}{2}$

Tensile capacity of plate $P_t = A_e . p_y$

Cl. 3.3.3. $A_e = K_e . A_{net} \not> A_g$.

$= 1.2 \times 2340 = 2808\,mm^2 < 3000\,mm^2$

$P_t = \frac{2808 \times 275}{10^3} = 772.2\,mm^2$

CURPE CONSULTANTS 46 Orburn Road Dunfield	Project: Beam Splices.	Job ref
	Part of structure: Beam splice design-bolted	Calc sheet no: C6 / 3 rev /
	Drawing ref; Calc by: R.A.S.; Date	Check by; Date

Ref	Calculations	Output
BS5950		

Since $F_t < P_t$ $\quad 515.3 < 772.2 \quad$ O.K.

Cover plate sat. in tension.

Top cover plate in compression is also adequate with lateral support provided. — *Flange cover plates sat.*

Bolts in web splice (i) web of beam.

Vertical shear $F_v = 300$ kN.

$\therefore$ Torsional moment $= F_v . e_1$
$= 300 \times 35 = 10500$ kNmm

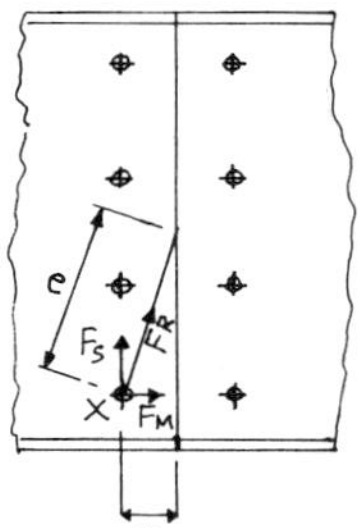

Maximum resultant force F_R occurs at top and bottom rows of bolts

Resultant force F_R on one bolt is the vectorial sum of:

F_s due to vertical shear;

F_M due to torsional moment.

Consider bolt x

$$F_s = \frac{F_v}{N} = \frac{300}{4} = 75 \text{ kN}$$

$$F_M = \frac{\text{Torsional moment} . Y . A_b}{I_b \text{ of one vert. column of bolts}}$$

$I_b = \sum A_b . Y^2 \qquad A_b$ = area of one bolt

CURPE CONSULTANTS 46 Orburn Road Dunfield	Project: Beam Splices	Job ref
	Part of structure: Beam splice design-bolted	Calc sheet no: C6 / 4 rev /
	Drawing ref; Calc by: R.A.S.; Date	Check by; Date

Ref	Calculations	Output
BS5950		

$$I_b = 2 . A_b . [Y_1^2 + Y_2^2]$$
$$= 2 . A_b . [62.5^2 + 187.5^2]$$

$$F_M = \frac{10500 \times 187.5 . A_b}{2 . A_b . [62.5^2 + 187.5^2]}$$

$F_M = 25.2$ kN.

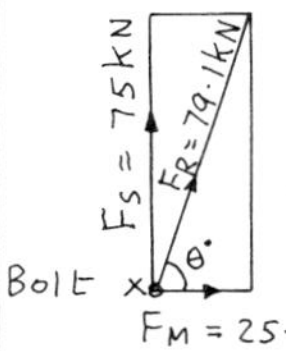

Vector force diagram

$$F_R = \sqrt{25.2^2 + 75^2} = 79.1 \text{ kN.}$$

Slip resistance of bolt $P_{SL} = 71.3$ kN

Bolt in double shear $= 2 P_{SL} = 71.3 \times 2 = 142.6$ kN

Bearing capacity $= d.t.p_{bg} \not> \frac{1}{3} e.t.p_{bg}$

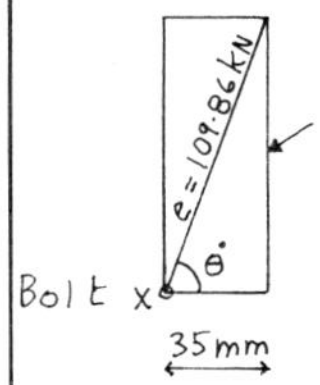

Edge of web

End distance e for web determined by proportion from vector force diagram

25.2 : 79.1 kN

35 : 109.86 mm

$$\therefore d.t.p_{bg} = \frac{20 \times 9.6 \times 825}{10^3} = 158.4$$

$$\not> \frac{1}{3} \times \frac{109.86 \times 9.6 \times 825}{10^3} = 290 \text{ kN}$$

$\therefore$ Bolt capacity = 142.6 kN

Since F_R < Bolt capacity

$79.1 < 142.6$ satisfactory so far

CURPE CONSULTANTS 46 Orburn Road Dunfield	Project: Beam Splices	Job ref:
	Part of structure: Beam splice-design-bolted	Calc sheet no: C6 / 5, rev: 1
	Drawing ref: — Calc by: R.A.S. Date: —	Check by: — Date: —

Ref	Calculations	Output
BS5950		

(2) Web plates.

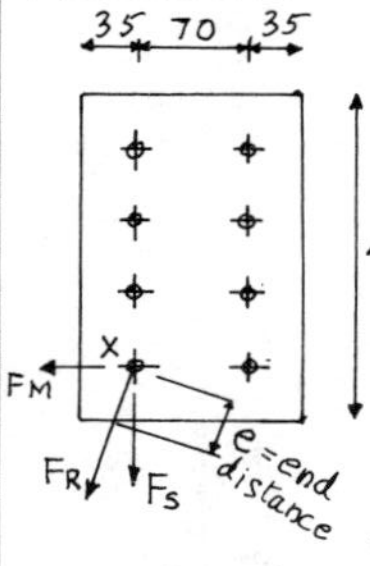

e = end distance

To find e by proportion

75 : 79.1

35 : 36.9

e = 36.9 mm

CL. 6.4.2.2 — Bearing capacity

$d.t.p_{bg} \not> \frac{1}{3} e.t.p_{bg}$

$d.t.p_{bg} = \frac{20 \times 10 \times 825}{10^3} = 165\,kN$

$\frac{1}{3} e.t.p_{bg} = \frac{1}{3} \times \frac{36.9 \times 10 \times 825}{10^3} = 101.4\,kN$

∴ Bearing cap. of one plate = 101.4 kN

" " " two plates = 202.8 kN

∴ Bearing cap. of web connection = 158.4 kN (i.e. the smaller of web and web plates)

Cap. of bolt = 142.6 kN (i.e. smaller of bearing cap and double shear slip resistance)

Since F_R < cap. of bolt   79.1 < 142.6 — **Output:** Web bolts sat.

Check cover web plates for shear + bending

Shear force to one plate $= \frac{F_v}{2} = \frac{300}{2} = 150\,kN$

Torsional moment to one plate $= \frac{150 \times 35}{10^3} = 5.25\,kNm$

CL. 4.2.3 — Shear cap. of one plate $P_v = 0.6\,p_y\,A_v$

CL. 4.2.3 (c) — $P_v = \frac{0.6 \times 275 \times 0.9 \times 10\,(445 - 4 \times 22)}{10^3}$

$= 530\,kN$

Since shear force 150 < 530   o.k. shear

Now check moment cap. Mc of the plate

CURPE CONSULTANTS 46 Orburn Road Dunfield	Project: Beam Splices	Job ref:
	Part of structure: Beam splice design-bolted	Calc sheet no: C6 / 6, rev: 1
	Drawing ref: — Calc by: R.A.S Date: —	Check by: — Date: —

Ref	Calculations	Output
BS5950		

Since 150 < 0.6Pv then

$M_c = p_y . Z$.

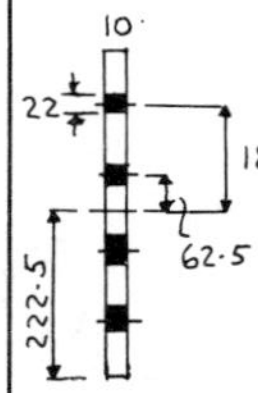

I of plate $= \frac{10 \times 445^3}{12} - 2(10 \times 22) 62.5^2 - 2(10 \times 22) 187.5^2$

$I = 73434270 - 1718750 - 15468750$

$\therefore I = 56246770\ mm^4$

$Z = \frac{I}{y} = \frac{56246770}{222.5} = 252794\ mm^3$

$M_c = p_y . Z = \frac{275 \times 252794}{10^6} = 69.5\,kNm$

Since torsional moment < Mc

5.25 < 69.5

Web plates are sat. for bending — **Output:** Web plates sat.

Check web of beam for shear

Net area = D.t − holes

$= (528.3 \times 9.6) - 4(22 \times 9.6) = 4266.88\ mm^2$

$P_v = 0.6\,p_y . A_v$

$= \frac{0.6 \times 275 \times 4266.8}{10^3} = 697\,kN > 300\,kN$ — **Output:** Beam web shear sat.

CURPE CONSULTANTS 46 Orburn Road Dunfield	Project Stanchion Splices		Job ref	
	Part of structure Stanchion splice - bolted		Calc sheet no C6 / 7	rev /
	Drawing ref	Calc by R.A.S.	Date	Check by	Date

Ref	Calculations	Output
BS5950		

Ex. 2. Stanchion Splice

Factored loading. Dead + Wind

Axial compressive force $F_c = 400\text{kN}$

Bending moment $M_x = 60\text{kNm}$

Shear force $F_v = 20\text{kN}$.

All M20 HSFG bolts.

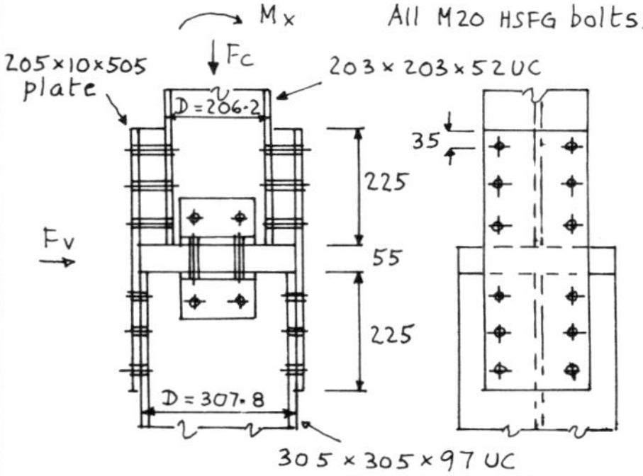

Mx
Fc
205×10×505 plate
203×203×52UC
D = 206·2
35
225
55
225
Fv
D = 307·8
305×305×97UC

Ends machined for direct bearing so check connection for tension due to moment.

For top section 203×203×52UC

Axial comp. stress $f_c = \frac{F_c}{A} = \frac{400\times10^3}{66\cdot4\times10^2} = 60\cdot2\,\text{N/mm}^2$

Bending stress $f_{bc} = f_{bt} = \frac{M}{Z_x} = \frac{60\times10^6}{510\times10^3} = 117\cdot6\,\text{N/mm}^2$

Since $f_{bt} > f_c$ tension is present

Design splice for full moment 60kNm i.e. conservatively.

CURPE CONSULTANTS 46 Orburn Road Dunfield	Project Stanchion Splices		Job ref	
	Part of structure Stanchion splice - bolted		Calc sheet no C6 / 8	rev /
	Drawing ref	Calc by R.A.S.	Date	Check by	Date

Ref	Calculations	Output
BS5950		

Flange cover plates

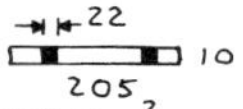

22
10
205

Gross area $= 205\times10 = 2050\,\text{mm}^2$

Net area $= 2050 - (2\times22\times10) = 1610\,\text{mm}^2$

Tensile force $F_t = \frac{M}{D+t_p} = \frac{60\times10^3}{(307\cdot8+10)} = 188\cdot8\,\text{kN}$

Tension capacity of plate $P_t = A_e.p_y$.

CL.3.3.3 $A_e = K_e$. net area $\not>$ gross area.

$= 1\cdot2\times1610 = 1932 < 2050$

$\therefore A_e = 1932\,\text{mm}^2$ $p_y = 275$

$P_t = \frac{1932\times275}{10^3} = 531\,\text{kN}$.

Since $F_t < P_t$ flange cover plate o.K. — Output: Flange cover plates sat.

Bolts in plates

Slip resistance 1 bolt $P_{SL} = 1\cdot1\,K_s.\,\mu.\,P_o$

CL. 6.4.2.1 $P_{SL} = 1\cdot1\times1\cdot0\times0\cdot45\times144 = 71\cdot3\,\text{kN}$

CL. 6.4.2.2 Bearing resistance 1 bolt $P_{bg} = d.t.p_{bg} \leq \frac{1}{3}e.t.p_{bg}$

$d.t.p_{bg} = \frac{20\times10\times825}{10^3} = 165\,\text{kN}$

$\frac{1}{3}\times\frac{35\times10\times825}{10^3} = 96\,\text{kN}$.

$\therefore P_{bg} = 96\,\text{kN}$

Slip resistance governs since $P_{SL} < P_{bg}$

Force at face of upper stanchion $= \frac{M}{D}$

$\therefore$ Force $= \frac{60\times10^3}{206\cdot2} = 291\,\text{kN}$

For 6 bolts capacity $= 6\times71\cdot3 = 428\,\text{kN}$.

Since $291 < 428$ bolts o.K.

Shear force taken by web cleat bolts. will be found satisfactory

Output: Flange plate bolts sat. Web cleats & bolts sat. Splice sat.

CURPE CONSULTANTS 46 Orburn Road Dunfield	Project: Steelwork Connections	Job ref
	Part of structure: Top and bottom cleat bolted connection.	Calc sheet no: C6 / 9, rev: /
	Drawing ref: — ; Calc by: R.A.S. ; Date: —	Check by: — ; Date: —

Ref BS5950	Calculations	Output
	Ex.3. <u>Top and bottom cleat connections</u>	

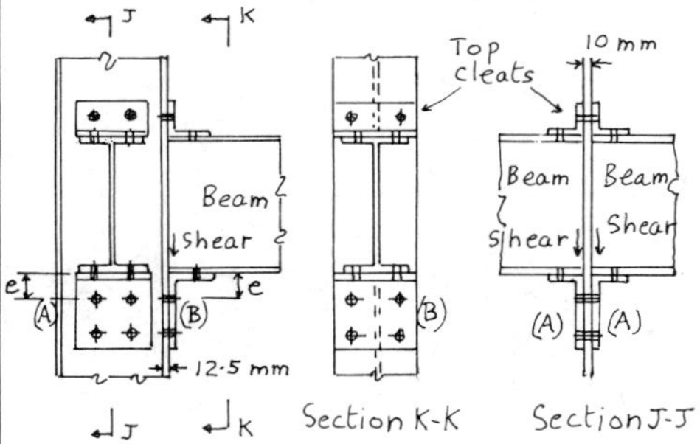

Ref BS5950	Calculations	Output
	Seating or bottom cleats (A) and (B) resist respective end shears with top cleats (or web cleats) steadying the top of the beam.	
	Cleat (B) 150 × 90 × 12 ⌐. End distance $e = 55$ mm. M20 bolts 4·6 grade	
	Check bolts for shear and bearing. Bolts in single shear.	
CL.6.3.2.	Shear $P_s = p_s . A_s = \frac{160 \times 245}{10^3} = 39\cdot2$ kN	$P_s = 39\cdot2$ kN
CL.6.3.3.2	Bearing. For bolt $P_{bb} = d.t.p_{bb} = \frac{20 \times 12 \times 435}{10^3} = 104\cdot4$ kN	$P_{bb} = 104\cdot4$ kN
CL.6.3.3.3	For conn. parts $P_{bs} = d.t.p_{bs} \leq \frac{1}{2} e.t.p_{bs}$	

CURPE CONSULTANTS 46 Orburn Road Dunfield	Project: Steelwork Connections	Job ref
	Part of structure: Top and bottom cleats bolted connections	Calc sheet no: C6 / 10, rev: /
	Drawing ref: — ; Calc by: RAS. ; Date: —	Check by: — ; Date: —

Ref BS5950	Calculations	Output
	$P_{bs} = \frac{20 \times 12 \times 460}{10^3} \leq \frac{1}{2} \times \frac{55 \times 12 \times 460}{10^3}$	
	$110\cdot4 \leq 151\cdot8$	$P_{bs} = 110\cdot4$ kN
	∴ Bearing capacity of one bolt = 104·4 kN	Bearing cap. 104·4 kN.
	∴ Capacity of one bolt = 39·2 kN.	
	For 4 bolts capacity = 4 × 39·2 kN = 156·8 kN	Bolts can resist end shear = 156·8 kN.
	Cleats (A) 150 × 90 × 12 ⌐. End distance $e = 55$ mm. M20 bolts 4·6 grade	
	Check bolts for shear and bearing. Bolts in double shear	
CL.6.3.2	Shear $P_s = 39\cdot2$ kN single shear $= 2 \times 39\cdot2 = 78\cdot4$ kN double shear	$P_s = 78\cdot4$ kN
CL 6.3.3.2	Bearing. For bolt $P_{bb} = d.t.p_{bb} = \frac{20 \times 10 \times 435}{10^3} = 87$ kN	$P_{bb} = 87$ kN
CL 6.3.3.3	For conn. parts $P_{bs} = d.t.p_{bs} \leq \frac{1}{2} e.t.p_{bs}$	
	$P_{bs} = \frac{20 \times 10 \times 460}{10^3} \leq \frac{1}{2} \times \frac{55 \times 10 \times 460}{10^3}$	
	$= 92 \leq 126\cdot5$	$P_{bs} = 92$ kN.
	∴ Bearing cap. of one bolt = 87 kN.	
	For 4 bolts capacity = 4 × 78·4 = 313·6 kN	
	Bolts can resist end shear to each beam $= \frac{313\cdot6}{2} = 156\cdot8$ kN.	Bolts can resist end shear = 156·8 kN per beam.

CURPE CONSULTANTS 46 Orburn Road Dunfield	Project: Steelwork Connections			Job ref	
	Part of structure: Eccentric bolted connection			Calc sheet no: C6 / 11 /	rev
	Drawing ref	Calc by: RAS.	Date	Check by	Date

Ref	Calculations	Output
BS5950		

Ex.4. Eccentric connection - bolted Shear & Torsion

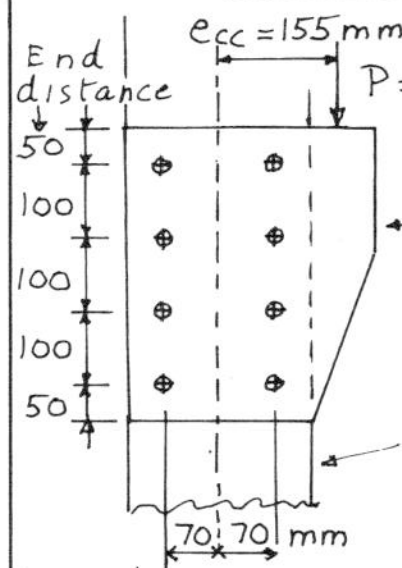

N = number of bolts

∴ Assume area of bolt as unity

Shear force $= 150\,kN$

Moment $= P.e = 150 \times 155$ kNmm

Vert. force due to shear $F_S = \frac{P}{N} = \frac{150}{8} = 18.75\,kN$ — Output: $F_S = 18.75\,kN$

Force due to moment $F_M = \frac{P.e.r}{\Sigma r^2}$

$r = \sqrt{70^2 + 150^2} = 165.53\,mm$

$\Sigma r^2 = \Sigma x^2 + \Sigma y^2$

$\Sigma x^2 = 8(1 \times x^2) = 8(1 \times 70^2) = 39200\,mm^4$

$\Sigma y^2 = 4(1 \times Y_1^2) + 4(1 \times Y_2^2)$
$= 4(1 \times 150^2) + 4(1 \times 50^2) = 100000\,mm^4$

$\Sigma r^2 = 39200 + 100000 = 139200\,mm^4$

$F_M = \frac{P.e.r}{\Sigma r^2} = \frac{150 \times 155 \times 165.53 \times 1}{139200} = 27.6\,kN$ — Output: $F_M = 27.6\,kN$

Resultant shear force on most severely stressed bolt can be found graphically or by calculation. — Output: Bolt X

CURPE CONSULTANTS 46 Orburn Road Dunfield	Project: Steelwork Connections			Job ref	
	Part of structure: Eccentric bolted connection			Calc sheet no: C6 / 12 /	rev
	Drawing ref	Calc by: RAS.	Date	Check by	Date

Ref	Calculations	Output
BS5950		

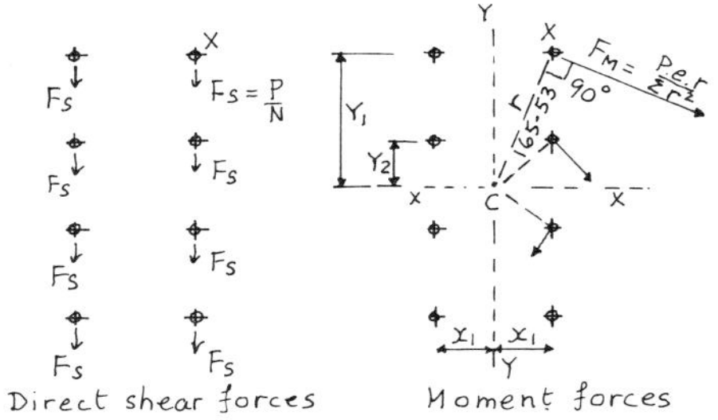

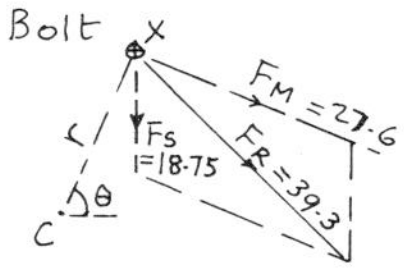

Graphical solution

Using formula $F_R = \sqrt{F_S^2 + F_M^2 + 2\,F_S.F_M.\cos\theta}$

$F_R = \sqrt{18.75^2 + 27.6^2 + (2 \times 18.75 \times 27.6 \times \frac{70}{165.73})}$

$F_R = 39.3\,kN$. — Output: $F_R = 39.3\,kN$

CL. 6.3.2. Shear capacity $P_S = p_s.A_s. = \frac{160 \times 245}{10^3} = 39.2\,kN$ — Output: $P_S = 39.2\,kN$

Bearing capacity lesser of P_{bb} or P_{bs}.

CL. 6.3.3.2 For bolt $P_{bb} = d.t.p_{bb} = \frac{20 \times 12.5 \times 435}{10^3} = 108.7\,kN$ — Output: $P_{bb} = 108.7\,kN$

CL. 6.3.3.3 For conn. ply $P_{bs} = d.t.p_{bs} \leq \frac{1}{2} e\,t\,p_{bs}$

$= \frac{20 \times 12.5 \times 460}{10^3} \leq \frac{1}{2} \times \frac{50 \times 12.5 \times 460}{10^3}$

$115 \leq 143.75$ — Output: $P_{bs} = 115\,kN$

∴ Bearing capacity 108.7 kN.

∴ Bolt capacity $= 39.2 < 39.3$ — Output: Slightly overstressed, assume OK or increase vert bolt centres.

CURPE CONSULTANTS 46 Orburn Road Dunfield	Project: Steelwork Connection	Job ref:
	Part of structure: Eccentric bolted connection	Calc sheet no: C6 / 13 / rev:
	Drawing ref: · Calc by: RAS · Date:	Check by: · Date:

Ref BS5950	Calculations	Output

Ex.5. Eccentric connection - bolted
Shear & Bending

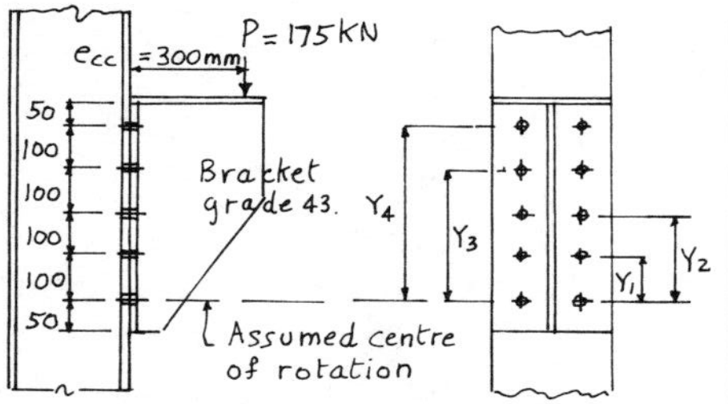

Bolts M20, 4·6 grade
N = number of bolts
Assume area of bolt as unity
Shear force = 175 kN
Moment = P.e = 175 × 300

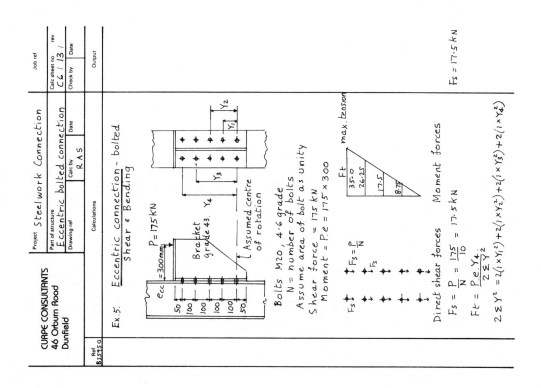

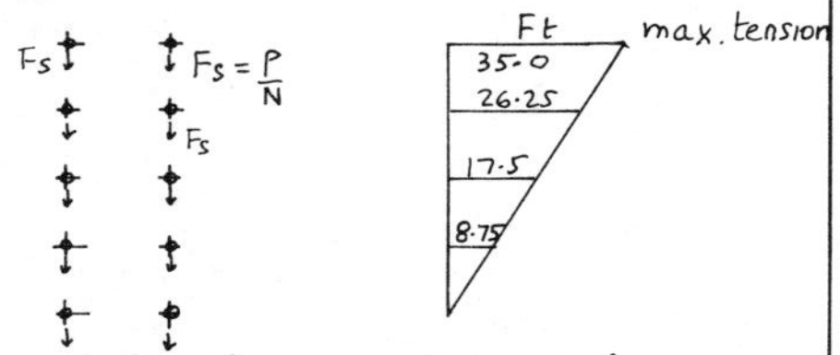

Direct shear forces   Moment forces

$F_s = \frac{P}{N} = \frac{175}{10} = 17{\cdot}5\,kN$ — Output: $F_s = 17{\cdot}5\,kN$

$F_t = \frac{P.e.Y_4}{2\Sigma Y^2}$

$2\Sigma Y^2 = 2(1\times Y_1^2)+2(1\times Y_2^2)+2(1\times Y_3^2)+2(1\times Y_4^2)$

CURPE CONSULTANTS 46 Orburn Road Dunfield	Project: Steelwork Connections	Job ref:
	Part of structure: Eccentric bolted connection	Calc sheet no: C6/ 14/ rev:
	Drawing ref: · Calc by: R.A.S. · Date:	Check by: · Date:

Ref BS5950	Calculations	Output
	$= 2(100^2)+2(200^2)+2(300^2)+2(400^2)$ $= 600000\ mm^4$	
	$F_t = \frac{175\times 300\times 400\times 1}{600000} = 35\,kN$	$F_t = 35\,kN$
CL.6.3.2.	Shear cap. of bolt $P_s = p_s.A_s = \frac{160\times 245}{10^3} = 39{\cdot}2\,kN$	$P_s = 39{\cdot}2\,kN$
	$P_s > F_s \quad 39{\cdot}2 > 17{\cdot}5$	Shear O.K.
CL.6.3.6.1	Tensile cap of bolt $P_t = p_t.A_t = \frac{195\times 245}{10^3} = 47{\cdot}8\,kN$	$P_t = 47{\cdot}8\,kN$
	$P_t > F_t \quad 47{\cdot}8 > 35{\cdot}0$	Tension O.K.
CL.6.3.6.3	Combined $\frac{F_s}{P_s} + \frac{F_t}{P_t} \leqslant 1{\cdot}4$	
	$\frac{17{\cdot}5}{39{\cdot}2} + \frac{35}{47{\cdot}8} = 1{\cdot}178 < 1{\cdot}4$	Combined stress O.K.

CURPE CONSULTANTS 46 Orburn Road Dunfield	Project: Steelwork Connections			Job ref	
	Part of structure: Eccentric welded connection			Calc sheet no: C6 / 15 /	rev
	Drawing ref	Calc by: R.A.S	Date	Check by	Date

Ref	Calculations	Output
BS 5950		

Ex.6. Eccentric connection - welded
Shear & Torsion.

P = 220 kN Factored.

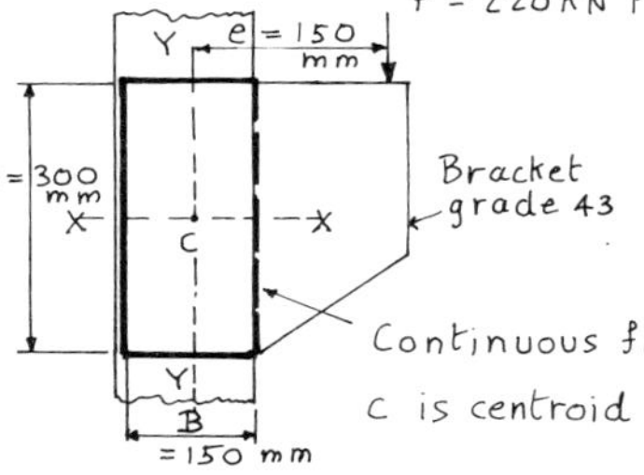

Continuous fillet weld.

C is centroid of welds.

Determine size of weld.

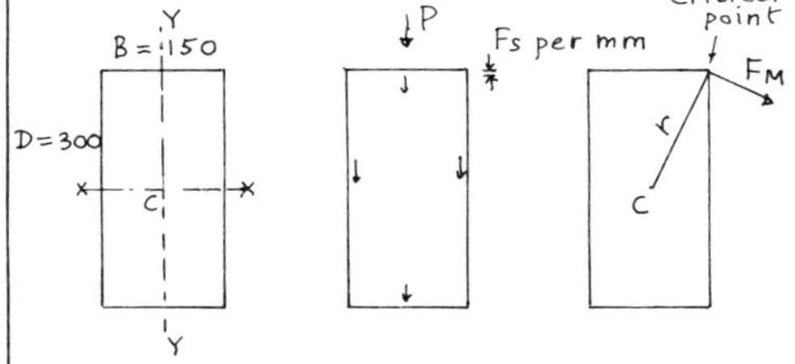

Welds. Direct shear forces. Torsional mom forces.

$F_s = \frac{P}{\text{length of weld}} = \frac{220 \times 10^3}{900} = 244.4 \text{ N/mm}$ — Output: $F_s = 244.4$ N/mm

$F_M = \frac{P.e.r}{I_p}$ $\quad I_p = I_{xx} + I_{yy}$

$I_{xx} = 2\left(\frac{1 \times 300^3}{12}\right) + 2(150 \times 1)150^2$

$= 11250000 \text{ mm}^4$

CURPE CONSULTANTS 46 Orburn Road Dunfield	Project			Job ref	
	Part of structure			Calc sheet no: C6 / 16 /	rev
	Drawing ref	Calc by	Date	Check by	Date

Ref	Calculations	Output
BS 5950		

$I_{YY} = 2\left(\frac{1 \times 150^3}{12}\right) + 2(300 \times 1)75^2$

$= 3937500 \text{ mm}^4$

$I_p = 11250000 + 3937500 = 15187500 \text{ mm}^4$ — Output: $I_p = 15187500 \text{ mm}^4$

$r = \sqrt{150^2 + 75^2} = 167.7 \text{ mm}$

$F_M = \frac{220 \times 10^3 \times 150 \times 167.7}{15187500} = 364.4 \text{ N/mm}$ — Output: $F_M = 364.4$ N/mm

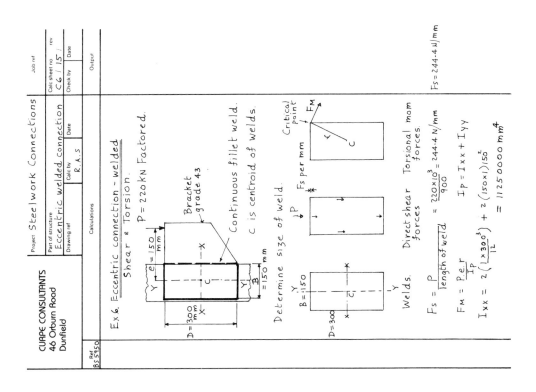

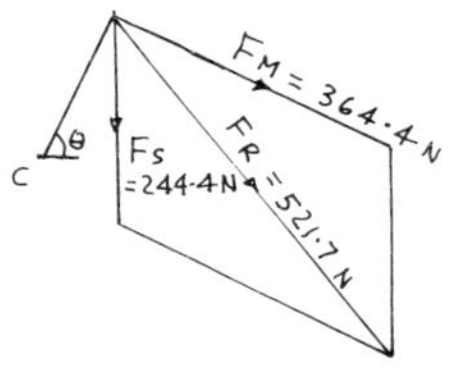

$F_R = \sqrt{F_s^2 + F_M^2 + 2 F_s . F_M . \cos\theta}$

$= \sqrt{244.4^2 + 364.4^2 + \left(2 \times 244.4 \times 364.4 \times \frac{75}{167.7}\right)}$

$= 521.7$ N/mm run of weld — Output: $F_R = 521.7$ N/mm

Cl. 6.6.5 — Strength of weld = 0.7 × leglength × pw

Table 36 — 521.7 = 0.7 × leg length × pw — Output: $p_w = 215 \text{ N/mm}^2$

∴ leg length (min) = 3.46 mm

Use 6 mm fillet weld — Output: Continuous 6 mm fillet weld. Sat.

CURPE CONSULTANTS
46 Orburn Road
Dunfield

Project: Steelwork Connections	Job ref
Part of structure: Eccentric welded connection	Calc sheet no: C6 / 17 / rev
Drawing ref / Calc by: R. A. S. / Date	Check by / Date

Ref: BS5950 | Calculations | Output

Ex 7. Eccentric connection - welded
Shear & Bending.

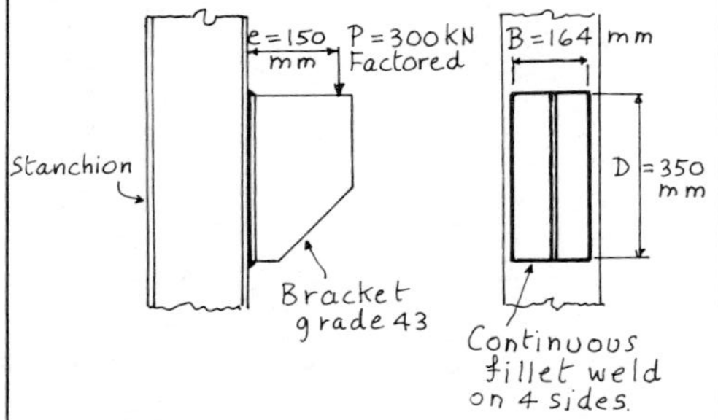

(a) Assume Neutral axis at level of centroid of welds.

$F_s = \dfrac{P}{\text{length of weld}}$

$F_s = \dfrac{300}{2(350+164)} = 0.292$ kN/mm run

I_{xx} of welds

$= 2\left(\dfrac{1 \times 350^3}{12}\right) + 2\left(164 \times 1 \times 175^2\right)$

$= 17190833 \text{ mm}^4$

$F_M = \dfrac{P.e.Y}{I_{xx}}$

$= \dfrac{300 \times 150 \times 175}{17190833}$

$= 0.458$ kN/mm run

F_s and F_M are combined vectorially to produce the resultant force F_R at the uppermost level of the welds.

CURPE CONSULTANTS
46 Orburn Road
Dunfield

Project: Steelwork Connections	Job ref
Part of structure: Eccentric welded connection	Calc sheet no: C6 / 18 / rev
Drawing ref / Calc by: R. A. S. / Date	Check by / Date

Ref: BS5950 | Calculations | Output

$F_R = \sqrt{0.292^2 + 0.458^2} = 0.543$ kN/mm

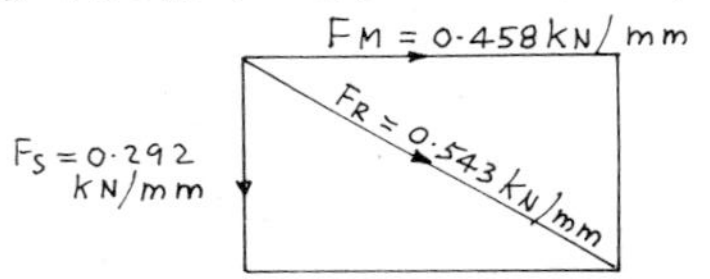

CL.6.6.5 Table 36 — Strength of weld = 0.7 × leg length × pw — Output: $p_w = 215 \text{ N/mm}^2$

$0.543 \times 10^3 = 0.7 \times \text{leg length} \times 215$

∴ leg length (min) = 3.6 mm

Use 6 mm fillet weld — Output: Continuous 6 mm fillet weld sat.

(b) Assume Neutral axis occurs at a distance of one sixth the depth of the weld measured from the bottom edge.

I_{NA} of welds

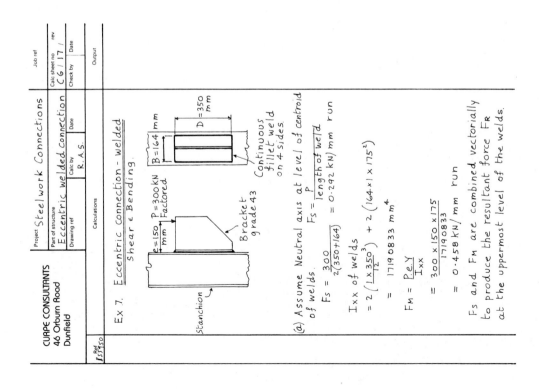

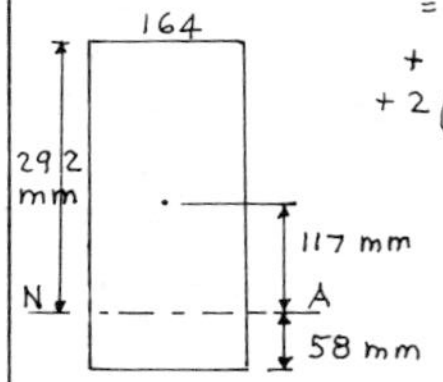

$= 164 \times 1 \times 292^2 = 13983296$

$+ 164 \times 1 \times 58^2 = 551696$

$+ 2\left(\dfrac{1 \times 350^3}{12} + 350 \times 1 \times 117^2\right) = 16728132$

$I_{NA} = 31263124$

$F_M = \dfrac{300 \times 150 \times 292}{31263124}$

$= 0.42$ kN/mm

$F_s = 0.292$ kN/mm

$F_R = \sqrt{0.292^2 + 0.42^2} = 0.512$ kN/mm

Use 6 mm fillet weld — Output: Continuous 6 mm fillet weld sat.

CURPE CONSULTANTS 46 Orburn Road Dunfield	Project: Steelwork Connections	Job ref
	Part of structure: Single angle welded connection	Calc sheet no: C6 / 19 / rev
	Drawing ref / Calc by: R.A.S / Date	Check by / Date

Ref BS5950	Calculations	Output
	Ex. 8. **Single angle welded connection**	

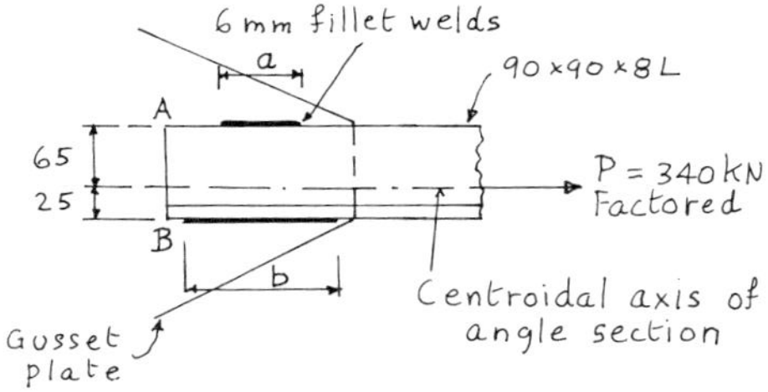

Ref BS5950	Calculations	Output
CL.6.6.5	Strength of fillet weld = 0·903 kN/mm run	
	Effective length req^d = $\frac{340}{0.903}$ = 376 mm	
	Balance welds about centroidal axis	
	Weld length a	
	Take moments about B	
	(0·903 × a × 90) = 340 × 25	
	a = 104·6 ≮ 90	
CL.6.6.5.2	Actual length = 104·6 + (2 × 6) = 116·6 mm	Say 117 mm
	Weld length b	
	Take moments about A	
	(0·903 × b × 90) = 340 × 65	
	b = 271·9 ≮ 90	
CL.6.6.5.2	Actual length = 271·9 + (2 × 6) = 283·9 mm	Say 284 mm

CURPE CONSULTANTS 46 Orburn Road Dunfield	Project: Steelwork Connections	Job ref
	Part of structure: Bolted web cleats connection	Calc sheet no: C6 / 20 / rev
	Drawing ref / Calc by: R.A.S. / Date	Check by / Date

Ref BS5950	Calculations	Output
	Ex. 9. **Bolted web cleats - Beam to Stanchion**	
	Vert. End Shear = 100 kN	

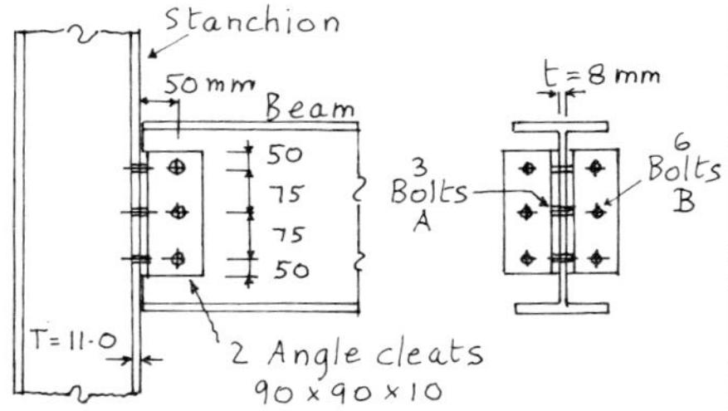

Ref BS5950	Calculations	Output
	M 20 bolts, 4·6 grade	
	Bolts A in double shear	
CL.6.3.2	Single shear capacity of one bolt P_s	
	$P_s = p_s . A_s$	
	$= \frac{160 \times 245}{10^3} = 39.2$ kN	P_s = 39·2 kN
	Double shear capacity of one bolt = 2P_s	
	= 2 × 39·2 = 78·4 kN	2P_s = 78·4 kN
	M = 100 × 50 = 5000 kN mm	

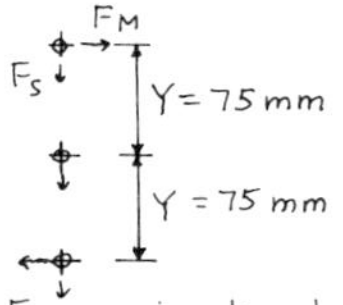

Ref BS5950	Calculations	Output
	I of bolts = ΣAY^2	
	I = 2 A(75^2) = 11250A mm^4	
	Z = $\frac{I}{Y}$ = $\frac{11250A}{75}$ = 150A mm^3	
	Force in top bolt:	
	Due to shear $F_S = \frac{100}{3}$ = 33·33 kN	F_S = 33·33 kN
	Due to moment $F_M = \frac{5000A}{150A}$ = 33·33 kN	F_M = 33·33 kN

CURPE CONSULTANTS 46 Orburn Road Dunfield	Project: Steelwork Connections			Job ref	
	Part of structure: Bolted web cleats connection.			Calc sheet no: C6 / 21 /	rev
	Drawing ref	Calc by: R.A.S.	Date	Check by	Date

Ref	Calculations	Output
	Resultant force $F_R = \sqrt{F_S^2 + F_M^2}$	
	$= \sqrt{33.33^2 + 33.33^2} = 47.13\ kN$	$F_R = 47.13\ kN$
	Since $F_R < 2P_s$ $47.13 < 78.4$ O.K.	Shear sat.
	Check bearing:	
Cl. 6.3.3.2	For bolt $P_{bb} = d.t.p_{bb} = \frac{20 \times 8 \times 435}{10^3} = 69.6\ kN$	$P_{bb} = 69.6\ kN$
Cl. 6.3.3.3	For conn. ply $P_{bs} = d.t.p_{bs} \leq \frac{1}{2} e.t.p_{bs}$	
	$= \frac{20 \times 8 \times 460}{10^3} \leq \frac{1}{2} \times \frac{50 \times 8 \times 460}{10^3}$	
	$= 73.6 \leq 92$	$P_{bs} = 73.6\ kN$
	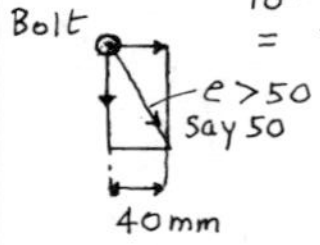	
	Cap. bearing lesser of P_{bb} and P_{bs}	
	Cap. bearing = 69.6 kN	Cap. bearing 69.6kN
	Since $F_R < P_{bb}$ $47.13 < 69.6$	Bearing sat. Bolts A sat.
	Bolts B in single shear	
Cl. 6.3.2	$P_s = p_s.A_s = 39.2\ kN$ as for A	$P_s = 39.2$
	For 6 bolts shear capacity $= 6 \times 39.2 = 235.2\ kN$	$6P_s = 235.2\ kN$
	Since $100 < 235.2$	Shear sat.
	Check bearing;	
Cl. 6.3.3.2	For bolt $P_{bb} = d.t.p_{bb} = \frac{20 \times 10 \times 435}{10^3} = 87\ kN$	$P_{bb} = 87\ kN$
Cl. 6.3.3.3	For conn. ply $P_{bs} = d.t.p_{bs} \leq \frac{1}{2} e.t.p_{bs}$ $e = 50$	
	$= \frac{20 \times 10 \times 460}{10^3} \leq \frac{1}{2} \times \frac{50 \times 10 \times 460}{10^3}$	
	$= 92 \leq 115$	$P_{bs} = 92\ kN$
	Cap. bearing lesser of P_{bb} and P_{bs}	Cap. bearing 87kN
	Cap. bearing 6 bolts $= 6 P_{bb} = 6 \times 87 = 522\ kN$	
	Since $522 > 100$	Bearing sat. Bolts B sat.

CURPE CONSULTANTS 46 Orburn Road Dunfield	Project: Steelwork Connections			Job ref	
	Part of structure: Flexible Welded end plate connection			Calc sheet no: C6 / 22 /	rev
	Drawing ref	Calc by: R.A.S.	Date	Check by	Date

Ref BS5950	Calculations	Output
	Ex. 10. Welded End Plate & bolted conn	
	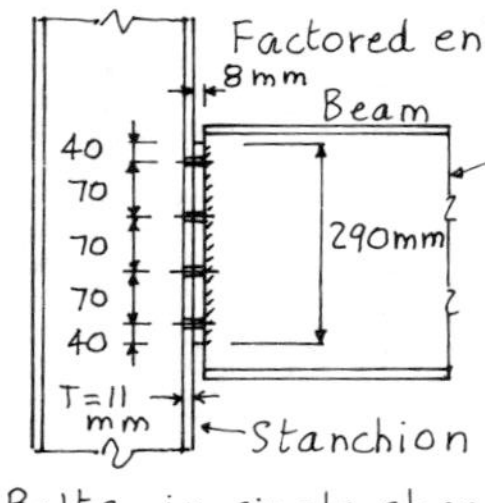	
	Factored end shear = 300kN.	
	Web t = 8 mm.	
	8 No. M20 bolts, 4.6 grade	
	Fillet weld both sides 6 mm.	
	End plate 8 mm thick	
	Bolts in single shear	
Cl 6.3.2	Shear $P_s = p_s.A_s$	
	$= \frac{160 \times 245}{10^3} = 39.2\ kN$	$P_s = 39.2\ kN$
	Shear cap. 8 bolts $= 8 \times 39.2 = 313.6\ kN$	Shear cap 8 bolts = 313.6kN
	Capacity in bearing bolt or conn. ply	
Cl. 6.3.3.2	For bolt $p_{bb} = d.t.p_{bb}$	
	$= \frac{20 \times 8 \times 435}{10^3} = 69.6\ kN$	$P_{bb} = 69.6\ kN$
Cl. 6.3.3.3	For conn. ply $P_{bs} = d.t.p_{bs} \leq \frac{1}{2}.e.t.p_{bs}$	
	$= \frac{20 \times 8 \times 460}{10^3} \leq \frac{1}{2} \times \frac{40 \times 8 \times 460}{10^3}$	
	$73.6 \leq 73.6$	$p_{bs} = 73.6\ kN$
	∴ Bearing cap = 69.6 kN	
	Bearing cap 8 bolts $= 8 \times 69.6 = 556.8\ kN$	Bear cap 8 bolts = 556.8kN
	∴ Strength of bolts controlled by shear	
	Since $313.6 > 300$ bolts sat.	Bolts sat.
Cl. 6.6.5.2	Welds Effective length $= 2[290 - (2 \times 6)]$	

CURPE CONSULTANTS 46 Orburn Road Dunfield	Project Steelwork Connections			Job ref	
	Part of structure Flexible Welded end plate connection			Calc sheet no C6 / 23 /	rev
	Drawing ref	Calc by R. A. S.	Date	Check by	Date

Ref BS5950	Calculations	Output
	$= 556$ mm	
CL.6.6.5 Table 36	Cap. of welds $= \dfrac{0.7 \times 6 \times 215 \times 556}{10^3} = 502$ kN	
	Since $502 > 300$ Welds sat	Welds sat.
	Shear cap. of end plate	
	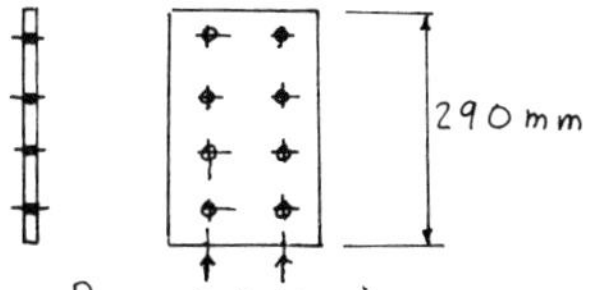	
CL.4.3.2.	$P_v = 0.6\ p_y . A_v$.	
	A_v for plate $= 0.9 A$	
	Accounting for holes.	
	Shear cap of plate	
	$= \dfrac{0.6 \times 275 \times 2 \times 0.9 \times 8(290-(4 \times 22))}{10^3}$	
	$= 480$ kN	
	Since $480 > 300$ shear sat.	End plate shear cap. sat.
	Shear cap. of beam web at plate	
CL.4.3.2.	Cap. of web $P_v = 0.6 p_y . A_v$	
	$A_v = 0.9 A$	
	$P_v = \dfrac{0.6 \times 275 \times 0.9 \times 290 \times 8}{10^3} = 344.5$ kN	
	Since $344.5 > 300$ web sat.	Shear cap. beam web sat.

7

Roof trusses and purlins

Roof trusses

Trusses are positioned at convenient centres along the building to provide support for the roof purlins. The minimum slope of the rafter depends on the particular type of roof cladding used, and should comply with that recommended in the manufacturer's specification. A smaller angle slope may provide problems arising from leakage at joints in the sheeting. For economical reasons, a height to ridge equal to one-fifth to one-quarter of the span is often selected. The framework is triangulated using standard arrangements of internal members for various spans of trusses.

Typical examples of symmetrical pitched and north light types are shown in Fig. 7.1.

Loads on trusses

There are three principal types of loading. These are:

1) Dead load due to the weight of sheeting, insulation, purlins, truss and ceiling.
2) Imposed load from BS 6399: Part 1.
3) Wind loads from CP3: Chapter V: Part 2. These loads are important in the case of relatively light roofs where roof suction due to wind can cause reversal of load in some truss members, particularly in the main tie.

The roof loading is applied to the truss by the purlins which are positioned along the rafter to suit the roof sheeting. The most economical rafter section can be achieved when purlin positions coincide with truss node points, since all members are subject to axial force only (Fig. 7.2(a)). When purlins are unavoidably positioned between node points, bending is produced in the rafter in addition to axial force. All other members are subjected to axial forces only (Fig. 7.2(b)).

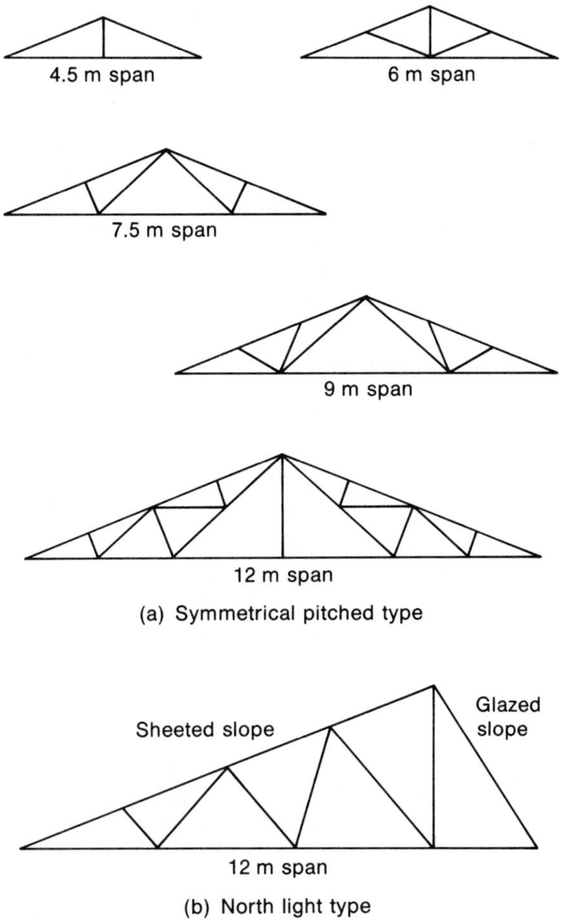

Fig. 7.1 *Roof truss layouts*

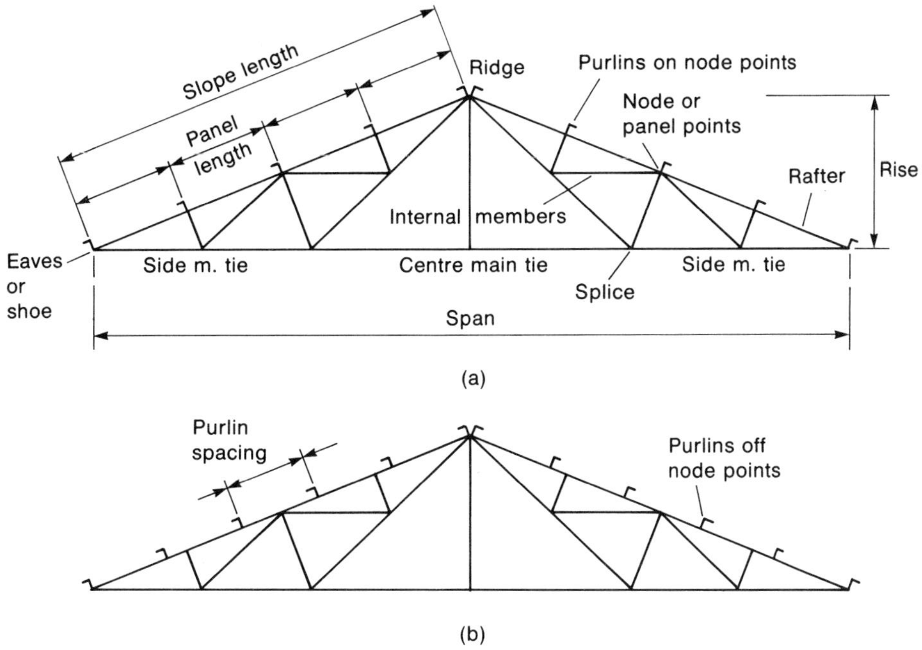

Fig. 7.2 *Purlins (a) on node points, (b) off node points*

Empirical design rules for members in trusses and lattice frames

To simplify the design of light trusses and frames, empirical rules are given in 4.10 of BS 5950. These rules are not applicable to heavy lattice construction which may involve detailed consideration of secondary stresses. Also these rules should not be adopted where fatigue is a critical factor. Secondary stresses caused by eccentricity at connections and moments resulting from rigid joints and truss deflection will be insignificant if the slenderness of the chord members in the plane of the truss is greater than 50, and that of most of the internal web members is greater than 100.

Assumptions made are as follows:

1) For the purpose of calculating axial forces in the members, the connections are pinned.
2) Account may be taken of the fixity of connections and the rigidity of adjacent members in determining effective lengths. The length of chord members may be taken as the distance between connection to the web members in plane and the distance between purlins or longitudinal ties out of plane.
3) Ties to chords and purlins should be connected properly to an adequate restraint system.
4) Local bending moment on the rafter may be taken as $\frac{WL}{6}$ if the exact positions of point loads are not known.
5) Purlins in trusses and light frames need not be checked for compressive stresses originating from their function as restraints.
6) Where sheeting is fixed directly to trusses in the absence of purlins, stability of the rafter should be considered. Restraint to the rafter should be assumed only in cases where the loading is mainly roof loading.

Typical members for trusses

Members are fabricated from standard steel sections such as:

1) angles;
2) tees;

3) channels;
4) rectangular hollow;
5) circular hollow.

The members are connected together at node points by:

1) gusset plates and bolts;
2) welding.

For hollow sections, welding is the more obvious choice, and for angles, bolting or welding may be considered. For welded construction, tee sections used for the rafter and main tie will facilitate the direct welding of members together at the node points.

Procedure for the analysis of trusses

Purlins may be located at node points or more often in practice between the node points.

1) Purlins located at node points

1) Determine the dead, imposed and wind loads (unfactored).
2) Use separate force diagrams to find the axial forces in members due to the unfactored dead and wind loads.
3) Tabulate forces due to unfactored dead, imposed and wind loads. Forces due to the imposed loads are found by proportion.
4) Apply load factors and tabulate resulting factored dead, imposed and wind forces.
5) Combine forces due to factored:

$$\text{dead + imposed} = 1.4D + 1.6I$$
$$\text{dead + wind} = 1.0D + 1.4W$$

2) Purlins not located at node points

Steps (1) to (5) are as above.

6) The rafter is then considered as a continuous beam loaded with the normal component of the unfactored dead purlin load, and the bending moments are determined using the method of moment distribution.

   Both ends of the rafter are considered as pinned for the given example.
7) Moments due to unfactored imposed loads are found by proportion.
8) Apply load factors and combine the factored dead and imposed bending moments.

Roof purlins

Purlins may be designed on the assumption that suitable cladding, adequately fixed, provides lateral restraint to an angle section, or to the face against which it is connected in the case of other sections.

Purlins may be butt jointed:

1) at all vertical supports (trusses, girders, etc.) so as to span only one bay;
2) preferably at alternate vertical supports and spanning as continuous members over two bays, with joints staggered in adjacent lines.

In all cases the deflection or purlins should be limited to suit the characteristics of the roof-cladding system adopted. Angle purlins are commonly used for smaller spans. For larger spans proprietary cold-rolled sections are often employed in preference to angles, because of their relative lightness and ease of handling during the fabrication, transport and erection stages of the steelwork.

Sag rods may be introduced particularly in long spans, to help stabilize the purlins during erection, reduce the deflections in the plane of the cladding, and assist the sheeting contractor. The code states that as one method of design, purlins formed of hot-rolled angles or hollow sections may be designed empirically in accordance with rules given in 4.12.4.3, BS 5950.

The general requirements for empirical design are that:

1) Unfactored loads are used.
2) Maximum span of purlins should be 6.5 m.
3) Where purlins are generally jointed to span only one bay, a minimum of two fasteners (bolts) should be used at each end connection.
4) Where purlins are generally jointed to span as continuous members over two bays, at

least one end should be connected by a minimum of two fasteners for any single bay member.

Specific rules for empirical design of purlins

1) Maximum roof slope 30°.
2) Loading on the purlin should be substantially uniformly distributed with not more than 10% due to other types of loading.
3) Minimum imposed load 0.75 kN/m^2.
4) Table 29 of BS 5950 gives empirical values of the required minimum Z, D and B for angle, CHS and RHS section purlins,

where Z = elastic modulus of the purlin about its axis parallel to the plane of the cladding,
D = dimension of the purlin perpendicular to the plane of the cladding,
B = dimension of the purlin parallel to the plane of the cladding.

In the case of an angle purlin, these values are given as:

$$Z \text{ minimum} = \frac{W_p L}{1800} \text{ cm}^3$$

$$D \text{ minimum} = \frac{L}{45} \text{ mm}$$

$$B \text{ minimum} = \frac{L}{60} \text{ mm}$$

where W_p = total unfactored load on one span of the purlin (kN) due to greater of (dead + imposed) or (wind − dead),
L = span (mm).

Where sag rods are introduced the spacing of these may be used to determine B.

Design of truss members

Members will be either compression members (struts) or tension members (ties).

Struts

Maximum slenderness
The maximum slenderness limits are:

Members resisting dead + imposed load	= 180
Members resisting wind load	= 250
Members acting as ties but subject to reversal of stress due to wind	= 350

These limits often control the size of lightly loaded members.

BS 5950 also states that deflection due to self weight should be checked for members whose slenderness exceeds 180. If this exceeds length/1000 then the effect of bending should be accounted for in design.

Limiting proportions for angles
To prevent local buckling, limiting width-to-thickness ratios for single angles and double angles with components separated are given in Table 7 of BS 5950.

For compact sections, $\frac{b}{t}$ and $\frac{d}{t}$ must not exceed 9.5ε.

For semi-compact sections, $\frac{b}{t}$ and $\frac{d}{t}$ must not exceed 15.0ε and, in addition, $\frac{(b+d)}{t}$ must not exceed 23ε where $\varepsilon = \sqrt{\left(\frac{275}{P_y}\right)}$.

Effective lengths and slenderness for rafters
The rafter or chord is normally a continuous member over a number of panels supported in its plane by the internal members of the truss, and by the roof purlins at right angles to the plane (see Fig. 7.3). In 4.10(e), BS 5950, the length of chord members is to be taken as:

1) Distance between connections to the web members in plane, i.e. panel length;
2) Spacing of purlins or longitudinal ties out of plane.

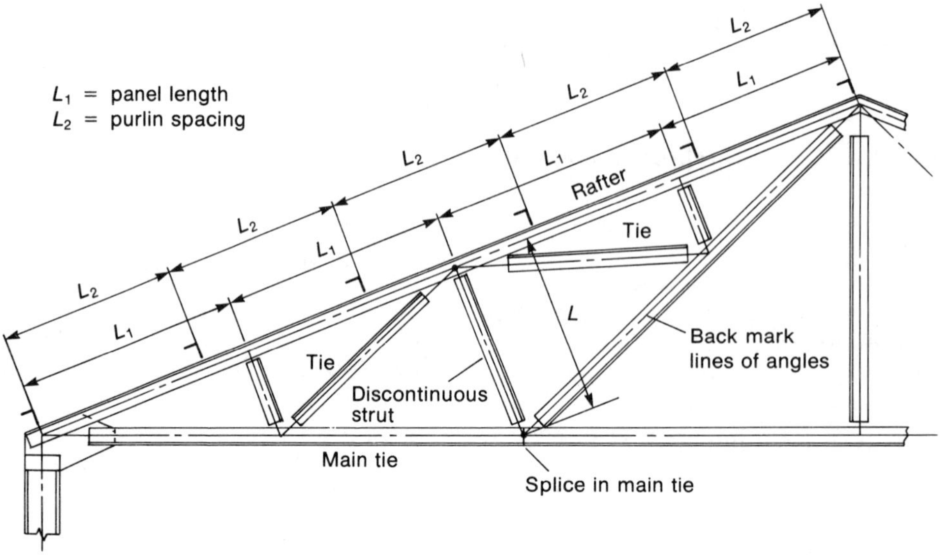

Fig. 7.3 *Truss members arrangement*

Effective lengths are determined in accordance with the conditions of restraint in the appropriate plane as given in Table 24 of BS 5950.

1) To calculate the slenderness λ for single angle rafters, use the maximum value of:

$$\frac{0.85L_1}{r_{vv}} \text{ or } \frac{L_2}{r_{vv}}$$

2) To calculate the slenderness λ for double angle rafters separated back-to-back by gusset plates, use the maximum value of:

$$\frac{0.85L_1}{r_{xx}} \text{ or } \frac{L_2}{r_{yy}}$$

Also reference should be made to 4.7.13.1(d), where the slenderness about the axis parallel to the connected surfaces λ_b should be calculated from 4.7.9(c), BS 5950:

$$\lambda_b = \sqrt{(\lambda_m^2 + \lambda_c^2)}$$

where $\lambda_m = \dfrac{L_2}{r_{yy}}$, i.e. the slenderness of the whole member about that axis.

λ_c = maximum slenderness of one angle (based on its minimum r value) between adjacent connections, and must not exceed 50.

Effective lengths and slenderness of discontinuous angle members

1) For single angle struts connected to a gusset or directly to another member at each end by two or more fasteners in line along the angle or by equivalent welding, the slenderness λ should be taken as the maximum value of:

$$\frac{0.85L}{r_{vv}} \text{ or } \frac{0.7L}{r_{aa}} + 30$$

where r_{aa} = radius of gyration about the axis parallel to the connected leg.

2) For double angle struts connected to both sides of a gusset by two or more fasteners in line along the angle or by equivalent welding, the slenderness λ should be taken as the maximum value of:

$$\frac{0.85L}{r_{xx}} \text{ or } \frac{0.85L}{r_{yy}} \text{ or } \frac{0.7L}{r_{xx}} + 30$$

3) For double angle struts connected directly to one or both sides of another member, the slenderness λ should be taken as the maximum value of:

$$\frac{1.0L}{r_{xx}} \text{ or } \frac{1.0L}{r_{yy}} \text{ or } \frac{0.7L}{r_{xx}} + 30$$

Also λ_b should be calculated as above and taken into account for double angle struts.

Compression resistance

For single angles or double angles with components separated, the compression resistance is given by:

$$P_c = A_g p_c$$

where A_g = gross area of section.

From Table 25 of BS 5950, for buckling about any axis, Table 27(c) is selected to obtain the compressive strength p_c.

If the section is slender then the design strength must be reduced.

Rafters subjected to axial load and moment

With the purlins located between the node points, the rafters are subjected to axial load and moment. The buckling resistance moment for single angles is given in 4.3.8, BS 5950 as:

$$M_b = 0.8 p_y Z \text{ for } \frac{L}{r_{vv}} \leqslant 100$$

$$= 0.7 p_y Z \text{ for } \frac{L}{r_{vv}} \leqslant 180$$

where Z = elastic modulus about the appropriate axis,
r_{vv} = radius of gyration about the weakest axis,
L = unrestrained length.

Conservatively for double angles acting together in bending, the rules given in 4.3.8 for single angles may be used taking the maximum of $\frac{L_1}{r_{xx}}$ and $\frac{L_2}{r_{yy}}$ as the criterion.

Load capacity check:

$$\frac{F}{A_g p_y} + \frac{M_x}{M_{cx}} + \frac{M_y}{M_{cy}} \ngtr 1$$

Overall buckling check with simplified approach:

$$\frac{F}{A_g p_c} + \frac{mM_x}{M_b} + \frac{mM_y}{p_y Z_y} \ngtr 1$$

M_x = applied moment about major axis,
M_y = applied moment about minor axis,
A_g = gross area of section,
M_{cx} = moment capacity about major axis in absence of axial load,
M_{cy} = moment capacity about minor axis in absence of axial load,
F = applied axial load,
p_c = compressive strength,
m = equivalent uniform moment factor (conservatively taken as 1.0),
M_b = buckling resistance moment capacity,
Z_y = elastic section modulus about minor axis,
p_y = design strength.

Ties

Clause 4.6 of BS 5950 covers the design of axially loaded tension members. The tension capacity P_t of a member is given by:

$$P_t = A_e P_y$$

where A_e = effective area of the section.

The net area of a section or element of a section is taken as its gross area less deductions for holes.

Effective areas

1) Single angles connected through one leg only are treated as axially loaded members, and the effective area given as:

$$A_e = a_1 + \left[a^2 \left(\frac{3a_1}{3a_1 + a_2} \right) \right]$$

where a_1 = net sectional area of the connected leg,
a_2 = sectional area of unconnected leg.

2) Double angles are also treated as axially loaded members using the net area if they are connected to both sides of a gusset or section, provided that the components are held longitudinally parallel and connected by bolts or welds in at least two places and held apart by packing pieces.

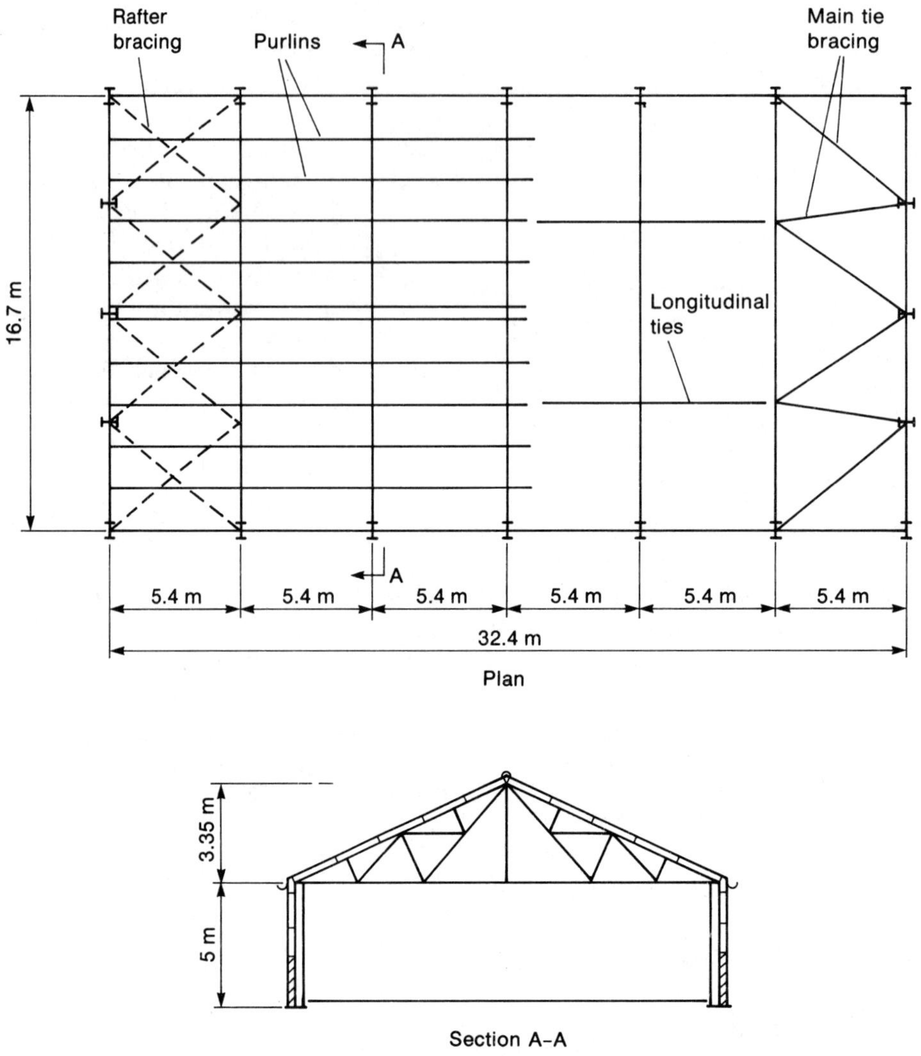

Rafter bracing
Purlins
A
Main tie bracing
16.7 m
Longitudinal ties
A
5.4 m
5.4 m
5.4 m
5.4 m
5.4 m
5.4 m
32.4 m
Plan
3.35 m
5 m
Section A–A

Fig. 7.4

CURPE CONSULTANTS 46 Orburn Road Dunfield	Project: Pitched roof structure		Job ref	
	Part of structure: Wind loading		Calc sheet no: C7 / 1	rev: /
	Drawing ref	Calc by: RAS	Date	Check by	Date

Ref CP3	Calculations	Output

Ex.1. Wind load on structure. Fig.7.4.

Table I — Basic wind speed $V = 45\ m/s$

Table 2 — Topography factor $S_1 = 1.0$

Ground roughness (3)

Class B loading

Height to top of roof $H = 8.35\,m$

Table 3 — Then for roof $S_2 = 0.71$

Height to top of wall $H = 5\,m$

Table 3 — Then for walls $S_2 = 0.65$

Fig. 2. — $S_3 = 1.0$

4.3(2) — Design wind speed $V_s = V.S_1.S_2.S_3.$

For roof $V_s = 45 \times 1 \times 0.71 \times 1 = 31.95$

For walls $V_s = 45 \times 1 \times 0.65 \times 1 = 29.25$

Table 4 — Dynamic pressure $q = K.V_s^2$

For roof $q = 0.613 \times 31.95^2 = 625\ N/m^2$

For walls $q = 0.613 \times 29.25^2 = 524\ N/m^2$

External pressure coefficients C_{pe}

Table 7 — (1) Walls

$\frac{h}{w} = \frac{5}{16.7} = 0.299 < \frac{1}{2}$

$\frac{\ell}{w} = \frac{32.4}{16.7} = 1.94 \qquad \frac{3}{2} < 1.94 < 4$

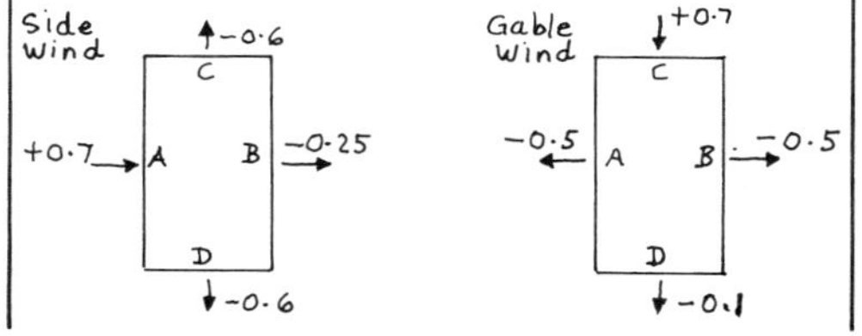

CURPE CONSULTANTS 46 Orburn Road Dunfield	Project: Pitched roof structure		Job ref	
	Part of structure: Wind loading		Calc sheet no: C7 / 2	rev: /
	Drawing ref	Calc by: R.A.S.	Date	Check by	Date

Ref CP3	Calculations	Output

Table 8 — (2) Roof

$\frac{h}{w} = \frac{5}{16.7} = 0.299 < \frac{1}{2}$

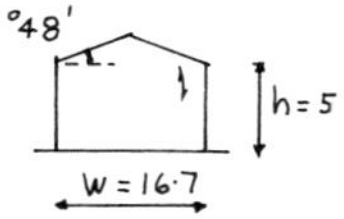

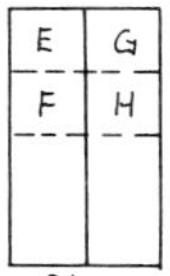

Plan.

	EF	GH
Side wind	−0.32	−0.4

	EG	FH
Gable wind	−0.7	−0.6

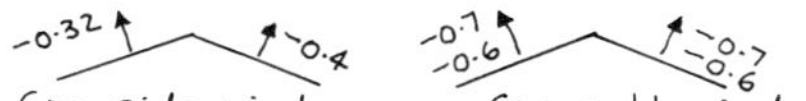

Cpe side wind   Cpe gable wind.

Combined Cpe and Cpi   Cases (a)(b)(c)(d).

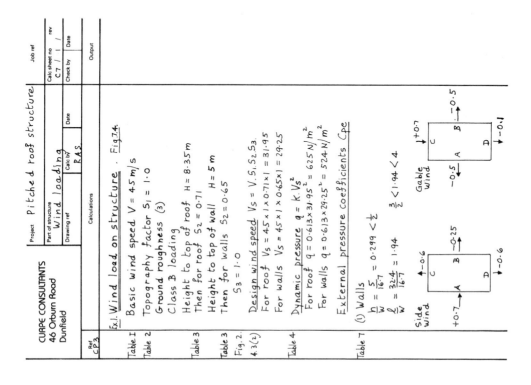

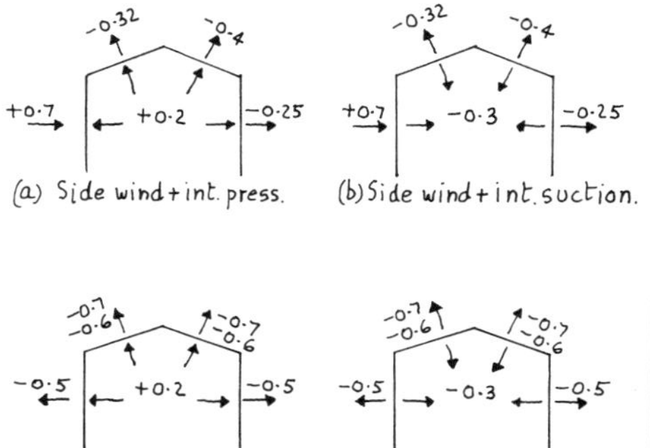

(a) Side wind + int. press.   (b) Side wind + int. suction.

(c) Gable wind + int. press.   (d) Gable wind + int. suction

CURPE CONSULTANTS 46 Orburn Road Dunfield	Project: Pitched roof structure	Job ref
	Part of structure: Roof purlins	Calc sheet no: C7 / 3 rev: 1
	Drawing ref: — Calc by: R.A.S. Date: —	Check by: — Date: —

Ref BS5950	Calculations	Output
CL.4.12.4.	Ex.2. **Design of Purlin** Fig. 7.4	

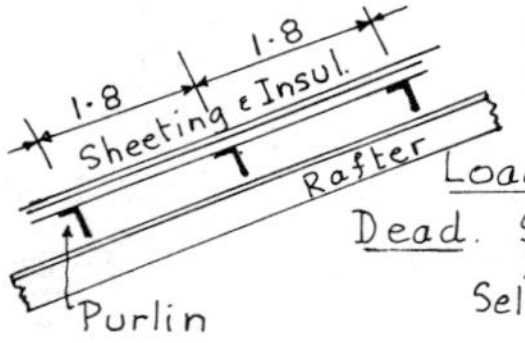

Span $L = 5400$ mm
Spacing = 1·8 m.

Loading

Dead. Sheeting 0·25
Insulation
Self wt purlin 0·10
(Slope) 0·35 kN/m^2

Ref	Calculations	Output
CL. 4.12.4.3	Imposed $0{\cdot}75\,kN/m^2$ (on plan) $= 0{\cdot}696\,kN/m^2$ (Slope)	
	Total unfactored load $= (0{\cdot}35 + 0{\cdot}696)\,1{\cdot}8 \times 5{\cdot}4$ $= 10{\cdot}167\,kN = W_p$.	
Table 29	Z_x reqd $= \frac{W_p . L}{1800} = \frac{10{\cdot}167 \times 5400}{1800} = 30{\cdot}5\,cm^3$	
"	Minimum $D = \frac{L}{45} = \frac{5400}{45} = 120$ mm	
"	" $B = \frac{L}{60} = \frac{5400}{60} = 90$ mm	
	$150 \times 90 \times 10$ ⌐ ($Z_x = 53{\cdot}3\,cm^3$)	USE $150 \times 90 \times 10$ ⌐
Table 29	Try using sag rods at mid span	or
	$B = \frac{2700}{60} = 45$ mm	
	$D = 120$ mm (as above)	
	$125 \times 75 \times 10$ ⌐ ($Z_x = 36{\cdot}5\,cm^3$)	USE $125 \times 75 \times 10$ ⌐ with sag rods mid-span.
	Cold rolled sections suitable – see Manufacturer's Catalogue for safe loads.	or Cold rolled sections with sag rods

CURPE CONSULTANTS 46 Orburn Road Dunfield	Project: Pitched roof structure	Job ref
	Part of structure: Roof truss design	Calc sheet no: C7 / 4 rev: 1
	Drawing ref: — Calc by: R.A.S. Date: —	Check by: — Date: —

Ref BS5950	Calculations	Output
	Ex.3. **Design of Roof Truss** Fig. 7.4.	

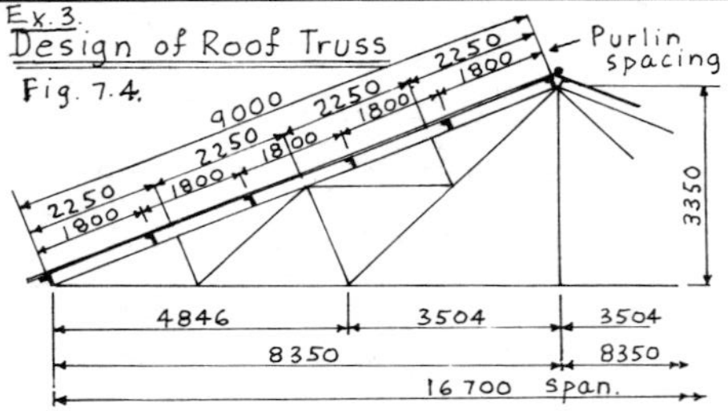

Loading data

Dead load: measured on slope.

Sheeting and insulation = 0·25
Purlin = 0·10
Truss = 0·10
0·45 kN/m^2

Imposed load $= 0{\cdot}75\,kN/m^2$ on plan corresponding to $0{\cdot}696\,kN/m^2$ on slope

Ref	Calculations	Output
CP3.	**Wind load**	
	Condition of maximum uplift occurs with longitudinal (gable) wind.	
	Dynamic pressure $q = 625\,N/m^2$	
	C_{pe} maximum $= -0{\cdot}7$ $C_{pi} = +0{\cdot}2$	

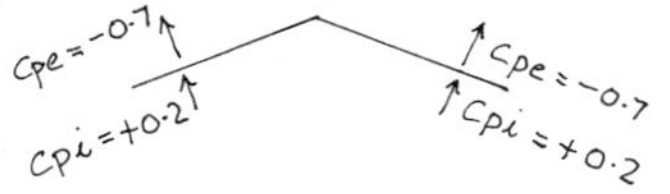

A horizontal bracing system and longitudinal ties support the main ties laterally against possible reversal of load

CURPE CONSULTANTS 46 Orburn Road Dunfield	Project: Pitched roof structure			Job ref	
	Part of structure: Roof truss design			Calc sheet no: C7 / 5	rev: /
	Drawing ref	Calc by: R.A.S.	Date	Check by	Date

Ref	Calculations	Output
	due to wind loading (with light roofs)	

<u>Truss loads</u>

<u>Dead</u> load per slope $= 0{\cdot}45 \times 9 \times 5{\cdot}4 = 21{\cdot}84$ kN

End panel point load $= \frac{21{\cdot}84}{8} = 2{\cdot}73$ kN

Internal panel point load $= \frac{21{\cdot}84}{4} = 5{\cdot}46$ kN

Forces in members due to dead load are found by means of a force diagram.

Forces due to <u>imposed</u> load are found by proportion, that is:

$$\text{Force due to dead load} \times \frac{0{\cdot}696}{0{\cdot}45}$$

Wind load per slope $= q.A.(C_{pe} - C_{pi})$

$$= \frac{625}{10^3} \times 9 \times 5{\cdot}4\left[-0{\cdot}7 - (0{\cdot}2)\right] = 27{\cdot}3 \text{ kN.}$$

End panel load $= \frac{27{\cdot}3}{8} = 3{\cdot}41$ kN

Internal panel point load $= \frac{27{\cdot}3}{4} = 6{\cdot}82$ kN

Forces in members due to wind are found by means of a force diagram.

Force diagrams for dead and wind loads are shown on the next sheets with suitable scales for dimensions and forces.

C7/6

All forces kN (unfactored)

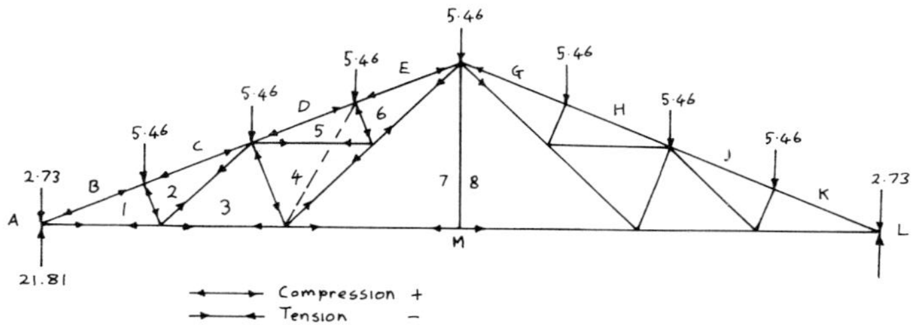

Frame Diagram

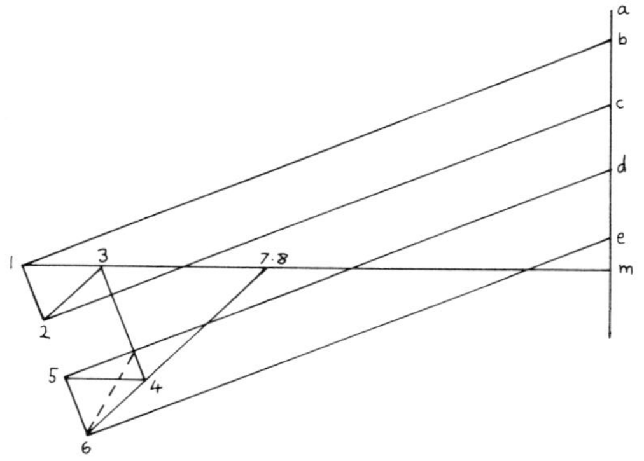

Force Diagram – Dead Loads

C7/7

All Forces kN (unfactored)

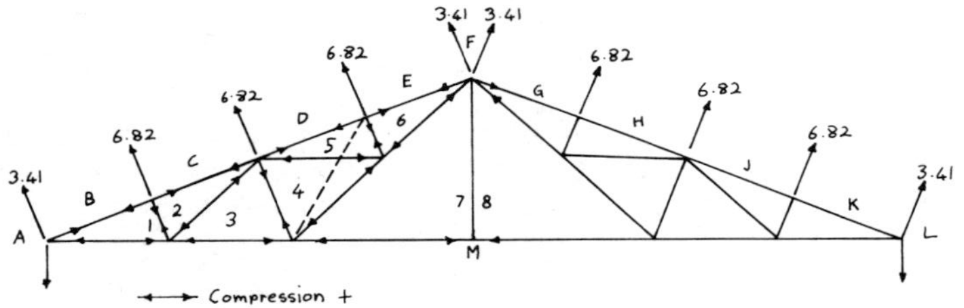

Frame Diagram

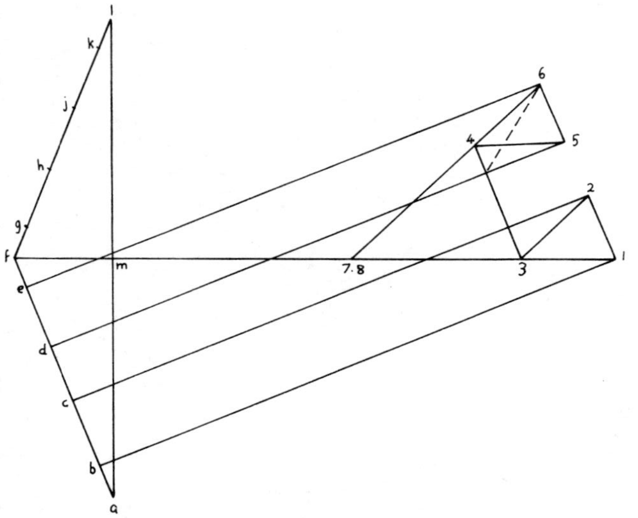

Force Diagram – wind Loads

CURPE CONSULTANTS 46 Orburn Road Dunfield	Project: Pitched roof structure			Job ref
	Part of structure: Roof truss design			Calc sheet no: C7 / 8 rev: /
	Drawing ref	Calc by: R.A.S.	Date	Check by / Date

Ref: BS5950 | Calculations | Output

Member	Dead (D)	Imposed (I)	Wind (W)	1.4(D)	1.6(I)	1.4(W)	1.4(D) + 1.6(I)	1.0(D) + 1.4(W)
B1	+51.5	+79.6	−58.0	+72.1	+127.4	−81.2	+199.5	−29.7
C2	+49.5	+76.5	−58.0	+69.3	+122.4	−81.2	+191.7	−31.7
D5	+47.5	+73.4	−58.0	+66.5	+117.4	−81.2	+183.9	−33.7
E6	+45.5	+70.3	−58.0	+63.7	+112.5	−81.2	+176.2	−35.7
M1	−48.0	−74.2	+52.0	−67.2	−118.7	+72.8	−185.9	+24.8
M3	−41.5	−64.1	+42.5	−58.1	−102.6	+59.5	−160.7	+18.0
M7	−27.2	−42.1	+26.4	−38.1	−67.4	+37.0	−105.5	+9.8
1-2	+5.0	+7.7	−6.8	+7.0	+12.3	−9.5	+19.3	−4.5
2-3	−6.5	−10.0	+10.0	−9.1	−16.0	+14.0	−25.1	+7.5
3-4	+10.0	+15.4	−13.0	+14.0	+24.6	−18.2	+38.6	−8.2
4-5	−6.5	−10.0	+10.0	−9.1	−16.0	+14.0	−25.1	+7.5
5-6	+5.0	+7.7	−6.8	+7.0	+12.3	−9.5	+19.3	−4.5
4-7	−13.5	−20.9	+17.5	−18.9	−33.4	+24.5	−52.3	+11.0
6-7	−20.0	−30.9	+27.0	−28.0	−49.4	+37.8	−77.4	+17.8
7-8	0	0	0	0	0	0	0	0

Forces in members + Compression − Tension

Moments in Rafters

Rafter analysed as a continuous beam:

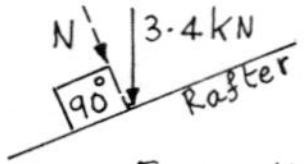

Dead load to intermediate purlin $= 0.35 \times 1.8 \times 5.4$
$= 3.4\,kN$

$$\text{Force } N = 3.4 \times \frac{8.35}{9} = 3.154 \text{ kN}$$

Unfactored loads

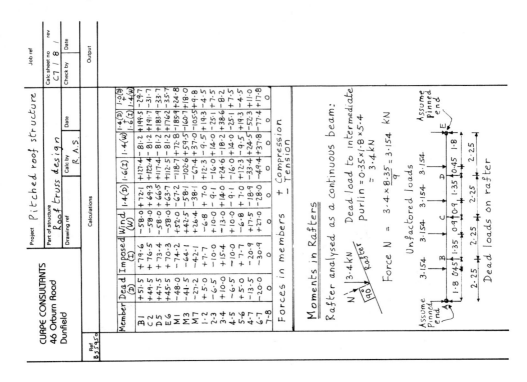

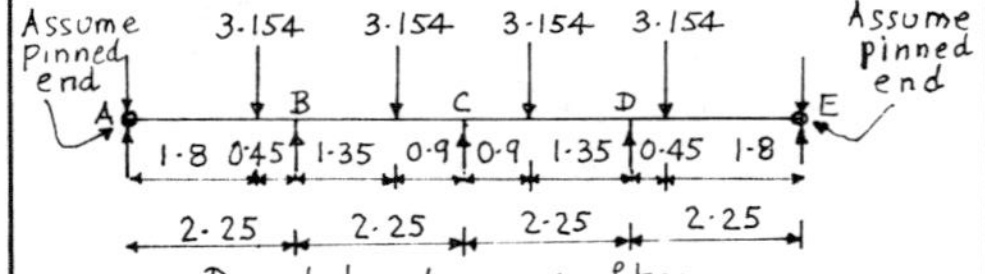

Dead loads on rafter

CURPE CONSULTANTS 46 Orburn Road Dunfield	Project: Pitched roof structure			Job ref
	Part of structure: Roof truss design			Calc sheet no: C7 / 9 / rev
	Drawing ref	Calc by: R A S.	Date	Check by / Date

Ref	Calculations	Output
BS5950		

Fixed End Moments

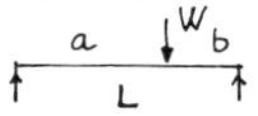

Span A-B

$M_{AB} = \frac{Wab^2}{L^2} = \frac{3\cdot154 \times 1\cdot8 \times 0\cdot45^2}{2\cdot25^2} = 0\cdot227$ kNm

$M_{BA} = \frac{Wa^2b}{L^2} = \frac{3\cdot154 \times 1\cdot8^2 \times 0\cdot45}{2\cdot25^2} = 0\cdot908$ kNm

Span B-C

$M_{BC} = \frac{3\cdot154 \times 1\cdot35 \times 0\cdot9^2}{2\cdot25^2} = 0\cdot681$ kNm

$M_{CB} = \frac{3\cdot154 \times 1\cdot35^2 \times 0\cdot9}{2\cdot25^2} = 1\cdot022$ kNm

Distribution factors

Joint B. $K_{BA} = 0\cdot75$, $K_{BC} = 1\cdot0$

$\therefore$ $D.F._{BA} = \frac{0\cdot75}{0\cdot75 + 1\cdot0} = 0\cdot4286$

$D.F._{BC} = \frac{1\cdot0}{0\cdot75 + 1\cdot0} = 0\cdot5714$

Joint C. $D.F._{CB} = D.F._{CD} = 0\cdot5$

	0·4286	0·5714	0·5	0·5	0·5714	0·4286	
−0·227	+0·908	−0·681	+1·022	−1·022	+0·681	−0·908	+0·227
+0·227	+0·113					−0·113	−0·227
0	+1·021	−0·681	+1·022	−1·022	+0·681	−1·021	0
	−0·146	−0·194	0	0	+0·194	+0·146	
			−0·097	+0·097			
0	+0·875	−0·875	+0·925	−0·925	+0·875	−0·875	0

Moment Distribution

Free Moments

Span A-B $M = \frac{Wab}{L} = \frac{3\cdot154 \times 1\cdot8 \times 0\cdot45}{2\cdot25} = 1\cdot135$ kNm

Span B-C $M = \frac{3\cdot154 \times 1\cdot35 \times 0\cdot9}{2\cdot25} = 1\cdot703$ kNm

CURPE CONSULTANTS 46 Orburn Road Dunfield	Project: Pitched roof structure			Job ref
	Part of structure: Roof truss design			Calc sheet no: C7 / 10 / rev
	Drawing ref	Calc by: R. A. S.	Date	Check by / Date

Ref	Calculations	Output
BS5950		

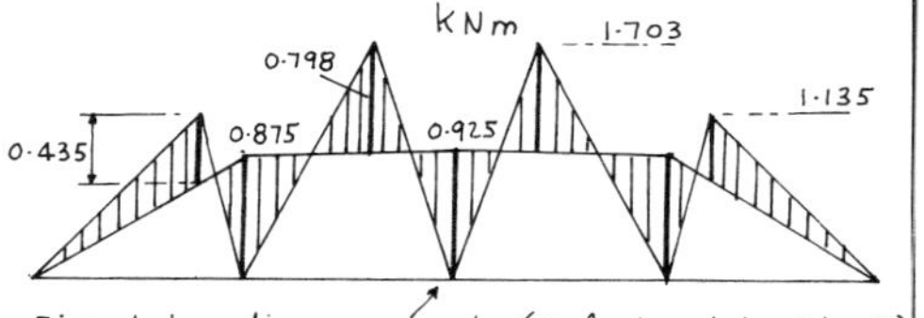

Final bending moments (Unfactored dead loads)

Imposed load to intermediate purlin

$= 0\cdot696 \times 1\cdot8 \times 5\cdot4 = 6\cdot76$ kN

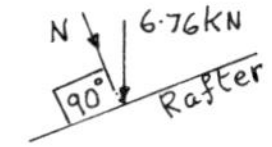

Force $N = 6\cdot76 \text{ kN} \times \frac{8\cdot35}{9} = 6\cdot27$ kN

Bending moments (by proportion) Unfactored

$0\cdot435 \times \frac{6\cdot27}{3\cdot154} = 0\cdot865$ kNm

$0\cdot875 \times \frac{6\cdot27}{3\cdot154} = 1\cdot739$ kNm

$0\cdot798 \times \frac{6\cdot27}{3\cdot154} = 1\cdot586$ kNm

$0\cdot925 \times \frac{6\cdot27}{3\cdot154} = 1\cdot839$ kNm

Combined Bending Moments due to factored dead + imposed loads.

$1\cdot4(0\cdot435) + 1\cdot6(0\cdot865) = 1\cdot99$ kNm

$1\cdot4(0\cdot875) + 1\cdot6(1\cdot739) = 4\cdot01$ kNm

$1\cdot4(0\cdot798) + 1\cdot6(1\cdot586) = 3\cdot65$ kNm

$1\cdot4(0\cdot925) + 1\cdot6(1\cdot839) = 4\cdot24$ kNm

CURPE CONSULTANTS 46 Orburn Road Dunfield	Project: Pitched roof structure	Job ref
	Part of structure: Roof truss design	Calc sheet no: C7 / 11 / rev
	Drawing ref / Calc by: R.A.S / Date	Check by / Date

Ref BS5950	Calculations	Output

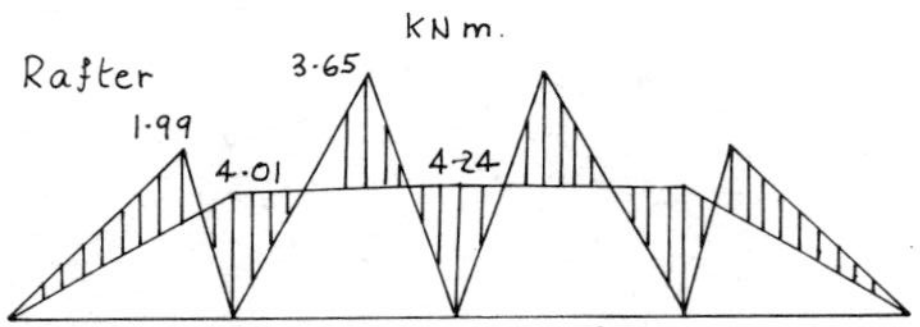

Combined Bending Moments - factored (D) + (I)

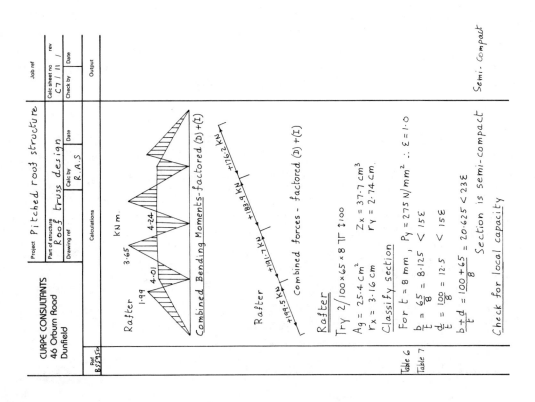

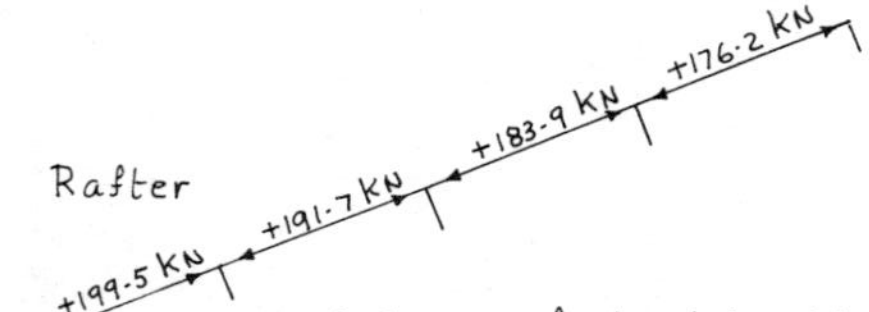

Combined forces - factored (D) + (I)

Rafter

Try 2/100×65×8 ⊤⊤ ↕100

$A_g = 25{\cdot}4\ cm^2$ $\quad Z_x = 37{\cdot}7\ cm^3$

$r_x = 3{\cdot}16\ cm$ $\quad r_y = 2{\cdot}74\ cm$

Classify section

Table 6 — For $t = 8\ mm$, $P_y = 275\ N/mm^2$ $\therefore \varepsilon = 1{\cdot}0$

Table 7 — $\frac{b}{t} = \frac{65}{8} = 8{\cdot}125 < 15\varepsilon$

$\frac{d}{t} = \frac{100}{8} = 12{\cdot}5 < 15\varepsilon$

$\frac{b+d}{t} = \frac{100+65}{8} = 20{\cdot}625 < 23\varepsilon$

Section is semi-compact — **Output:** Semi-Compact

Check for local capacity

CURPE CONSULTANTS 46 Orburn Road Dunfield	Project: Pitched roof structure	Job ref
	Part of structure: Roof truss design	Calc sheet no: C7 / 12 / rev
	Drawing ref / Calc by: RAS / Date	Check by / Date

Ref	Calculations	Output

CL. 4.8.3.2.

$$\frac{F}{A_g.p_y} + \frac{M_x}{M_{cx}} + \frac{M_y}{M_{cy}} \not> 1{\cdot}0$$

$$A_g.p_y = \frac{25{\cdot}4 \times 10^2 \times 275}{10^3} = 698{\cdot}5\ KN.$$

$$M_{cx} = p_y.Z_x = \frac{275 \times 37{\cdot}7 \times 10^3}{10^6} = 10{\cdot}367\ KNm.$$

At point B

$F = 199{\cdot}5\ KN$ $\qquad M_x = 4{\cdot}01\ KNm$

$$\frac{199{\cdot}5}{698{\cdot}5} + \frac{4{\cdot}01}{10{\cdot}367} + 0 = 0{\cdot}2856 + 0{\cdot}3868$$

$= 0{\cdot}6724 < 1{\cdot}0$ Satisfactory — **Output:** Rafter Local capacity Sat.

At point C

$F = 191{\cdot}7\ KN$ $\qquad M_x = 4{\cdot}24\ KNm$

$$\frac{191{\cdot}7}{698{\cdot}5} + \frac{4{\cdot}24}{10{\cdot}367} + 0 = 0{\cdot}2744 + 0{\cdot}409$$

$= 0{\cdot}6834 < 1{\cdot}0$ satisfactory

Check for overall buckling

CL. 4.8.3.3.1 — Using simplified apprach

$$\frac{F}{A_g.p_c} + \frac{m.M_x}{M_b} + \frac{m.M_y}{p_y.Z_y} \not> 1{\cdot}0$$

Slenderness:

Tables 25, 27(c)

$$\lambda_x = \frac{0{\cdot}85L_1}{r_x} = \frac{0{\cdot}85 \times 2{\cdot}25 \times 10^3}{3{\cdot}16 \times 10} = 60{\cdot}5 \not> 180$$

For $p_y = 275$ $\qquad p_{cx} = 200\ N/mm^2$

$$\lambda_y = \frac{1{\cdot}0 L_2}{r_y} = \frac{1{\cdot}0 \times 1{\cdot}8 \times 10^3}{2{\cdot}74 \times 10} = 65{\cdot}7 \not> 180$$

For $p_y = 275$ $\qquad p_{cy} = 189{\cdot}96\ N/mm^2$

$\therefore p_c = 189{\cdot}96\ N/mm^2$ the lesser value

CL. 4.7.4 — Comp. resistance $P_c = A_g.p_c$

$$= \frac{25{\cdot}4 \times 10^2 \times 189{\cdot}96}{10^3} = 482{\cdot}5\ KN$$

$$M_{cx} = p_y.Z_x = \frac{275 \times 37{\cdot}7 \times 10^3}{10^6} = 10{\cdot}367\ KNm.$$

CURPE CONSULTANTS 46 Orburn Road Dunfield	Project: Pitched roof structure	Job ref
	Part of structure: Roof truss design	Calc sheet no: C7/ 13 / rev
	Drawing ref / Calc by: R.A.S. / Date	Check by / Date

Ref BS5950	Calculations	Output
CL.4.3.8	Since $\lambda = 65.7 < 100$ then $M_b = 0.8\,p_y.Z$	
	$M_b = \frac{0.8 \times 275 \times 37.7 \times 10^3}{10^6} = 8.294$ kNm	
	$\frac{F}{A_g.p_c} + \frac{m.M_x}{M_b} + \frac{m.M_y}{p_y.Z_y} \not> 1.0$	
	B-1 $\frac{199.5}{482.5} + \frac{1.0 \times 4.01}{8.294} + 0 = 0.896 < 1.0$	Sat. so far but see re-check below
	C-2 $\frac{191.7}{482.5} + \frac{1.0 \times 4.24}{8.294} + 0 = 0.9085 < 1.0$	
	Now check slenderness of one angle component between adjacent connections	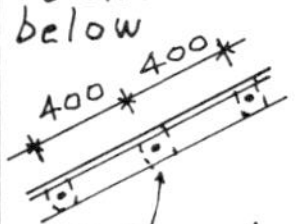
	Spacing of single bolts connecting angles together say $50t = 50 \times 8 = 400$ mm	
4.7.9(c)	$\lambda_b = \sqrt{\lambda_m^2 + \lambda_c^2}$	
	$\lambda_m = \frac{L_2}{r_y} = 65.7$ from previous calcs	
	$\lambda_c = \frac{400}{r_v} = \frac{400}{1.4 \times 10} = 28.5 < 50$	
	$\lambda_b = \sqrt{65.7^2 + 28.5^2} = 71.5$	
	$\lambda_x = 60.5$ from previous calcs	
	For $p_y = 275$ $p_c = 179$ N/mm² for $\lambda = 71.5$	
	Since $\lambda < 100$ then $M_b = 8.294$ kNm as before	
	$\therefore P_c = A_g.p_c = \frac{25.4 \times 10^2 \times 179}{10^3} = 454.6$ kN	
	$\frac{F}{A_g.p_c} + \frac{m.M_x}{M_b} + \frac{m.M_y}{p_y.Z_y} \not> 1.0$	
	B-1 $\frac{199.5}{454.6} + \frac{1.0 \times 4.01}{8.294} + 0 = 0.9223 < 1.0$	Rafter Use 2/100×65×8 ⊤ connected together as shown
	C-2 $\frac{191.7}{454.6} + \frac{1.0 \times 4.24}{8.294} + 0 = 0.9328 < 1.0$	

CURPE CONSULTANTS 46 Orburn Road Dunfield	Project: Pitched roof structure	Job ref
	Part of structure: Roof truss design	Calc sheet no: C7 / 14 / rev
	Drawing ref / Calc by: R.A.S. / Date	Check by / Date

Ref BS5950	Calculations	Output
	Main Tie	
	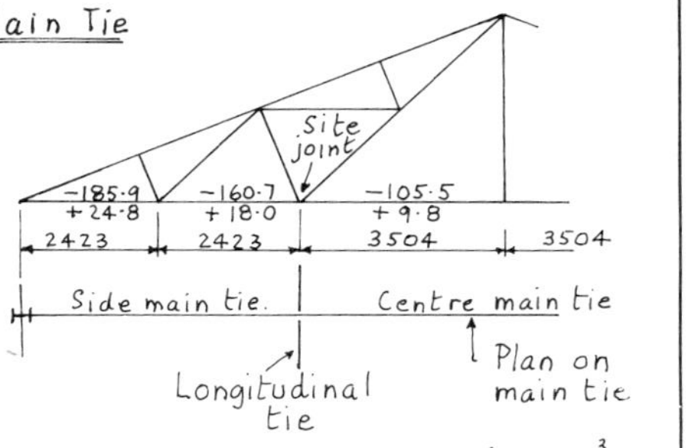	
	Side main tie: M1–M3. Ae reqd $= \frac{185.9 \times 10^3}{275} = 676$	
	Max. forces $F = -185.9$ kN Tension (D+I)	
	$F = +24.8$ kN Compression (D+W)	
	Try 80×60×8L with long leg connected (vert)	
CL. 4.7.3.2	As strut $\lambda_y = \frac{L_{EY}}{r_y} = \frac{2 \times 2423}{1.73 \times 10} = 280 \not> 350$	
	$\lambda_v = \frac{L_{EV}}{r_v} = \frac{2423}{1.27 \times 10} = 191 \not> 350$	
Table 27(c)	$p_c = 23$ N/mm² using larger λ	
CL.4.7.4	$P_c = A_g.p_c = \frac{10.6 \times 10^2 \times 23}{10^3} = 24.38$ kN	
	Since $24.8 > 24.38$ Inadequate	80×60×8L Inadequate
	Try 80×80×8L. Section is semi-compact.	Semi-compact
CL. 4.7.3.2	As strut $\lambda_y = \frac{L_{EY}}{r_y} = \frac{2 \times 2423}{2.43 \times 10} = 199.4 \not> 350$	
	$\lambda_v = \frac{L_{EV}}{r_v} = \frac{2423}{1.56 \times 10} = 155.3 \not> 350$	
Table 27(c)	$p_c = 42.1$ N/mm² using larger λ	
CL.4.7.4	$P_c = A_g.p_c = \frac{12.3 \times 10^2 \times 42.1}{10^3} = 51.7$ kN	
	Since $24.8 < 51.7$, Adequate as strut	80×80×8L OK as strut

CURPE CONSULTANTS
46 Orburn Road
Dunfield

Project: Pitched roof structure			Job ref
Part of structure: Roof truss design			Calc sheet no: C7 / 15 / rev
Drawing ref	Calc by: RAS	Date	Check by / Date

Ref BS5950	Calculations	Output
	Check as tie:	
	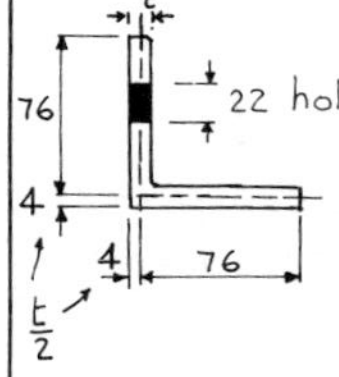	
	$a_1 = (76-22)\,8 = 432\,mm^2$	
	$a_2 = (76)\,8 = 608\,mm^2$	
	$A_e = a_1 + \left[a_2\left(\frac{3a_1}{3a_1+a_2}\right)\right]$	
	$= 432 + \left[608\left(\frac{3\times432}{1296+608}\right)\right]$	
	$= 432 + 413.8 = 845.8\,mm^2$	
	Tension capacity $= P_t = A_e.p_y$	
	$= \frac{845.8\times275}{10^3} = 232.6\,kN$	o.k. as tie, strut
	Since $185.9 < 232.6$, Adequate as tie	Use 80×80×8 L
	Centre main tie M7 – M8. A_e reqd $= \frac{105.5\times10^3}{275} = 383$	
	Max forces $F = -105.5\,kN$ Tension (D+I)	
	$F = +9.8\,kN$ Compression (D+W)	
	Try 80×80×8 L	
CL. 4.7.3.2	As strut $\lambda_y = \frac{L_{EY}}{r_Y} = \frac{2\times3504}{2.43\times10} = 288.4 \not> 350$	
	$\lambda_v = \frac{L_{EV}}{r_V} = \frac{3504}{1.56\times10} = 224.6 \not> 350$	
Table 27(c)	$p_c = 21.3\,N/mm^2$ using larger λ	
CL. 4.7.4	$P_c = A_g.p_c = \frac{12.3\times10^2\times21.3}{10^3} = 26.2\,kN$	o.k. as strut, tie
	Since $9.8 < 26.2$, Adequate as strut	Use
	Since $105.5 < 232.6$, " " tie	80×80×8 L
	Alternative section for main tie (using double angles) A_e reqd $= \frac{185.9\times10^3}{275} = 676\,mm^2$	
	Try 2/50×50×6 JL separated by gussets.	
	M1 – M3	
CL. 4.7.3.2	As strut $\lambda_x = \frac{L_{EX}}{r_X} = \frac{2423}{1.5\times10} = 161.5 \not> 350$	

CURPE CONSULTANTS
46 Orburn Road
Dunfield

Project: Pitched roof structure			Job ref
Part of structure: Roof truss design			Calc sheet no: C7 / 16 / rev
Drawing ref	Calc by: RAS	Date	Check by / Date

Ref BS5950	Calculations	Output
	$\lambda_y = \frac{L_{EY}}{r_Y} = \frac{4846}{2.38\times10} = 203.6 \not> 350$	
Table 27c	$p_c = 40.6\,N/mm^2$ using larger λ	
	$P_c = A_g.p_c = \frac{11.4\times10^2\times40.6}{10^3} = 46.3\,kN > 24.8$	o.k. as strut
	As tie	
	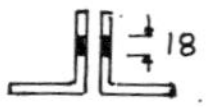	
CL. 4.6.3.3	Effective area = gross area − holes	
	$A_e = (11.4\times10^2) - (2\times18\times6)$	
	$= 924\,mm^2$	o.k. as tie
CL. 4.6.1.	Tension capacity $P_t = A_e.p_y$	Alternatively
	$= \frac{924\times275}{10^3} = 254\,kN > 185.9$	Use 2/50×50×6 JL
	M7 – M8. $F = -105.5$, $F = +9.8$. A_e reqd $= \frac{105.5\times10^3}{275} = 383$	
	As strut $\lambda_x = \frac{L_{EX}}{r_X} = \frac{3504}{1.5\times10} = 233.6 \not> 350$	
	$\lambda_y = \frac{L_{EY}}{r_Y} = \frac{2\times3504}{2.38\times10} = 294.4 \not> 350$	
	$p_c = 20.5\,N/mm^2$ using larger λ	
CL. 4.7.4	$P_c = A_g.p_c = \frac{11.4\times10^2\times20.5}{10^3} = 23.4\,kN > 9.8$ Adequate	o.k. as strut, tie
	As tie $P_t = 254\,kN > 105.5$ Adequate	Alternatively Use 2/50×50×6 JL
	Member 3-4 Discontinuous strut, $L = 1.8\,m$	
	$F = +38.6\,kN$ Compression (D+I)	
	$F = -8.2\,kN$ Tension (D+W)	
	Try 50×50×6 L	
Table 7	Section classification $\frac{b}{t} = \frac{d}{t} = \frac{50}{6} = 8.33$	
	$\varepsilon = 1.0$ ∴ Section not slender	
Table 28	$\lambda = \frac{0.85L}{r_{vv}} = \frac{0.85\times1.8\times10^3}{0.968\times10} = 158 \not> 180$	
	Also $\lambda = \frac{0.7L}{r_{aa}} + 30 = \frac{0.7\times1.8\times10^3}{1.5\times10} + 30 = 114 \not> 180$	
Table 27(c)	$p_c = 62.6\,N/mm^2$ using larger λ	

CURPE CONSULTANTS 46 Orburn Road Dunfield	Project: Pitched roof structure	Job ref
	Part of structure: Roof truss design	Calc sheet no: C7 / 17 / rev
	Drawing ref: — Calc by: RAS Date: —	Check by: — Date: —

Ref BS5950	Calculations	Output
CL.4.7.4	$P_c = A_g \cdot p_c = \frac{5.69 \times 10^2 \times 62.6}{10^3} = 35.6$ kN.	
	Since $38.6 > 35.6$ section is inadequate	50×50×6 L Inadequate
	Try 60×60×6 L	
Table 7	Section classification $\frac{b}{t} = \frac{d}{t} = \frac{60}{6} = 10 < 15\varepsilon$	
	$\frac{b+d}{t} = \frac{60+60}{6} = 20 < 23\varepsilon$ section is semi-compact	Semi-compact.
Table 28	$\lambda = \frac{0.85L}{r_{vv}} = \frac{0.85 \times 1.8 \times 10^3}{1.17 \times 10} = 130.8$	
	$\lambda = \frac{0.7L}{r_{aa}} + 30 = \frac{0.7 \times 1.8 \times 10^3}{1.82 \times 10} + 30 = 99.2$	
Table 27c	$p_c = 85$ N/mm^2 using larger λ	
	$P_c = A_g \cdot p_c = \frac{6.91 \times 10^2 \times 85}{10^3} = 58.7$ kN	
	Since $38.6 < 58.7$ Section is adequate	Use 60×60×6L
	Members 1-2, 5-6 Discontinuous strut	
	L = 0.9 m. Try 50×50×6 L (as minimum)	
	F = +19.3 Compression (D + I)	
	F = −4.5 Tension (D + W)	
	Section not slender from previous calculation	
Table 28	Strut $\lambda = \frac{0.85L}{r_{vv}} = \frac{0.85 \times 0.9 \times 10^3}{0.968 \times 10} = 79 \not> 180$	
	$\lambda = \frac{0.7L}{r_{aa}} + 30 = \frac{0.7 \times 0.9 \times 10^3}{1.5 \times 10} + 30 = 72 \not> 180$	
Table 27c	$p_c = 163$ N/mm^2 using larger λ	
	$P_c = A_g \cdot p_c = \frac{5.69 \times 10^2 \times 163}{10^3} = 92.7$ kN	
	Since $19.3 < 92.7$ section adequate	O.K. as strut
	Tie 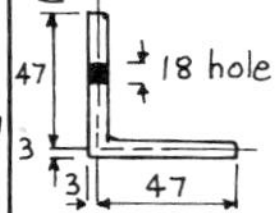	
	$a_1 = (47 - 18)6 = 174$ mm^2	
	$a_2 = (47)6 = 282$ mm^2	
CL.4.6.3.1	$A_e = a_1 + \left[a_2\left(\frac{3a_1}{3a_1 + a_2}\right)\right]$	
	$= 174 + \left[282\left(\frac{3 \times 174}{522 + 282}\right)\right]$	

CURPE CONSULTANTS 46 Orburn Road Dunfield	Project: Pitched roof structure	Job ref
	Part of structure: Roof truss design	Calc sheet no: C7 / 18 / rev
	Drawing ref: — Calc by: RAS. Date: —	Check by: — Date: —

Ref BS5950	Calculations	Output
	$A_e = 357$ mm^2	
CL.4.6.1	$P_t = A_e \cdot p_y = \frac{357 \times 275}{10^3} = 98.2$ kN	o.K. as tie Use 50×50×6 L
	Since $4.5 < 98.2$ kN section adequate	
	Members 2-3, 4-5	
	F = −25.1 kN Tension (D + I)	
	F = +7.5 kN Compression (D + W)	
	L = 2423 mm	
	Try 50×50×6 L (as minimum)	
	As tie:	
	P_t = 98.2 kN from above	
	Since $25.1 < 98.2$, section adequate as tie.	o.K. as tie
	As strut: Section not slender	
	$\lambda = \frac{0.85L}{r_{vv}} = \frac{0.85 \times 2423}{0.968 \times 10} = 212.8 \not> 350$	
	$\lambda = \frac{0.7L}{r_{aa}} + 30 = \frac{0.7 \times 2423}{1.5 \times 10} + 30 = 143 \not> 350$	
Table 27c	$p_c = 37.1$ N/mm^2 using larger λ	
	$P_c = \frac{5.69 \times 10^2 \times 37.1}{10^3} = 21.1$ kN	
	Since $7.5 < 21.1$, section adequate as strut	o.K. as strut Use 50×50×6 L
	Member 7-8	
	F = 0 L = 3350 mm	
	This member restrains centre main tie at mid point so design for slenderness 180 max.	
	Minimum $r_{vv} = \frac{0.85L}{180} = \frac{3350}{180} \times 0.85 = 15.8$ mm	
	80×80×6 L, r_{vv} = 15.7 mm, consider O.K.	Use 80×80×6 L
	Member 4-7, 6-7	
	A_e reqd $= \frac{77.4 \times 10^3}{275} = 281.4$ mm^2	

CURPE CONSULTANTS 46 Orburn Road Dunfield	Project: Pitched roof structure		Job ref	
	Part of structure: Roof truss design		Calc sheet no: C7 / 19	rev: 1
	Drawing ref	Calc by: RAS / Date	Check by	Date

Ref BS5950	Calculations	Output

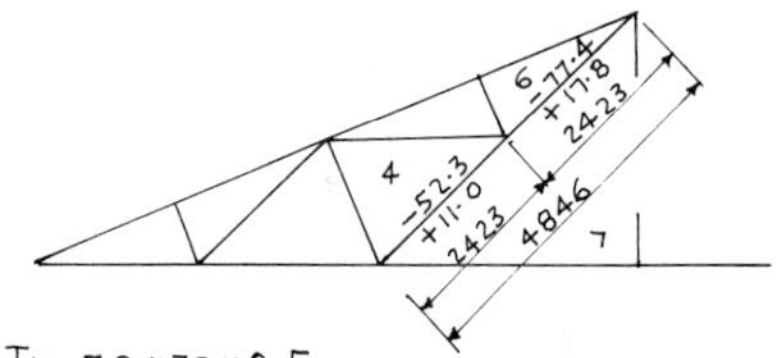

Try 70 × 70 × 8 ⌐

Member 6-7 as strut

Table 28 — $\lambda = \dfrac{0.85L}{r_{vv}} = \dfrac{0.85 \times 2423}{1.36 \times 10} = 151.4 \not> 350$

$\lambda = \dfrac{0.7L}{r_{aa}} + 30 = \dfrac{0.7 \times 2423}{2.11 \times 10} + 30 = 110.3 \not> 350$

Overall length member as strut

Table 28 — $\lambda = \dfrac{0.85L}{r_{YY}} = \dfrac{0.85 \times 4846}{2.11 \times 10} = 195$

$\lambda = \dfrac{0.7L}{r_{aa}} + 30 = \dfrac{0.7 \times 4846}{2.11 \times 10} + 30 = 190.8$

Table 27c — $p_c = 43\ \text{N/mm}^2$ for $\lambda = 195$

CL. 4.7.4 — $P_c = A_g \cdot p_c = \dfrac{10.6 \times 10^2 \times 43}{10^3} = 45.5\,\text{kN}$

Since 17.8 < 45.5, section adequate as strut — **Output:** o.K. as strut

As tie:

Connected to gusset plate M20 bolt, 22 dia hole

Effective area $A_e = 704\ \text{mm}^2$

$P_t = 194\,\text{kN}$

Since 77.4 < 194, section adequate as tie — **Output:** o.K. as tie. use 70×70×8 ⌐

8

Composite construction

General requirements

This chapter will consider the technique of design in which two major structural materials, concrete and steel, work together integrally to form composite beam sections.

Floors and roofs in buildings are commonly made of *in situ* reinforced concrete slabs supported by reinforced concrete beams, or alternatively by structural steel beams. The monolithic nature of the concrete slabs and beams makes it possible to design the internal beams as T-beams and the perimeter beams as L-beams. Where steel beams are employed it used to be normal practice to design the beams generally as simply supported, to carry the concrete slab, imposed load and finishes.

With the introduction of mechanical shear connectors, it is possible to connect the concrete slab to the steel beam so as to obtain T-beams or L-beam action, so long used to advantage in the design of reinforced concrete structures. The slab supports the applied loads such as imposed load and finishes, and transfers these loads to the supporting beams. It also acts as the top flange of a composite T- or L-beam section.

For beams, composite action may be applied to:

1) a concrete slab resting on and anchored to the top flange of a steel beam by means of mechanical shear connectors which are welded to the steel beam at calculated spacing;
2) slabs with permanent formwork consisting of:
   a) prestressed concrete planks resting on the top flange of a steel beam with *in situ* concrete topping embedding the shear connectors;
   b) profiled steel sheet decking fixed to the steel beam, with *in situ* concrete topping embedding the shear connectors. Bond may be provided at the steel–concrete interface by, for example, dimples formed at manufacture on the steel sheets.
3) totally concrete-encased steel beams, the concrete being cast integrally with the concrete slab. No shear connectors are necessary in this case if certain conditions are met.

For illustrations, see Fig. 8.1.

Only simply supported composite beam sections, with the concrete slab resting on and anchored to the top flange of a steel beam by means of stud-headed shear connectors, as at (1) above, will be considered here and designed in accordance with CP 117: Part 1.

Effective width of concrete compression flange

CP 117: Part 1 presents rules to determine how much width of the concrete slab should be considered as effectively working with the steel beam for composite action. For a T-beam section, the effective width of the concrete flange must not be greater than the least of the following:

1) one-third of the effective span of the beam,

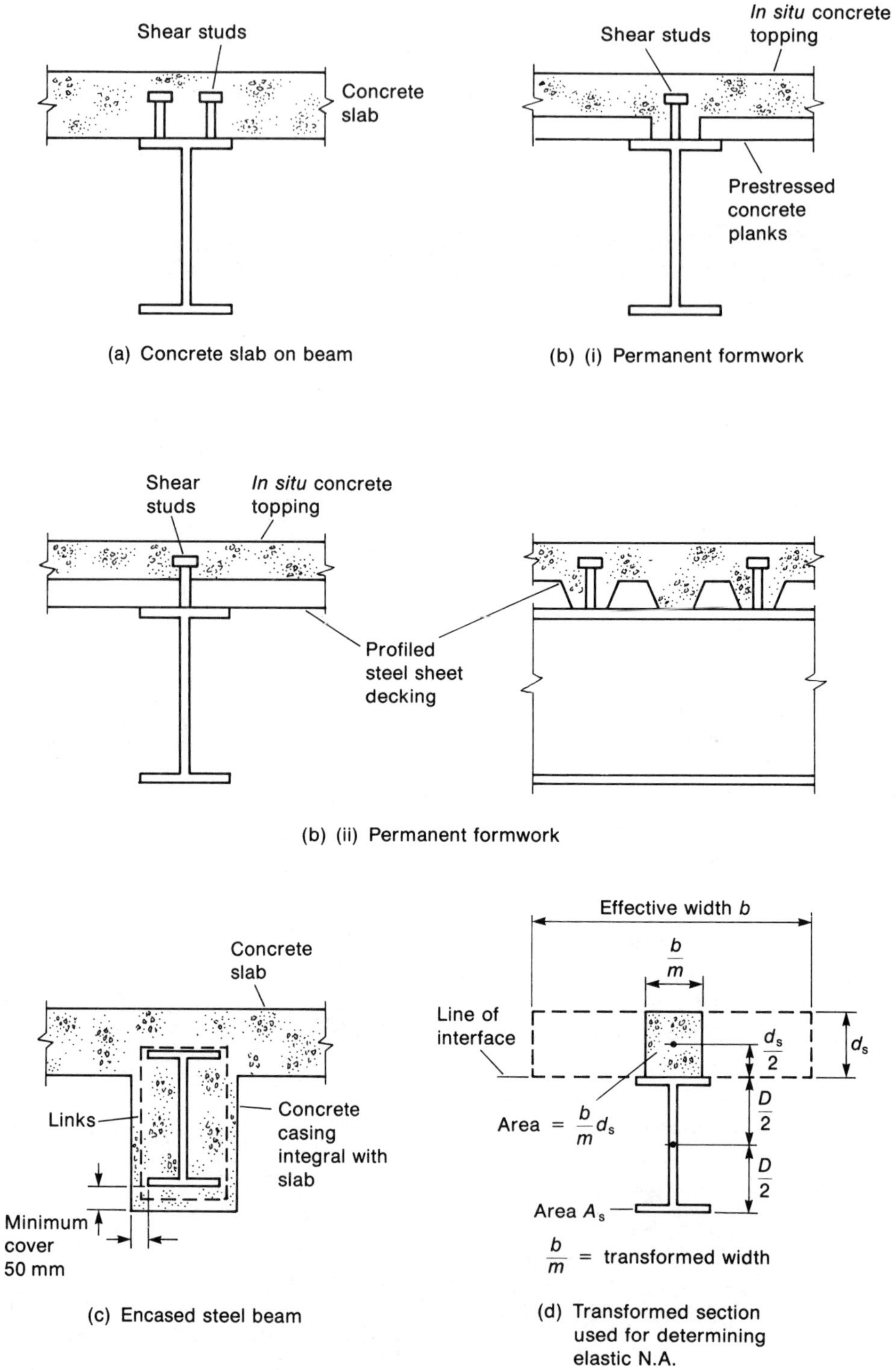
Shear studs
Concrete slab
(a) Concrete slab on beam
Shear studs
In situ concrete topping
Prestressed concrete planks
(b) (i) Permanent formwork
Shear studs
In situ concrete topping
Profiled steel sheet decking
(b) (ii) Permanent formwork
Concrete slab
Links
Concrete casing integral with slab
Minimum cover 50 mm
(c) Encased steel beam
Effective width b
$\frac{b}{m}$
Line of interface
$\frac{d_s}{2}$
d_s
Area $= \frac{b}{m} d_s$
$\frac{D}{2}$
$\frac{D}{2}$
Area A_s
$\frac{b}{m}$ = transformed width
(d) Transformed section used for determining elastic N.A.

Fig. 8.1

2) the centres of the steel beams,
3) the width of the support plus twelve times the slab thickness.

Shear connection

To ensure that composite action takes place, the entire horizontal shear at the interface between the slab and the steel beam can be transferred by some form of shear connection.

For totally concrete-encased steel beams, longitudinal shear failure occurs in the concrete slab rather than at the surface perimeter of the steel beam. The design of such beams is for good reasons limited to the elastic method, referred to later on in the chapter.

For uncased steel beams, the most practicable form of shear connection is some form of dowel welded to the top flange and eventually embedded in the concrete slab when the floor or roof is cast. Again, longitudinal shear failure occurs in the slab.

Shear connectors

The functions of shear studs are to transfer the entire horizontal shear at the interface between slab and steel beam, and to prevent the tendency for separation of the two materials, i.e. prevent uplift of the slab.

The most commonly used connector is the headed stud which is available for selection in various diameters and heights. The design values P_c of connectors for different concrete strengths are given in CP 117: Part 1. Shear connectors are always designed by the Load Factor Method, irrespective of whether the Elastic or the Load Factor design method is adopted.

The number (N) of shear connectors required in the zone between the sections of zero and maximum moment to transfer the total horizontal compressive force in the concrete F_{cc} can be calculated from:

$$N = \frac{F_{cc}}{P_c}$$

where P_c = design value of one shear connector.

In the absence of any heavy concentrated loads on the beam, these N connectors are spaced evenly over the half span of a simply supported beam with the other half span similar. The maximum spacing, however, should not be greater than:

1) 4 × slab thickness;
2) 600 mm.

The minimum overall height of headed studs should be 50 mm, and they should project 25 mm minimum into the compression zone of the slab.

CP 117 also requires that the shear force Q in kN/m run should not exceed certain values and a minimum amount of transverse steel in the bottom of the slab is required.

Deflection

Deflection should be calculated on the basis of a fully composite section using a modular ratio of:

15 for imposed loads; 30 for dead loads:

with the concrete in tension being ignored.

Deflection of composite beams takes place in two stages:

1) at construction where the steel beam only supports loads,
2) in service when the whole composite beam is effective.

Construction

Design is influenced to a large extent by the method of construction adopted, i.e. propped or unpropped construction.

a) *Propped construction.* In propped construction the steel beam is supported temporarily until the slab has become composite. This ensures that the composite section will take all loading on the composite beam section. Continuous support to the beam is not required, and propping at the quarter-span and mid-span points will generally be considered as adequate. These props are kept in place until the concrete has attained a minimum specified cube strength.

b) *Unpropped construction.* In unpropped construction the dead load acting before composite action has been achieved is taken by the steel beam alone. This load arises at the construction stage from self-weight of the concrete, the shuttering and the steel beam.

Design of concrete slab

The slab may be designed independently of composite action in accordance with BS 8110 or CP 114, i.e. stresses caused by composite action need not be added to the bending stresses in the slab.

Design of composite beam sections

Elastic or Load Factor methods may be adopted.

1) *Elastic.* Elastic stresses should be calculated on the basis of a fully composite section using a modular ratio of 15 with concrete assumed to have no tensile strength. If unpropped, the steel beam alone has to carry the initial construction stage loads, and is designed in accordance with BS 449.

When the concrete has achieved the specified strength and the beam is effectively a composite section, stresses are induced in both concrete and steel from applied imposed loads, finishes and partitions. The stresses in the steel beam are added to the initial stresses, the resulting stresses not to exceed the permissible stresses in BS 449. The stress in the concrete should not exceed the permissible stress in CP 114.

2) *Load Factor.* A load factor of 1.75 may be used for the design of composite beam sections. In determining the ultimate moment of resistance of the section, the stresses to be taken are:

a) specified yield strength of the steel Y_s,
b) 4/9 of the specified characteristic strength of the concrete f_{cu}.

For unpropped beams, the steel section should comply with BS 449 during the construction stage. In no case should the total elastic stress in the steel beam under working loads exceed 0.9 Y_s, or the stress in the concrete exceed $1/3 f_{cu}$. Ultimate moment M must not exceed ultimate moment of resistance of the composite section M_r.

Calculations of sectional properties of the composite section

It is necessary to know the sectional properties of the composite beam in order to calculate bending stresses and deflection under loading. Elastic and load factor values are required.

1) Elastic

The position of the elastic neutral axis (N.A.) is determined solely by the geometrical properties of the composite section and is independent of the stresses developed. The elastic neutral axis can occur:

1) within the depth of the concrete slab,
2) below the slab and in the steel beam.

The depth to the neutral axis can be found by assuming initially that the N.A. is in the steel beam and that the full depth of the concrete is effective. Moments of the areas of the transformed section are taken about the interface of the steel and concrete. If the moment of area of the steel beam is greater than that of the transformed concrete area, then the elastic N.A. will be in the steel beam, and if less, will be in the slab. The principle is that the N.A. must be on the same side as the larger moment of area. Refer to Fig. 8.1(d).

Referring to Fig. 8.1(d) and taking moments of areas about the interface,

$$\frac{b}{m} d_s \frac{d_s}{2} \quad \text{for the concrete}$$

$$A_s \frac{D}{2} \quad \text{for the steel.}$$

Then if $A_s \dfrac{D}{2} > \dfrac{bd_s^2}{2m}$ the N.A. is in the steel beam,

and if $A_s \dfrac{D}{2} < \dfrac{bd_s^2}{2m}$ the N.A. is in the concrete slab.

Both cases are now considered.

Case 1: Elastic N.A. within the concrete slab (Fig. 8.2).

To find the depth to the N.A. take moments of areas about the top of the slab.

$$\frac{bd_e^2}{2} + mA_s d_g = (bd_e + mA_s)d_e$$

$$bd_e^2 + 2mA_s d_g = 2bd_e^2 + 2mA_s d_e$$

$$bd_e^2 = 2mA_s d_g - 2mA_s d_e$$

$$bd_e^2 = 2mrbd_g^2 - 2mrbd_g d_e$$

$$d_e^2 = 2mrd_g^2 - 2mrd_g d_e$$

$$d_e^2 = 2mrd_g d_e - 2mrd_g^2 = 0$$

This is a quadratic equation to be solved with

$$a = 1; \quad b = 2mrd_g; \quad c = -2mrd_g^2$$

$$d_e = \frac{-2mrd_g \pm \sqrt{[4m^2r^2d_g^2 - (4 \times 1 \times -2mrd_g^2)]}}{2}$$

$$= d_g[\sqrt{\{mr(2 + mr)\}} - mr]$$

Moment of Inertia or Sccond Moment of Area:

$$I_g = \frac{bd_e^3}{3m} + I_s + A_s(d_g - d_e)^2 \text{ steel units}$$

Section Modulus:

For bottom flange of steel beam $Z_{st} = \dfrac{I_g}{d_b - d_e}$ steel units

For top of slab $Z_{cc} = \dfrac{mI_g}{d_e}$ concrete units

Case 2: Elastic N.A. within steel beam (Fig. 8.3).

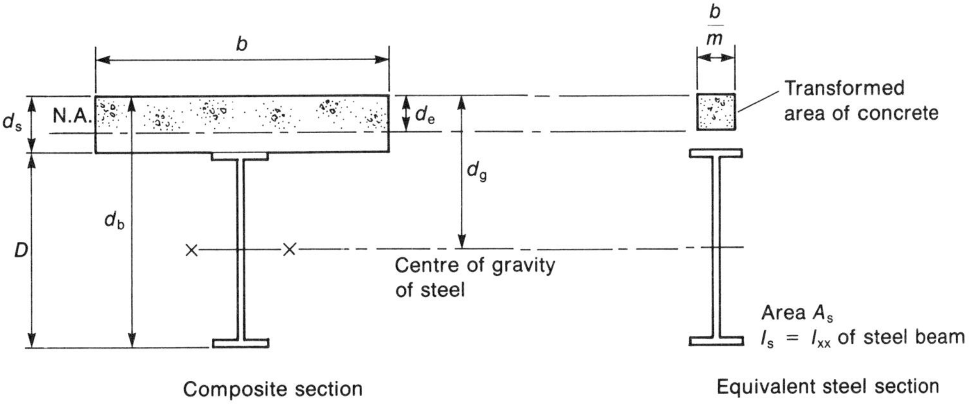

Fig. 8.2

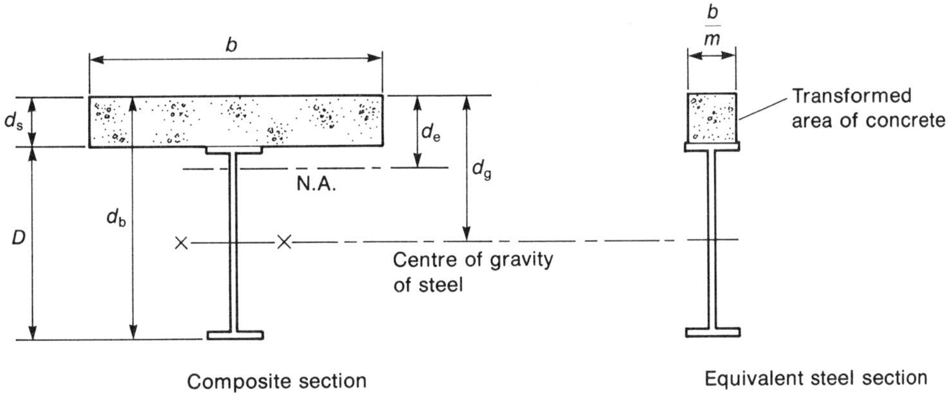

Fig. 8.3

To find depth to N.A. take moments of areas about the top of the slab:

$$\frac{bd_s^2}{2} + mA_s d_g = (bd_s + mA_s)d_e$$

$$bd_s^2 + 2mA_s d_g = 2bd_s d_e + 2mA_s d_e$$

$$d_e = \frac{bd_s^2 + 2mA_s d_g}{2bd_s + 2mA_s}$$

Dividing across by $2m$, then

$$d_e = \frac{\dfrac{bd_s^2}{2m} + A_s d_g}{\dfrac{bd_s}{m} + A_s}$$

Moment of Inertia or Second Moment of Area:

$$I_g = \frac{bd_s^3}{12m} + \frac{bd_s}{m}\left(d_e - \frac{d_s}{2}\right)^2 + I_s + A_s(d_g - d_e)^2$$

steel units

Section Modulus:

For bottom flange of steel beam $Z_{st} = \dfrac{I_g}{d_b - d_e}$

steel units

For top of slab $Z_{cc} = \dfrac{mI_g}{d_e}$

concrete units

2) Load factor

The position of the plastic neutral axis (N.A.) is determined not only by the geometrical properties of the composite section but also by the stresses at ultimate load where it is assumed that the steel is stressed to the yield stress Y_s and the concrete in compression to $\frac{4}{9}f_{cu}$.

The area of concrete below the plastic neutral axis is cracked and is therefore unstressed. The plastic neutral axis can occur:

1) within the depth of the concrete slab,
2) below the slab and in the steel beam, i.e. flange or web areas.

Case 1: Plastic N.A. within concrete slab (Fig. 8.4).

This occurs when $\alpha A_s \leqslant bd_s$, where $\alpha = \frac{9}{4}\dfrac{Y_s}{f_{cu}}$

Only the part of the slab above the N.A. is in compression.

Compressive force in concrete at ultimate load, F_{cc} = stress × area.

$$F_{cc} = \tfrac{4}{9}f_{cu}bd_n$$

Tensile force in steel F_{st} = stress × area

$$= Y_s A_s$$

Under ultimate load conditions $F_{cc} = F_{st}$

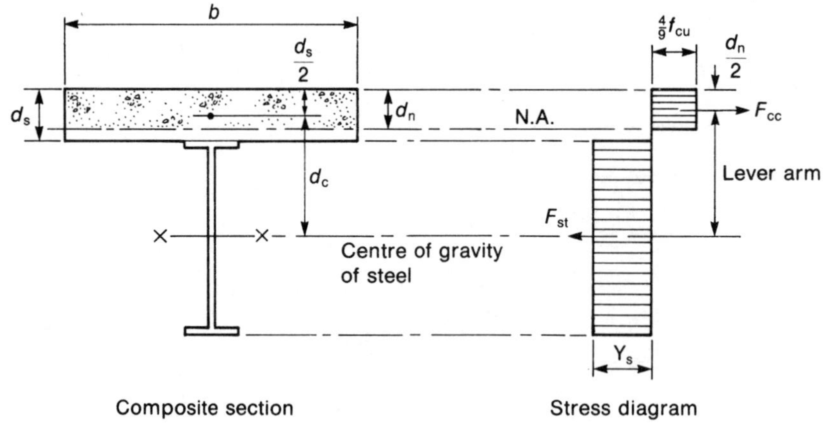

Fig. 8.4

Hence $\frac{4}{9} f_{cu} b d_n = Y_s A_s$

$$d_n = \frac{9}{4} \frac{Y_s A_s}{f_{cu} b} \text{ or } \frac{\alpha A_s}{b}$$

Ultimate moment of resistance M_r taking moments about F_{cc},

$$M_r = A_s Y_s \left(d_c + \frac{d_s}{2} - \frac{d_n}{2} \right) \text{ in terms of steel.}$$

$$\text{Also } M_r = \tfrac{4}{9} f_{cu} b d_n \left(d_c + \frac{d_s}{2} - \frac{d_n}{2} \right) \text{ in terms of concrete.}$$

Case 2: a) Plastic N.A. within the steel beam flange (Fig. 8.5).

This occurs when $bd_s < \alpha A_s < (bd_s + 2\alpha A_f)$ where $A_f = b_f t_f$.

All the concrete slab and part of the steel flange are in compression. The derivation of the formula is simplified by adding equal but opposite forces to the steel beam above the N.A., giving an equal stress distribution.

$$F_E = F_{cc} + 2F_{sc} \qquad F_{cc} = \tfrac{4}{9} f_{cu} b d_s$$

$$A_s Y_s = \tfrac{4}{9} f_{cu} b d_s + 2b_f(d_n - d_s) Y_s$$

$$A_s Y_s = \tfrac{4}{9} f_{cu} b d_s + 2b_f d_n Y_s - 2b_f d_s Y_s$$

$$A_s Y_s - \tfrac{4}{9} f_{cu} b d_s + 2b_f d_s Y_s = 2b_f d_n Y_s$$

$$d_n = \frac{A_s Y_s}{2b_f Y_s} - \frac{\frac{4}{9} f_{cu} b d_s}{2b_f Y_s} + \frac{2b_f d_s Y_s}{2b_f Y_s}$$

$$d_n = \frac{A_s}{2b_f} - \frac{\frac{4}{9}}{2} \frac{f_{cu} b d_s}{b_f Y_s} + d_s$$

$$\text{or } d_s + \frac{\alpha A_s - bd_s}{2b_f \alpha}$$

Ultimate moment of resistance M_r taking moments about F_{cc}:

$$M_r = A_s Y_s d_c - 2b_f(d_n - d_s) Y_s \left(\frac{d_s}{2} + \frac{d_n - d_s}{2} \right)$$

$$M_r = A_s Y_s d_c - 2b_f Y_s (d_n - d_s) \frac{d_n}{2}$$

$$M_r = A_s Y_s d_c - b_f Y_s d_n (d_n - d_s)$$

$$M_r = Y_s [A_s d_c - b_f d_n (d_n - d_s)]$$

Case 2: b) Plastic N.A. within the steel beam web (Fig. 8.6).

This occurs when $\alpha(A_s - 2A_f) > bd_s$

This case is similar to the previous in its analysis and

$$d_n = d_s + t_f + \frac{\alpha(A_s - 2A_f) - bd_s}{2\alpha t_w}$$

$$M_r = Y_s [A_s d_c - A_f(d_s + t_f) - t_w(d_n + t_f)(d_n - d_s - t_f)]$$

$$F_{cc} = \tfrac{4}{9} f_{cu} b d_s$$

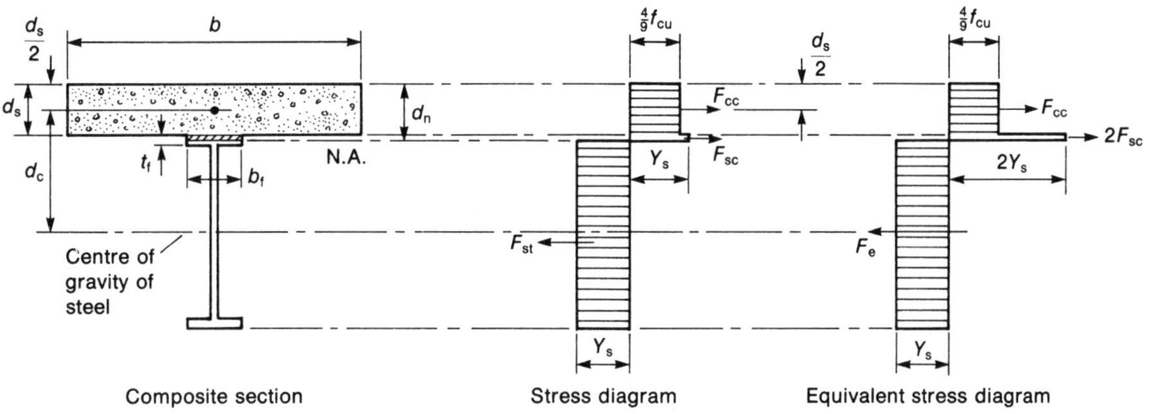

Fig. 8.5

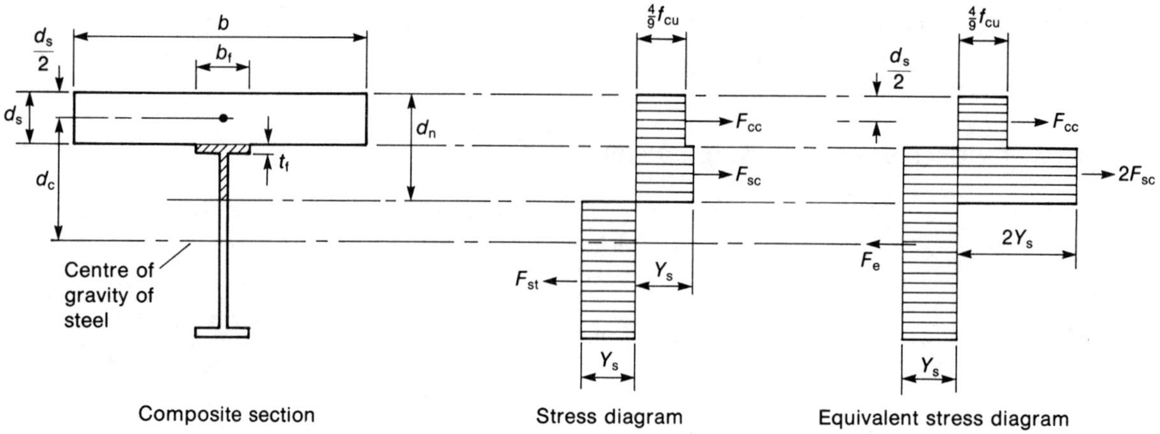
b
b_f
$\frac{d_s}{2}$
d_s
d_c
t_f
d_n
Centre of gravity of steel
$\frac{4}{9}f_{cu}$
F_{cc}
F_{sc}
F_{st}
Y_s
$\frac{4}{9}f_{cu}$
$\frac{d_s}{2}$
F_{cc}
$2F_{sc}$
$2Y_s$
F_e
Y_s
Composite section
Stress diagram
Equivalent stress diagram

Fig. 8.6

CURPE CONSULTANTS 46 Orburn Road Dunfield	Project: Composite Beams	Job ref
	Part of structure: Design- elastic method	Calc sheet no: C8 / 1 rev: 1
	Drawing ref: — Calc by: R.A.S. Date: —	Check by: — Date: —

Ref: CP117:1 | Calculations | Output

Ex. 1. Elastic Method

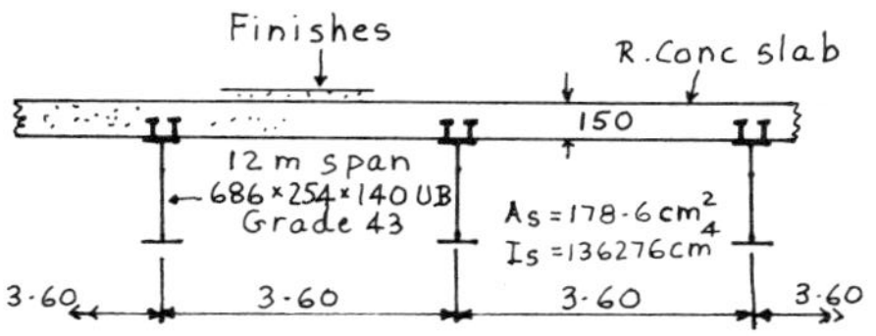

Section

$f_{cu} = 25\ N/mm^2$ $\quad m = 15$

$Y_s = 250\ N/mm^2$

Loading: Imposed $7{\cdot}0\ kN/m^2$; Finishes $0{\cdot}95$ "; Conc. slab $3{\cdot}6$ "

6. Effective width of flange

(a) $\frac{1}{3}$ span $= \frac{1}{3} \times 12 \times 10^3 = 4000\ mm$

(b) c/c ribs $= 3600\ mm$

(c) $b_r + 12 d_s = 253{\cdot}7 + (12 \times 150) = 2053{\cdot}7\ mm$

$\therefore\ b = 2053{\cdot}7\ mm$ — Output: $b = 2053{\cdot}7\ mm$

Properties of composite section

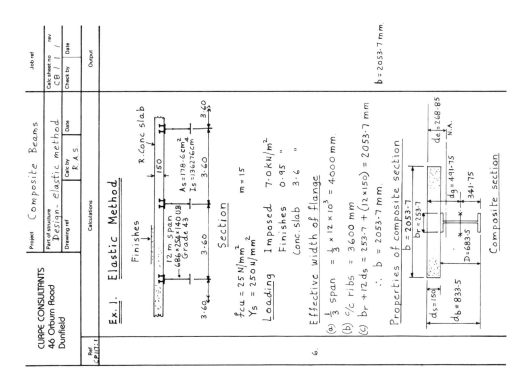

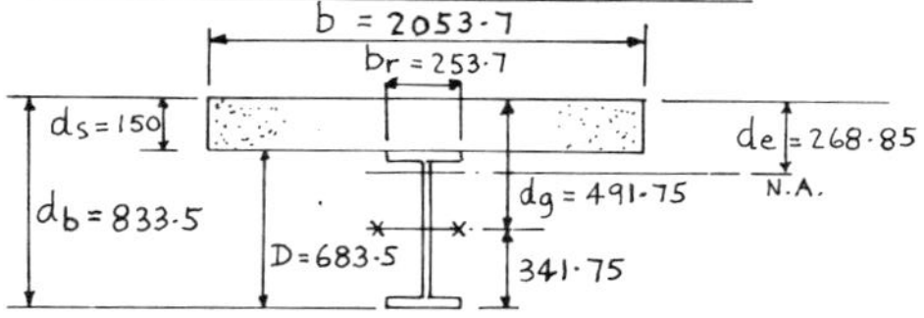

Composite section

CURPE CONSULTANTS 46 Orburn Road Dunfield	Project: Composite Beams	Job ref
	Part of structure: Design- elastic method	Calc sheet no: C8 / 2 rev: 1
	Drawing ref: — Calc by: R.A.S. Date: —	Check by: — Date: —

Ref: CP117:1 | Calculations | Output

Determine position of elastic N.A.

Moments of areas about interface:

$\frac{b}{m} \cdot d_s \cdot \frac{d_s}{2} = \frac{2053{\cdot}7}{15} \times \frac{150^2}{2} = 1540275$ (conc)

$A_s \cdot \frac{D}{2} = 178{\cdot}6 \times 10^2 \times 341{\cdot}75 = 6103655$ (steel)

Since $6103655 > 1540275$, N.A. lies in steel.

Taking moments about top of slab

$\frac{b d_s^2}{2} + m \cdot A_s \cdot d_g = (b \cdot d_s + m \cdot A_s)\, d_e$

$\frac{2053{\cdot}7 \times 150^2}{2} + (15 \times 178{\cdot}6 \times 10^2 \times 491{\cdot}75)$

$= [(2053{\cdot}7 \times 150) + (15 \times 178{\cdot}6 \times 10^2)]\, d_e$

$d_e = 268{\cdot}85\ mm$ — Output: $d_e = 268{\cdot}85\ mm$

Second moment of area or Moment of Inertia I_g

$I_g = \frac{b \cdot d_s^3}{12 m} + \frac{b \cdot d_s}{m}\left(d_e - \frac{d_s}{2}\right)^2 + I_s + A_s(d_g - d_e)^2$ steel units

$= \frac{2053{\cdot}7 \times 150^3}{12 \times 15} + \frac{2053{\cdot}7 \times 150}{15}\left(268{\cdot}85 - \frac{150}{2}\right)^2$

$+ (136276 \times 10^4) + 178{\cdot}6 \times 10^2 (491{\cdot}75 - 268{\cdot}85)^2$

$= 3060366174\ mm^4$ steel units — Output: $I_g = 3060366174\ mm^4$ Steel units

Section modulus

Concrete $\quad Z_{cc} = m \cdot \frac{I_g}{d_e} = \frac{15 \times 3060366174}{268{\cdot}85}$

$= 170747590\ mm^3$ conc. units — Output: $Z_{cc} = 170747590\ mm^3$ Conc. units

Steel $\quad Z_{st} = \frac{I_g}{d_b - d_e} = \frac{3060366174}{833{\cdot}5 - 268{\cdot}85}$

$= 5419934\ mm^3$ steel units — Output: $Z_{st} = 5419934\ mm^3$ steel units.

CURPE CONSULTANTS 46 Orburn Road Dunfield	Project: Composite Beams	Job ref:
	Part of structure: Design - elastic method	Calc sheet no: C8 / 3 rev: 1
	Drawing ref: — Calc by: RAS Date:	Check by: Date:

Ref	Calculations	Output

Ref: CP117:1

Design for no propping

Stage I Stress in steel beam due to construction loads, i.e. wet conc. and self weight of beam.

Loading : 150 mm. conc. slab = $3 \cdot 60\ kN/m^2$

Load $= 3 \cdot 60 \times 12 \times 3 \cdot 60 = 155 \cdot 52$ kN

Self wt steel beam $= \dfrac{140 \times 12 \times 9 \cdot 81}{10^3} = 16 \cdot 48$ kN

Total W = 172·00 kN

$M = \dfrac{WL}{8} = \dfrac{172 \times 12}{8} = 258$ kNm

Bending stresses : in steel beam

$f_{bc} = f_{bt} = \dfrac{M}{Z} = \dfrac{258 \times 10^6}{3988 \times 10^3} = 64 \cdot 7\ N/mm^2$

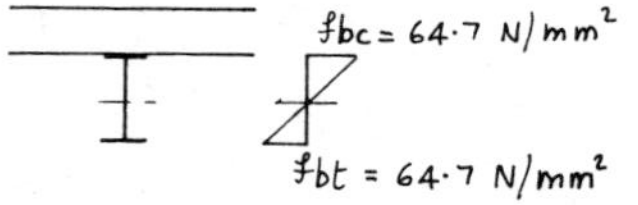

Output: $f_{bc} = f_{bt} = 64 \cdot 7\ N/mm^2$

Determine allowable bending stress with ends of beam connected to stanchions by top & bottom flange cleats, $\ell = 0 \cdot 85 L$

Then $\dfrac{\ell}{r_Y} = \dfrac{0 \cdot 85 \times 12 \times 10^3}{5 \cdot 38 \times 10} = 190$ $\quad \dfrac{D}{T} = 36$

$P_{bc} = 68\ N/mm^2$ (Table 3a, BS449)

Since $f_{bc} < P_{bc}$ and $f_{bt} < P_{bt}$ beam satisfactory at this stage.

Output: $P_{bc} = 68\ N/mm^2$ No propping Stage I Construction Bending sat.

CURPE CONSULTANTS 46 Orburn Road Dunfield	Project: Composite Beams	Job ref:
	Part of structure: Design - elastic method	Calc sheet no: C8 / 4 rev: 1
	Drawing ref: — Calc by: R.A.S. Date:	Check by: Date:

Ref	Calculations	Output

Ref: CP117:1

Stage 2 Stresses in composite section due to applied loads i.e. imposed + finishes.

Loading : Finishes 0·95
Imposed 7·00
$7 \cdot 95\ kN/m^2$

Load $= 7 \cdot 95 \times 12 \times 3 \cdot 60 = 343 \cdot 4$ kN.

$M = \dfrac{WL}{8} = \dfrac{343 \cdot 4 \times 12}{8} = 515 \cdot 10$ kNm

Bending stresses in composite section

$f_{cc} = \dfrac{M}{Z_{cc}} = \dfrac{515 \cdot 10 \times 10^6}{170747590} = 3 \cdot 016\ N/mm^2$ Conc. units.

$f_{st} = \dfrac{M}{Z_{st}} = \dfrac{515 \cdot 10 \times 10^6}{5419934} = 95 \cdot 038\ N/mm^2$ steel units

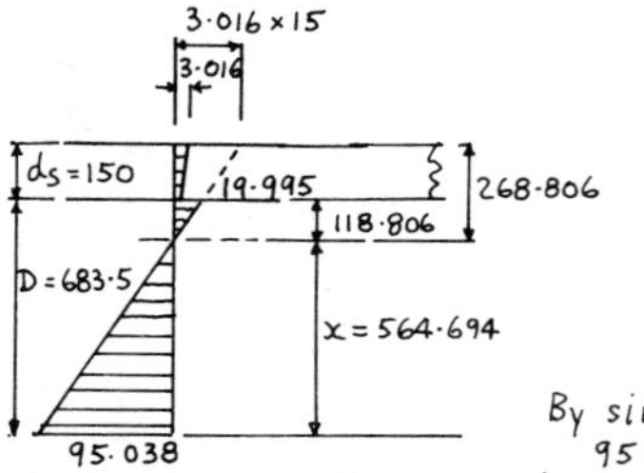

Stresses composite section.

By similar triangles
$95 \cdot 038 : x$
$(3 \cdot 016 \times 15) : 833 \cdot 5 - x$
$x = 564 \cdot 694$ mm

f_{bc} at top of steel $= \dfrac{3 \cdot 016 \times 15 \times 118 \cdot 806}{268 \cdot 806} = 19 \cdot 995\ N/mm^2$

or $\dfrac{95 \cdot 038 \times 118 \cdot 806}{564 \cdot 694} = 19 \cdot 995\ N/mm^2$ check

CURPE CONSULTANTS 46 Orburn Road Dunfield	Project: Composite Beams	Job ref
	Part of structure: Design - elastic method	Calc sheet no: CB / 5 / rev 1
	Drawing ref: — / Calc by: R.A.S. / Date: —	Check by: — / Date: —

Ref	Calculations	Output
CP117:1		

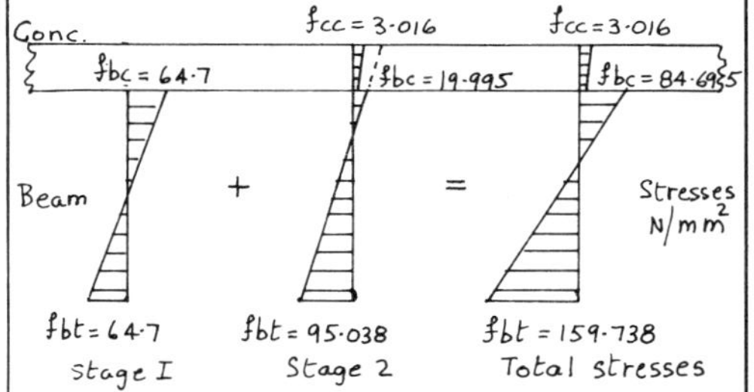

$P_{bc} = P_{bt} = 165\ N/mm^2$ Table 2, BS449.

Since $f_{bc} < P_{bc}$ and $f_{bt} < P_{bt}$

Steel section satisfactory for bending. — *Output:* No propping Stage 2. Bending sat.

Design for propping

Loading : 150 mm conc slab = 3.60
Finishes = 0.95
Imposed = 7.00
11.55 kN/m²

Load = 11.55 × 12 × 3.6 = 498.96 kN
Self weight of beam = 16.48 kN
515.44 kN

$$M = \frac{WL}{8} = \frac{515.44 \times 12}{8} = 773.10\ kNm.$$

Stresses in composite section

$$f_{cc} = \frac{M}{Z_{cc}} = \frac{773.10 \times 10^6}{170747590} = 4.527\ N/mm^2$$

$$f_{st} = \frac{M}{Z_{st}} = \frac{773.10 \times 10^6}{5419934} = 142.64\ N/mm^2$$

By propping, steel stress reduced, concrete stress increased. Beam satisfactory for bending. — *Output:* Propping. Bending sat.

CURPE CONSULTANTS 46 Orburn Road Dunfield	Project: Composite Beams	Job ref
	Part of structure: Design - load factor method	Calc sheet no: CB / 6 / rev 1
	Drawing ref: — / Calc by: R.A.S / Date: —	Check by: — / Date: —

Ref	Calculations	Output
CP117:1		

Ex. 2. Load Factor Method

Imposed load + finishes = 10 kN/m²

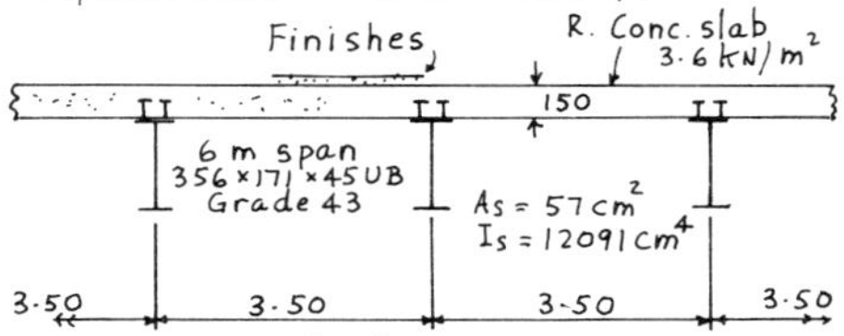

Section

$f_{cu} = 25\ N/mm^2$ Load factor = 1.75
$Y_s = 250\ N/mm^2$ m = 15

6. **Effective width of flange**

(a) $\frac{1}{3}$ span $= \frac{1}{3} \times 8 \times 10^3 = 2666$ mm

(b) c/c ribs = 3500 mm

(c) $b_r + 12 d_s = 171 + (12 \times 150) = 1971$ mm

∴ b = 1971 mm. — *Output:* b = 1971 mm.

Properties of composite section

(1) Elastic

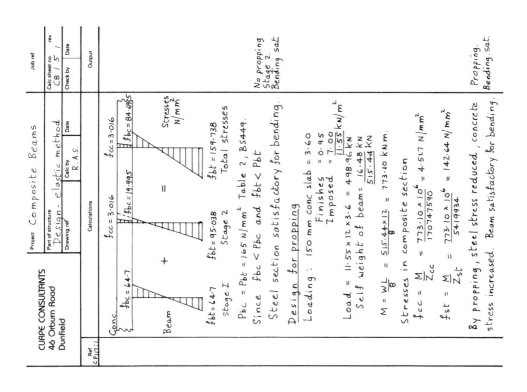

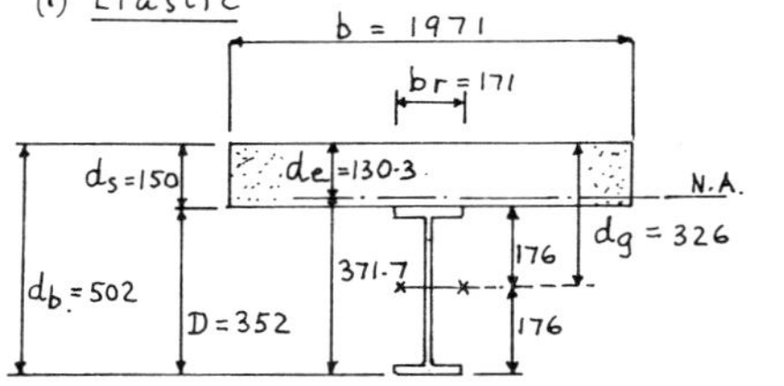

Composite section

CURPE CONSULTANTS 46 Orburn Road Dunfield	Project: Composite Beams			Job ref	
	Part of structure: Design - load factor method			Calc sheet no: C8 / 7	rev: /
	Drawing ref	Calc by: R.A.S.	Date	Check by	Date

Ref	Calculations	Output
CP117:1		

Determine position of elastic N.A.

Moments of areas about interface :

$\frac{b}{m}.d_s.\frac{d_s}{2} = \frac{1971}{15} \times \frac{150^2}{2} = 1478250$ (conc)

$A_s.\frac{D}{2} = 57 \times 10^2 \times 176 = 1003200$ (steel)

Since 1478250 > 1003200, N.A. lies in conc.

Now $r = \frac{A_s}{b.d_g} = \frac{57 \times 10^2}{1971 \times 326} = 0.008871.$

$d_e = d_g\left[\sqrt{mr(2+mr)} - mr\right]$

$= 326\left[\sqrt{15 \times 0.00871(2 + 15 \times 0.008871)} - (15 \times 0.008871)\right]$

$= 326\,(0.39969) = 130.3$ mm — **Output:** $d_e = 130.3$ mm.

Second moment of area or Moment of Inertia I_g

$I_g = \frac{b.d_e^3}{3m} + I_s + A_s(d_g - d_e)^2$ steel units.

$= \frac{1971 \times 130.3^3}{3 \times 15} + (12091 \times 10^4) + 57 \times 10^2(326 - 130.3)^2$

$= 96896336 + 120910000 + 218301390$

$= 436107726$ mm^4 steel units. — **Output:** $I_g = 436107726$ mm^4. Steel units

Section Modulus

$Z_{st} = \frac{I_g}{d_b - d_e} = \frac{436107726}{502 - 130.3} = 1173278$ mm^3 steel units — **Output:** $Z_{st} = 1173278$ mm^3. Steel units.

$Z_{cc} = \frac{m.I_g}{d_e} = \frac{15 \times 436107726}{130.3} = 50204265$ mm^3 conc. units. — **Output:** $Z_{cc} = 50204265$ mm^3. Conc units.

These properties enable elastic stresses to be calculated.

CURPE CONSULTANTS 46 Orburn Road Dunfield	Project: Composite Beams			Job ref	
	Part of structure: Design - load factor method			Calc sheet no: C8 / 8	rev: /
	Drawing ref	Calc by: R.A.S.	Date	Check by	Date

Ref	Calculations	Output
CP117:1		

Properties of composite section (contd)

(2) Ultimate

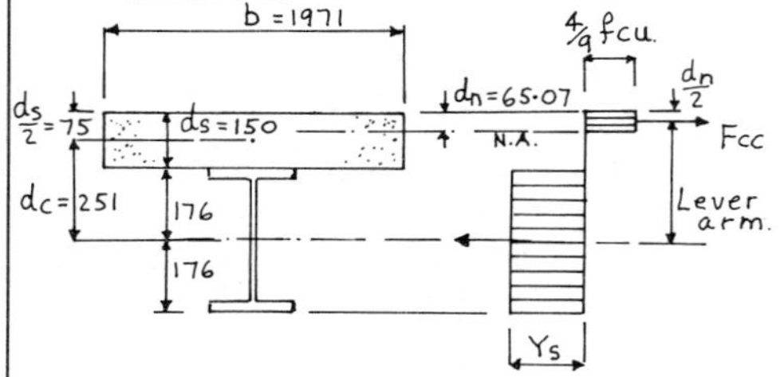

$\alpha = \frac{9}{4}.\frac{Y_s}{f_{cu}} = \frac{9}{4} \times \frac{250}{25} = 22.5$

A4. a — Try Case I for position of plastic N.A.

$\alpha.A_s \leqslant b.d_s$

$\alpha.A_s = 22.5 \times 57 \times 10^2 = 128250$

$b.d_s = 1971 \times 150 = 295650$

Since $\alpha.A_s < b.d_s$ plastic N.A. in slab.

Compressive force in conc $F_{cc} = \frac{4}{9}f_{cu}.b.d_n$

Tensile force in steel $F_{st} = Y_s.A_s$.

$F_{cc} = F_{st}$

Hence $\frac{4}{9}.f_{cu}.b.d_n. = Y_s.A_s.$

$d_n = \frac{9.Y_s.A_s}{4.f_{cu}.b} = \frac{9 \times 250 \times 57 \times 10^2}{4 \times 25 \times 1971} = 65.07$ mm — **Output:** $d_n = 65.07$ mm.

Ultimate moment of resistance M_r =

$A_s.Y_s\left(d_c + \frac{d_s}{2} - \frac{d_n}{2}\right)$

$M_r = (57 \times 10^2)\,250\left(251 + 75 - \frac{65.07}{2}\right)\frac{1}{10^6} = 418$ kNm — **Output:** $M_r = 418$ kNm

CURPE CONSULTANTS 46 Orburn Road Dunfield	Project: Composite Beams			Job ref	
	Part of structure: Design - load factor method			Calc sheet no: C8 / 9	rev: 1
	Drawing ref	Calc by: R.A.S.	Date	Check by	Date

Ref	Calculations	Output
CP117:1.	Also $M_r = \frac{4}{9} \cdot f_{cu} . b . d_n \left(d_c + \frac{d_s}{2} - \frac{d_n}{2}\right)$ $= \frac{4}{9} \times 25 \times 1971 \times 65.07 \left(251 + 75 - \frac{65.07}{2}\right) \frac{1}{10^6}$ $= 418$ kNm. (Check).	
	<u>Design for no propping</u> (a) Elastic stresses in steel beam under constructions loads *i.e.* wet conc and self weight of beam.	
	Wt of conc. slab $= 3.6 \times 3.5 \times 6 = 75.6$ kN. Self wt of beam $= \frac{45 \times 6 \times 9.81}{10^3} = 2.6$ kN Total $= 78.2$ kN.	
	$M = \frac{WL}{8} = \frac{78.2 \times 6}{8} = 58.65$ kNm. Bending stresses $f_{bc} = f_{bt} = \frac{M}{Z} = \frac{58.65 \times 10^6}{686.9 \times 10^3}$ $= 84.16$ N/mm^2	
	Allowable bending stress P_{bc} With $\ell = 0.85L$, then $\frac{\ell}{r_y} = \frac{0.85 \times 6 \times 10^3}{3.78 \times 10} = 134.9$ $\frac{D}{T} = 36.2$ $\therefore P_{bc} = 120.5$ N/mm^2 (Table 3a) Since $f_{bc} < P_{bc}$, steel beam satisfactory at this stage.	No propping Construction Bending sat.
	(b) Additional elastic stresses from imposed load and finishes on composite section. Imposed load + finishes $= 10 \times 6 \times 3.5 = 210$ kN. $M = \frac{WL}{8} = \frac{210 \times 6}{8} = 157.5$ kNm.	

CURPE CONSULTANTS 46 Orburn Road Dunfield	Project: Composite Beams			Job ref	
	Part of structure: Design - load factor method			Calc sheet no: C8 / 10	rev: 1
	Drawing ref	Calc by: R.A.S.	Date	Check by	Date

Ref	Calculations	Output
CP117:1	$f_{st} = \frac{M}{Z_{st}} = \frac{157.5 \times 10^6}{1173278} = 134.24$ N/mm^2 $f_{cc} = \frac{M}{Z_{cc}} = \frac{157.5 \times 10^6}{50204265} = 3.14$ N/mm^2	
	(c) Total elastic stress in steel beam $= 84.16 + 134.24 = 218.4$ N/mm^2 Since $218.4 < 0.9\,Y_s$ satisfactory Also total elastic stress in concrete $= 3.14$ N/mm^2 Since $3.14 < \frac{1}{3} f_{cu}$ satisfactory $\therefore$ Elastic stresses under working loads satisfactory.	Elastic stresses Working loads Bending sat.
	(d) Ultimate bending moment $\ngtr M_r$, the ultimate moment of resistance. Total load $W = 78.2 + 210 = 288.2$ kN Ultimate bending moment $= 1.75 \frac{WL}{8}$ $= \frac{1.75 \times 288.2 \times 6}{8} = 378.26$ kNm Since $378.26 < 418$ satisfactory	$M < M_r$ Bending sat.
	<u>Design for propping</u> (a) Elastic stresses in composite section under total loading. Total load $W = 288.2$ kN. $M = \frac{WL}{8} = \frac{288.2 \times 6}{8} = 216.15$ kNm. $f_{st} = \frac{M}{Z_{st}} = \frac{216.15 \times 10^6}{1173278} = 184.22$ N/mm^2 Since $184.22 < 0.9\,Y_s$ satisfactory	

CURPE CONSULTANTS 46 Orburn Road Dunfield	Project: Composite Beams			Job ref	
	Part of structure: Design - load factor method			Calc sheet no: C8 / 11.	rev: 1
	Drawing ref	Calc by: R.A.S.	Date	Check by	Date

Ref	Calculations	Output
CP117:1		
	$f_{cc} = \frac{M}{Z_{cc}} = \frac{216 \cdot 15 \times 10^6}{50204265} = 4 \cdot 305\ N/mm^2$	Propping Elastic stresses Bending sat.
	Since $4 \cdot 305 < \frac{1}{3} f_{cu}$ satisfactory	
	(b) Total load W = 288·2 kN	
	Ultimate bending moment $= 1 \cdot 75 \frac{WL}{8}$	
	$= \frac{1 \cdot 75 \times 288 \cdot 2 \times 6}{8} = 378 \cdot 26$ kNm	
	Since $378 \cdot 26 < 418$ satisfactory	M < Mr Bending sat.
7.	Shear Connectors	
App. B	With plastic N.A. within the conc. slab	
	$N = \frac{F_{cc}}{P_c} = \frac{F_{st}}{P_c}$	
	$\frac{F_{st}}{P_c} = \frac{A_s . Y_s}{P_c} = \frac{57 \times 10^2 \times 250}{10^3 \times 78} = 18 \cdot 27$	
	where design strength of a shear stud connector is given as 78 kN	
	Use 20 steel connectors per half span of beam i.e. 10 pairs. Other half span same.	Use 10 pairs per half span.
	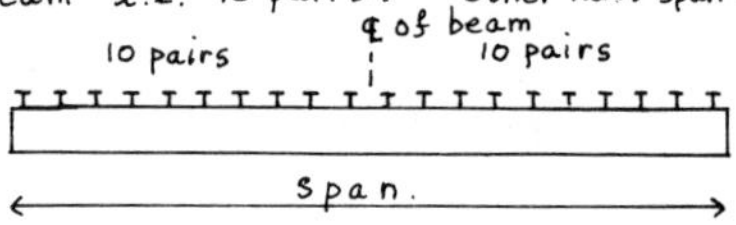	
8.	Deflection	
	Deflection should be calculated on the basis of a fully composite section, using m = 15 for Imposed loading and 30 for the dead load, conc. in tension being ignored.	

9

Brickwork and blockwork

General requirements

BS 5628: Part 1 gives recommendations for the structural design of unreinforced masonry. The philosophy is based on two requirements:

1) The design strength of all load-bearing members in a structure should have an adequate margin of safety over the ultimate strength of that member.
2) An adequate margin of safety should be provided against the structure becoming unserviceable by considerations other than strength.

For ultimate strength, the loads and material strengths used in the design process are based on the results of laboratory tests and statistical considerations. These are termed characteristic loads and characteristic material strengths.

For ultimate limit state condition, characteristic loads and material strengths are each modified by using partial safety factors:

i) γ_f for loads,
ii) γ_m for material strengths.

Design requirement

To ensure an adequate margin of safety against ultimate limit state being reached, then:

$$\text{Design strength} \geqslant \text{design load}$$

Design strength

$$\text{Design strength} = \frac{\text{characteristic strength}}{\gamma_m}$$

where γ_m, the partial safety factor for materials, is related to the quality control exercised during manufacture and construction. Values are given in Table 4 of BS 5628: Part 1, where normal and special categories are recognized.

Characteristic compressive strength f_k

Values of f_k in terms of brick, block and mortar strength are presented in Table 2 of BS 5628: Part 1. For compressive loading:

$$\text{Design strength} = \frac{f_k}{\gamma_m}$$

Characteristic loads

The following characteristic loads should be used in design. Characteristic dead load G_k and imposed load Q_k are taken as the loads defined in BS 6399: Part 1: 1984.

Design loads

$$\begin{aligned}\text{Design load} &= \text{characteristic load} \\ &\quad \times \text{partial safety factor} \\ &= (G_k \text{ or } Q_k)\gamma_f\end{aligned}$$

where γ_f is a partial safety factor for loads.

Values are given in Clause 22 of BS 5628: Part 1.

In the case of combined dead and imposed loads:

$$\begin{aligned}\text{Design dead load} &= 0.9G_k \text{ or } 1.4G_k \\ \text{Design imposed load} &= 1.6Q_k\end{aligned}$$

The numerical values 0.9, 1.4 and 1.6 are the γ_f factors. Where alternatives are shown, that producing the more severe condition should be selected. For example, in the case of a free-standing wall, the self-weight of the wall will be the only restoring force to resist overturning moments, so γ_f should be taken as 0.9.

Compressive strengths of units

For bricks, these values range from a minimum of 5 to a maximum of 100 N/mm^2, and for blocks from 2.8 to 35 N/mm^2, although some strengths may not be readily available. Values of corresponding characteristic compressive strengths for designated grades of mortar are presented in Table 2 of BS 5628: Part 1.

Mortar. Mortar fills the joints, evening out slight difference in size and shape of the bricks or blocks, and provides a uniform bed enabling loads to be distributed uniformly. It also helps to prevent the penetration of air and water through the joints in a wall. Mortar consists of an aggregate and binder. The aggregate is sand, and the binder a mixture of lime and cement, or cement with the addition of a plasticiser to assist workability. A good mortar should possess necessary workability in its fluid state and, on setting and hardening, obtain adequate strength and good bonding characteristics for structural stability and durability.

Identification of a particular mix proportion for a type of mortar is now given by the term 'mortar designation'. Four designations are shown as (i), (ii), (iii) and (iv) in order of decreasing strength, but increasing ability to accommodate movement caused by settlement, temperature and moisture changes. The choice of mortar mix will depend on strength requirements, the type of brick, the degree of exposure and the time of the year for the construction of the walls.

Modification factors

Brickwork

For standard format bricks, the value of f_k is modified if relevant, to allow for the effects of narrow brick walls or small plan area, as follows:

Narrow brick walls. Where solid brick walls or the loaded inner leaf of a cavity wall are constructed in standard format bricks with a thickness equal to the width of a single brick, the values of f_k may be multiplied by 1.15.

Small plan area. Where the horizontal cross-section of a loaded wall or column is less than 0.2 m^2, the value of f_k should be multiplied by the factor $(0.7 + 1.5A)$, where A = horizontal loaded cross-sectional area of the wall or column (m^2).

Blockwork

For blockwork, the increased size of individual blocks means that fewer joints are required than for an equivalent area of wall built with standard bricks. This results in an overall increased compressive strength for the wall. The characteristic compressive strength depends on the shape factor and on whether the units are solid or hollow.

The shape factor is obtained by dividing the height of the block by its least horizontal dimension. As for brickwork, the value of f_k is modified, if relevant, to allow for the effects of small plan area.

Design for vertical loading

The basic principle can be expressed as:

Design vertical loading <
design vertical load resistance

Walls and columns are subjected to vertical loads arising from self-weight of masonry and any supported floors, roof, beams, etc. A low squat wall will fail by crushing of the material. Walls of a given thickness and material strength tend to fail by buckling at lower loads as their height increases.

Eccentricity of loading which is often the normal condition also reduces the load-carrying capacity. The eccentricity is usually expressed as a fraction of the wall thickness. It is not necessary to consider the effects of eccentricity if:

$$e_x < 0.05t$$

where e_x = the eccentricity at the top of the wall,
t = the thickness of the wall.

A load transmitted to a wall by a single floor or roof is assumed to act on one-third of the length of the bearing area measured from the loaded edge, implying a triangular-shaped pressure distribution. Where a uniform floor is continuous over a wall, each side of the floor may be taken as being supported individually on half the total bearing area.

Slenderness ratio

This is defined as:

$$\frac{\text{effective height (or length)}}{\text{effective thickness}} = \frac{h_{ef} \text{ or } l_{ef}}{t_{ef}}$$

The effective length is used where this gives the lesser slenderness ratio.

In order to avoid the possibility of sudden buckling failure occurring without warning, a limiting slenderness ratio of 27 is adopted, except where very thin units (less than 90 mm thick) are used in the construction of buildings higher than two storeys, where it is limited to 20. Allowance is made for the effects of slenderness ratio and eccentricity by applying β, a capacity reduction factor from Table 7 of BS 5628: Part 1.

Lateral support

A lateral support may be provided along either a horizontal or a vertical line, depending on whether the slenderness ratio is based on a vertical or horizontal dimension. Horizontal lateral supports in the form of suitable floors or roofs provide either 'simple' or 'enhanced' resistance to lateral movement of the wall. Examples are shown in Fig. 9.1.

Vertical lateral supports in the form of intersecting or return walls again provide 'simple' or 'enhanced' resistances. Examples are as shown in Fig. 9.2.

Effective height

The effective height is related to the lateral support provided along the horizontal planes. The effective height of a wall may be taken as:

1) 0.75 times the clear distance between lateral supports which provide 'enhanced' resistance, or
2) the clear distance between lateral supports which provide 'simple' resistance.

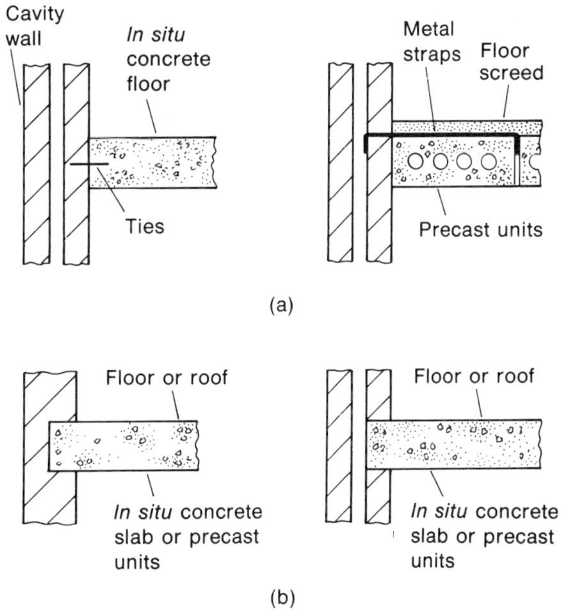

Fig. 9.1 *Horizontal lateral supports. (a) simple resistance, (b) enhanced resistance*

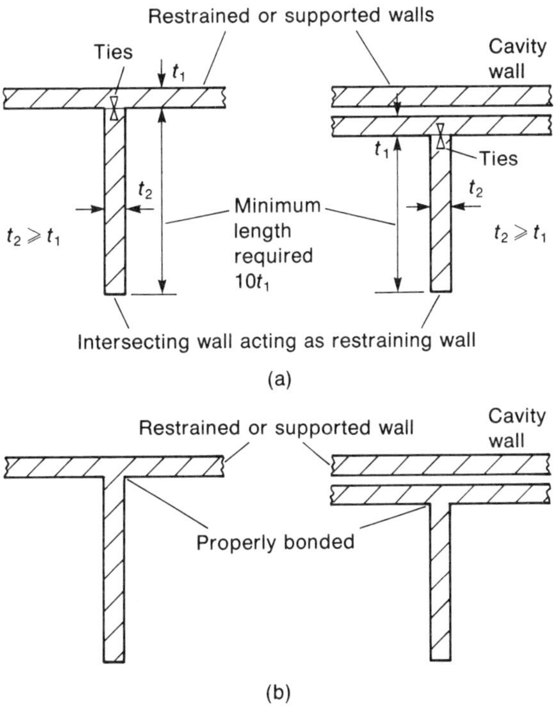

Fig. 9.2 *Vertical lateral supports. (a) simple resistance, (b) enhanced resistance*

Effective length

The effective length is related to the lateral support provided along the vertical planes.

The effective length of a wall may be taken as:

1) 0.75 times the clear distance between vertical lateral supports which provide 'enhanced' resistance, or
2) the clear distance between lateral supports which provide 'simple' resistance.

Effective thickness

Case 1: Walls not stiffened by piers or intersecting walls.

For a single leaf wall:

$$t_{ef} = \text{actual thickness } t.$$

For a cavity wall, t_{ef} is the greatest of:

1) $\frac{2}{3}(t_1 + t_2)$; or
2) t_1; or
3) t_2

where t_1 = actual thickness of one leaf,
t_2 = actual thickness of the other leaf.

Case 2: Walls stiffened by piers or intersecting walls.

The effective thickness t_{ef} of a wall or leaf of a cavity wall is:

$$t_{ef} = tK$$

where t = actual thickness of wall or leaf,
K = appropriate stiffness coefficient from Table 5 of BS 5628: Part 1.

For a wall stiffened by intersecting walls, K may be determined from Table 5 on the assumption that the intersecting walls are equivalent to piers of width equal to the thickness of the intersecting wall and thickness equal to three times the thickness of the stiffened wall.

Design vertical load resistance of walls

The design vertical load resistance of a wall per unit length is given by:

$$\frac{\beta t f_K}{\gamma_m}$$

where β = capacity reduction factor from Table 7 of BS 5628: Part 1,
f_K = characteristic compressive strength of the masonry,
γ_m = partial safety factor for the material,
t = thickness of wall.

Cavity walls

These are constructed with two leaves of masonry tied together at specified intervals by adequate cavity ties. Where one leaf only supports the load from roof or floor, the load-bearing capacity of the wall should be based on the horizontal cross-section of that leaf only, although the stiffening effect of the other leaf can be taken into account when calculating the slenderness ratio.

Tensile strength

Tensile strength is governed by the tensile bond at the brick/mortar interface. It is not generally acceptable to rely on masonry in direct tension but, at the designer's discretion, half the values given in Table 3 of BS 5628: Part 1 may be per-when wind uplift on roofs is transmitted to masonry walls. In no circumstances may combined direct and flexural tensile stresses exceed Table 3 values.

Table 2. Characteristic compressive strength of masonry, f_k, in N/mm^2

(a) Constructed with standard format bricks

Mortar designation	*Compressive strength of unit (N/mm^2)*								
	5	*10*	*15*	*20*	*27.5*	*35*	*50*	*70*	*100*
(i)	2.5	4.4	6.0	7.4	9.2	11.4	15.0	19.2	24.0
(ii)	2.5	4.2	5.3	6.4	7.9	9.4	12.2	15.1	18.1
(iii)	2.5	4.1	5.0	5.8	7.1	8.5	10.6	13.1	15.5
(iv)	2.2	3.5	4.4	5.2	6.2	7.3	9.0	10.8	12.7

(b) Constructed with blocks having a ratio of height to least horizontal dimension of 0.6

Mortar designation	*Compressive strength of unit (N/mm^2)*							
	2.8	*3.5*	*5.0*	*7.0*	*10*	*15*	*20*	*35 or greater*
(i)	1.4	1.7	2.5	3.4	4.4	6.0	7.4	11.4
(ii)	1.4	1.7	2.5	3.2	4.2	5.3	6.4	9.4
(iii)	1.4	1.7	2.5	3.2	4.1	5.0	5.8	8.5
(iv)	1.4	1.7	2.2	2.8	3.5	4.4	5.2	7.3

(c) Constructed with hollow blocks having a ratio of height to least horizontal dimension of between 2.0 and 4.0

Mortar designation	*Compressive strength of unit (N/mm^2)*							
	2.8	*3.5*	*5.0*	*7.0*	*10*	*15*	*20*	*35 or greater*
(i)	2.8	3.5	5.0	5.7	6.1	6.8	7.5	11.4
(ii)	2.8	3.5	5.0	5.5	5.7	6.1	6.5	9.4
(ii)	2.8	3.5	5.0	5.4	5.5	5.7	5.9	8.5
(iv)	2.8	3.5	4.4	4.8	4.9	5.1	5.3	7.3

(d) Constructed from solid concrete blocks having a ratio of height to least horizontal dimension of between 2.0 and 4.0

Mortar designation	*Compressive strength of unit (N/mm^2)*							
	2.8	*3.5*	*5.0*	*7.0*	*10*	*15*	*20*	*35 or greater*
(i)	2.8	3.5	5.0	6.8	8.8	12.0	14.8	22.8
(ii)	2.8	3.5	5.0	6.4	8.4	10.6	12.8	18.8
(iii)	2.8	3.5	5.0	6.4	8.2	10.0	11.6	17.0
(iv)	2.8	3.5	4.4	5.6	7.0	8.8	10.4	14.6

Table 4. Partial safety factors for material strength, γ

		Category of construction control	
		Special	*Normal*
Category of manufacturing control of structural units	Special	2.5	3.1
	Normal	2.8	3.5

Table 7. Capacity reduction factor, β

Slenderness ratio h_{ef}/t_{ef}	*Eccentricity at top of wall,* e_x			
	*Up to 0.05*t *(see note 1)*	*0.1*t	*0.2*t	*0.3*t
0	1.00	0.88	0.66	0.44
6	1.00	0.88	0.66	0.44
8	1.00	0.88	0.66	0.44
10	0.97	0.88	0.66	0.44
12	0.93	0.87	0.66	0.44
14	0.89	0.83	0.66	0.44
16	0.83	0.77	0.64	0.44
18	0.77	0.70	0.57	0.44
20	0.70	0.64	0.51	0.37
22	0.62	0.56	0.43	0.30
24	0.53	0.47	0.34	
26	0.45	0.38		
27	0.40	0.33		

NOTE 1. It is not necessary to consider the effects of eccentricities up to and including 0.05t.
NOTE 2. Linear interpolation between eccentricities and slenderness ratios is permitted.
NOTE 3. The derivation of β is given in appendix B of BS 5628: Part 1.

Tables 2, 4 and 7 of BS 5628: Part 1, used in the following worked examples.

CURPE CONSULTANTS 46 Orburn Road Dunfield	Project: BRICKWORK & BLOCKWORK Part of structure: Solid brick wall Drawing ref: / Calc by: RAS / Date:	Job ref: Calc sheet no: C9 / 1 / rev: 1 Check by: / Date:

Ref BS5628	Calculations	Output
	Ex. 1. Solid Wall- axial loading	
	Roof Dead loading $5\ kN/m^2$	
	Imposed " $1.5\ kN/m^2$	
	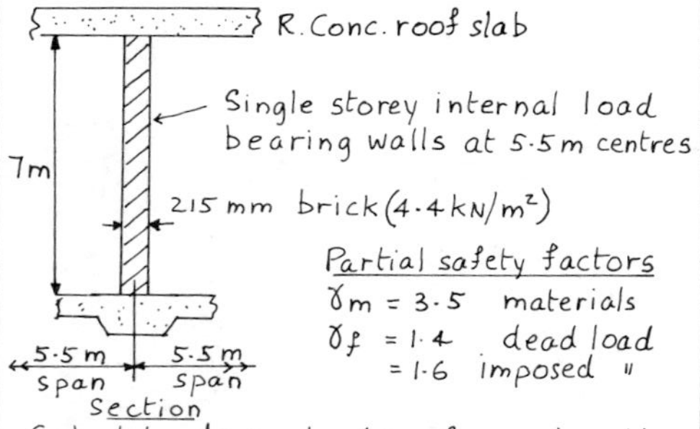	
CL. 27 Table 4	Partial safety factors	
	$\gamma_m = 3.5$ materials	
CL. 22.	$\gamma_f = 1.4$ dead load	
	$= 1.6$ imposed "	
	Calculate design load n_w for 1 m length	
	Characteristic dead load G_k:	
	roof $= \frac{5.5 + 5.5}{2} \times 5 \times 1 = 27.5\ kN$	
	wall $= 7 \times 4.4 \times 1 = 30.8\ kN$	
	$G_k = 58.3\ kN$	$G_k = 58.3\ kN$
	Characteristic imposed load Q_k:	
	roof $= \frac{5.5 + 5.5}{2} \times 1.5 \times 1 = 8.25\ kN$	$Q_k = 8.25\ kN$
	Design load =	
	$n_w = 1.4\ G_k + 1.6\ Q_k$	
	$= 1.4(58.3) + 1.6(8.25) = 94.82\ kN/m$	$n_w = 94.82\ kN$
	Slenderness ratio and β	
CL. 28.3.1.1.	With enhanced resistance $h_{ef} = 0.75h$	
CL. 28.4.1.	Slenderness ratio $= \frac{h_{ef}}{t_{ef}} = \frac{0.75 \times 7 \times 10^3}{215}$	
CL. 28.1.	$= 24.5 < 27$	

CURPE CONSULTANTS 46 Orburn Road Dunfield	Project: BRICKWORK & BLOCKWORK Part of structure: Solid block wall. Drawing ref: / Calc by: R. A. S. / Date:	Job ref: Calc sheet no: C9 / 2 / rev: 1 Check by: / Date:

Ref BS5628	Calculations	Output
Table 7	Hence from Table 7, $\beta = 0.51$	$\beta = 0.51$
	Calculate design vertical load resistance of wall	
CL. 32.2.1	$\frac{\beta . t . f_k}{\gamma_m}$ for 1 m length	
	$= \frac{0.51 \times 215 \times 10^3 . f_k}{3.5}$ N/m	
	Equate design load and resistance	
	$94.82 \times 10^3 = \frac{0.51 \times 215 \times 10^3\ f_k}{3.5}$	
	f_k required $= 3.02\ N/mm^2$	f_k reqd $= 3.02$
	Select brick strength and mortar	
Table 2 a	From Table 2(a) using mortar designation (iv) a brick of compressive strength $10\ N/mm^2$ gives $f_k = 3.5\ N/mm^2$, satisfactory	Brick $10\ N/mm^2$ mortar (iv) $f_k = 3.5$ sat.
	Ex. 2. Solid Block Wall - axial loading	
	Roof Dead $5\ kN/m^2$	
	Imposed $1.5\ kN/m^2$	
	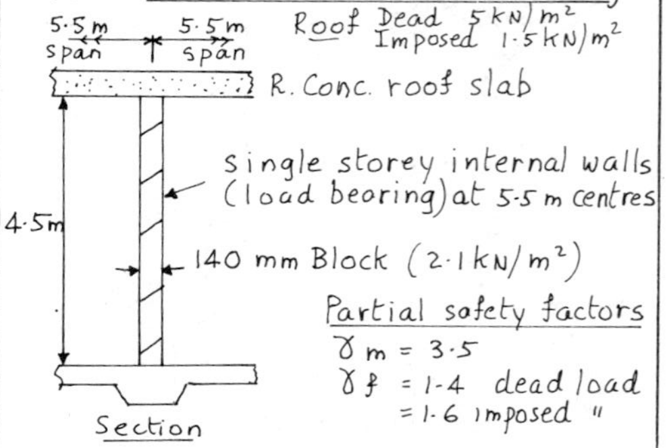	
CL. 27.	Partial safety factors	
Table 4	$\gamma_m = 3.5$	
CL. 22.	$\gamma_f = 1.4$ dead load	
	$= 1.6$ imposed "	

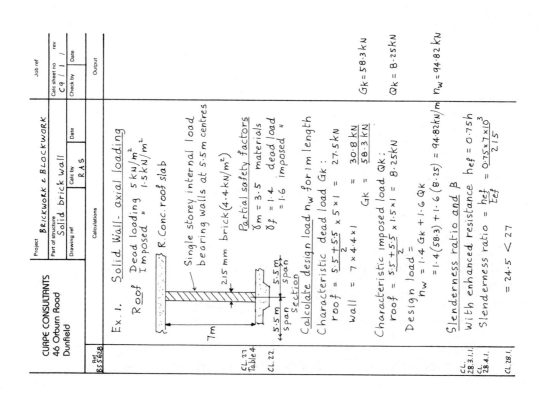

CURPE CONSULTANTS 46 Orburn Road Dunfield	Project: BRICKWORK & BLOCKWORK	Job ref
	Part of structure: Solid block wall	Calc sheet no: C9 / 3 rev: /
	Drawing ref; Calc by: R.A.S; Date	Check by; Date

Ref BS5628	Calculations	Output
	Calculate design load n_w for 1 m length	
	Gk. roof $= \frac{5.5+5.5}{2} \times 5 \times 1 = 27.5$	
	wall $= 2.1 \times 4.5 \times 1 = 9.45$	
	Gk $= 36.95$ kN	Gk = 36.95 kN
	Qk. roof $= \frac{5.5+5.5}{2} \times 1.5 \times 1 = 8.25$ kN	Qk = 8.25 kN
	Design load $n_w = 1.4\,Gk + 1.6\,Qk$	
	$= 1.4(36.95) + 1.6(8.25) = 64.9$ kN	$n_w = 64.9$ kN
	Slenderness ratio and β	
CL. 28	Slenderness ratio $= \frac{h_{ef}}{t_{ef}} = \frac{0.75 \times 4.5 \times 10^3}{140}$	
CL. 28.1.	$= 24 < 27$	
Table 7	Hence from Table 7, $\beta = 0.53$	$\beta = 0.53$
	Calculate design vertical load resistance of wall	
CL. 32.2.1	$\frac{\beta . t . f_k}{\gamma_m}$ for 1 m length	
	$= \frac{0.53 \times 140 \times 10^3 f_k}{3.5}$	
	Equate design load and resistance	
	$64.9 \times 10^3 = \frac{0.53 \times 140 \times 10^3 f_k}{3.5}$	
	f_k required $= 3.06$ N/mm^2	f_k reqd = 3.06
	Select block strength and mortar	
Table 2 d	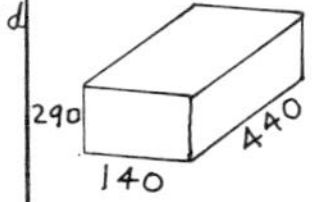 Shape factor $= \frac{290}{140} = 2.07$	
	Table 2(d) mortar (IV) Solid block of comp. strength 3.5 N/mm^2 gives $f_k = 3.5$ N/mm^2 satisfactory	Solid conc block 440×290×140 mortar (IV) Comp. str. 3.5 sat.

CURPE CONSULTANTS 46 Orburn Road Dunfield	Project: BRICKWORK & BLOCKWORK	Job ref
	Part of structure: Solid wall - eccentric loading	Calc sheet no: C9 / 4 rev: /
	Drawing ref; Calc by: R.A.S.; Date	Check by; Date

Ref BS5628	Calculations	Output
	Ex.3. Solid wall - eccentric loading	
	Roof: Dead loading = 7 kN/m^2; Imposed loading (storage) = 10 kN/m^2	
	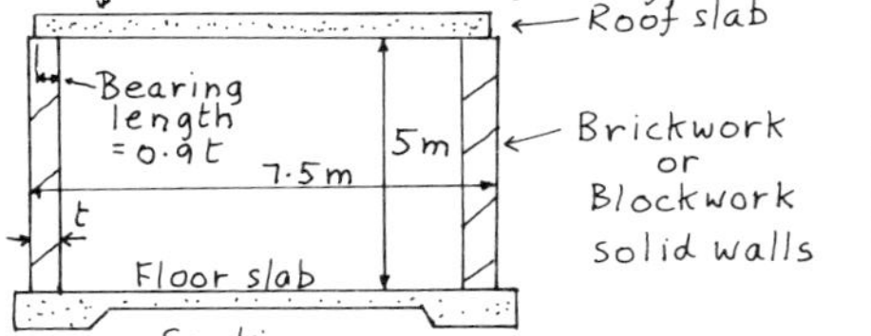	
	Section	
CL. 27.	Partial safety factors	
Table 4	$\gamma = 3.5$	
CL. 22.	$\gamma_f = 1.4$ dead load	
	$= 1.6$ imposed load	
	Self weight of walls	
	4.5 kN/m^2 for 215 mm thick brickwork	
	3.0 kN/m^2 " 190 mm " blockwork	
	(a) Design of brick wall	
	Try 215 mm thick	
	Calculate design load n_w for 1 m length	
	Gk. Slab $= \frac{7.5}{2} \times 7 \times 1 = 26.25$	
	Wall $= 4.5 \times 5 \times 1 = 22.50$	
	Gk $= 48.75$ kN/m	Gk = 48.75 kN
	Qk. imposed $= \frac{7.5}{2} \times 10 \times 1 = 37.5$ kN/m	Qk = 37.5 kN.
	Design load $n_w = 1.4\,Gk + 1.6\,Qk$	
	$= 1.4(48.75) + 1.6(37.5)$	
	$= 128.25$ kN/m	$n_w = 128.25$ kN.
CL. 28.1.	Slenderness ratio and β	

CURPE CONSULTANTS 46 Orburn Road Dunfield	Project: BRICKWORK & BLOCKWORK Part of structure: Solid wall - eccentric loading Drawing ref: · Calc by: R.A.S. · Date:	Job ref: Calc sheet no: C9 / 5 · rev: 1 Check by: · Date:
Ref BS5628	**Calculations**	**Output**
	Slenderness ratio $= \frac{h_{ef}}{t_{ef}} = \frac{0.75 \times 5 \times 10^3}{215}$	
	$= 17.5 < 27$	
	$e_x = \frac{t}{2} - \frac{0.9t}{3} = 0.5t - 0.3t = 0.2t$	
CL. 31.	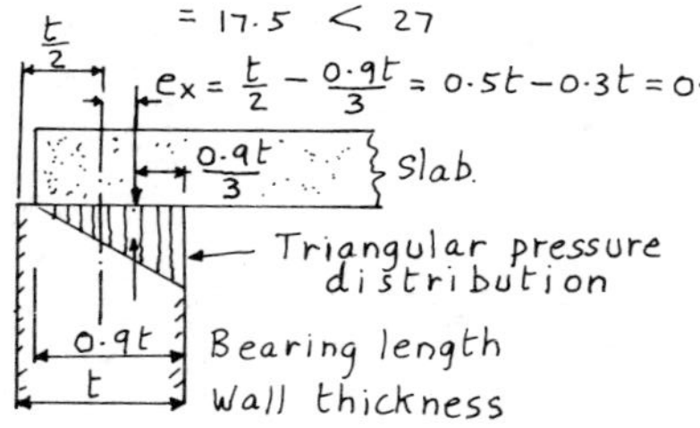	
Table 7	From Table 7, with S.R. = 17.5 and $e_x = 0.2t$ by interpolation, $\beta = 0.58$	$\beta = 0.58$
	Equate design load and resistance of wall	
	$128.25 \times 10^3 = \frac{0.58 \times 215 \times f_k \times 10^3}{3.5}$	
	f_k required $= 3.6\ N/mm^2$	f_k reqd = 3.6
Table 2a	From Table 2(a) using mortar (iii) with brick comp. strength $10\ N/mm^2$ gives $f_k = 4.1\ N/mm^2$, satisfactory	Brick Comp. strength $10\ N/mm^2$ sat. mortar (iii)
	(b) Design of block wall	
	Try 190 mm thick solid blocks.	
	Calculate design load n_w for 1m length	
	Gk. Slab $= \frac{7.5}{2} \times 7 \times 1 = 26.25$	
	Wall $= 3 \times 5 \times 1 = 15.00$	
	Gk $= 41.25\ kN/m$	Gk = 41.25
	Qk. Imposed $= \frac{7.5}{2} \times 10 \times 1 = 37.5\ kN/m$	Qk = 37.5

CURPE CONSULTANTS 46 Orburn Road Dunfield	Project: BRICKWORK & BLOCKWORK Part of structure: Solid wall - eccentric loading Drawing ref: · Calc by: R.A.S. · Date:	Job ref: Calc sheet no: C9 / 6 · rev: 1 Check by: · Date:
Ref BS5628	**Calculations**	**Output**
	Design load $n_w = 1.4(41.25) + 1.6(37.5)$	
	$= 117.75\ kN/m$	$n_w = 117.75$
	Slenderness ratio and β	
CL. 28.1	Slenderness ratio $= \frac{0.75 \times 5 \times 10^3}{190}$	
	$= 19.74 < 27$	
	Eccentricity $e_x = 0.2t$	
Table 7	From Table 7, with S.R = 19.74 and $e_x = 0.2t$ by interpolation $\beta = 0.517$	$\beta = 0.517$
	Equate design load and resistance of wall	
	$117.75 \times 10^3 = \frac{0.517 \times 190 \times f_k \times 10^3}{3.5}$	
	f_k required $= 4.19\ N/mm^2$	f_k reqd = 4.19
	Select block size, strength and mortar	
CL. 23.	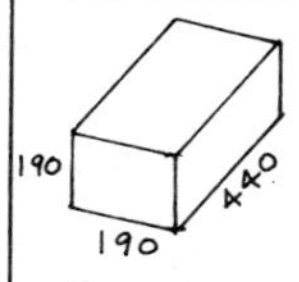 Shape factor $= \frac{190}{190} = 1.0$	
	Try solid blocks with compressive strength $10\ N/mm^2$ set in mortar (iv)	
Table 2b " 2d	From Table 2(b), $f_k = 3.5$ From " 2(d), $f_k = 7.0$	
	By interpolation, $f_k = 4.5\ N/mm^2$	$f_k = 4.5$
	Since $4.19 < 4.5$, satisfactory	Solid conc. block 440×190×190 mortar (iv) Comp. str. 10.0 sat.

CURPE CONSULTANTS 46 Orburn Road Dunfield	Project: BRICKWORK & BLOCKWORK			Job ref	
	Part of structure: Cavity wall			Calc sheet no: C9 / 7	rev: 1
	Drawing ref	Calc by: R.A.S.	Date	Check by	Date

Ref BS5628	Calculations	Output
	Ex.4. Cavity Wall.	
	Roof: Dead loading = 7 kN/m² ; Imposed " = 1·5 kN/m²	

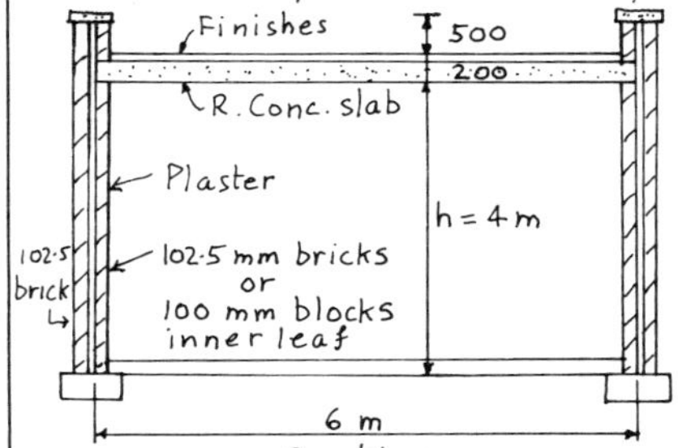

Section

Ref	Calculations	Output
CL. 27	Partial safety factors	
Table 4	$\gamma_m = 3 \cdot 5$	
CL. 22	$\gamma_f = 1 \cdot 4$ dead load; $= 1 \cdot 6$ imposed load	
	Self weight of wall : Inner leaf	
	102·5 brick + plaster = 2·5 kN/m²	
	100 block + plaster = 1·8 kN/m²	
CL. 29.1.1	Outer leaf is self supporting	
	(a) Design for bricks 102·5 mm thick.	
	Gk. Roof slab + finishes = $\frac{6}{2} \times 7 \times 1$ = 21·00	
	Wall, inner leaf (4+0·5) 2·5 × 1 = 11·25	
	Gk = 32·25	Gk = 32·25
	Qk. Imposed = $\frac{6}{2} \times 1 \cdot 5 \times 1$ = 4·5 kN/m	Qk = 4·5
	Design load on wall $n_w = 1 \cdot 4 G_k + 1 \cdot 6 Q_k$	
	$= 1 \cdot 4 (32 \cdot 25) + 1 \cdot 6 (4 \cdot 5) = 52 \cdot 35$ kN/m	$n_w = 52 \cdot 35$

CURPE CONSULTANTS 46 Orburn Road Dunfield	Project: BRICKWORK & BLOCKWORK			Job ref	
	Part of structure: Cavity wall.			Calc sheet no: C9 / 8	rev: 1
	Drawing ref	Calc by: R.A.S.	Date	Check by	Date

Ref BS5628	Calculations	Output
	Slenderness ratio and β	
CL. 28.3.1.1.	$h_{ef} = 0 \cdot 75 h$	
CL 28.4.1.	$t_{ef} = \frac{2}{3}(102 \cdot 5 + 102 \cdot 5) = 136 \cdot 66$ mm	
CL. 28.1.	Slenderness ratio $= \frac{h_{ef}}{t_{ef}} = \frac{0 \cdot 75 \times 4 \times 10^3}{136 \cdot 66}$	
	$= 22 < 27$	

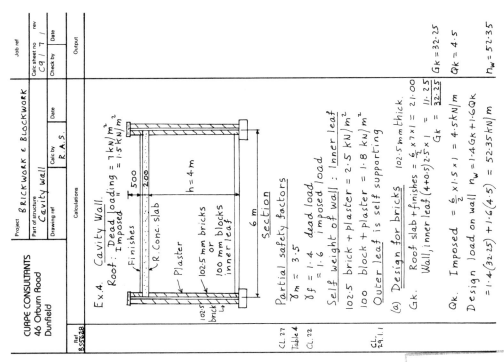

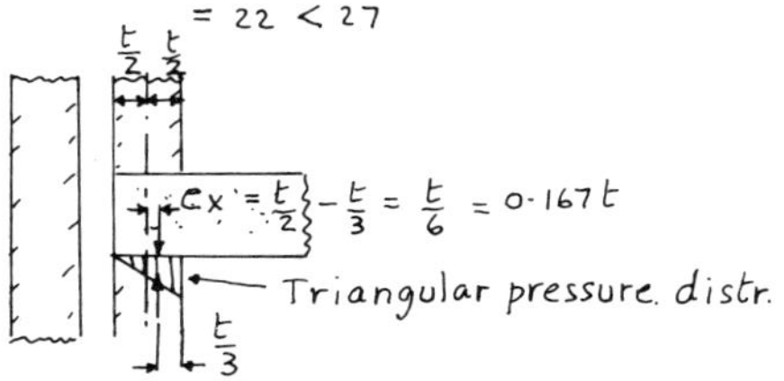

Ref	Calculations	Output
CL. 31.	$e_x = \frac{t}{2} - \frac{t}{3} = \frac{t}{6} = 0 \cdot 167 t$	
	From Table 7, with S.R = 22 and $e_x = 0 \cdot 167 t$	
Table 7	$\beta = 0 \cdot 473$	$\beta = 0 \cdot 473$
	Equate design load and resistance of wall	
CL. 23.1.2	$52 \cdot 35 \times 10^3 = \frac{0 \cdot 473 \times 102 \cdot 5 (f_k \times 1 \cdot 15) 10^3}{3 \cdot 5}$	
	f_k required = 3·28 N/mm²	f_k reqd = 3·28
	Select brick strength and mortar	
Table 2a	From Table 2(a) using mortar (iv), brick with comp. strength 10 N/mm² gives f_k = 3·5 N/mm², satisfactory	Bricks comp. str 10 N/mm², (iv) mortar, f_k = 3·5
	(b) Design for blocks 100 mm thick	
	Gk. Roof slab + finishes = $\frac{6}{2} \times 7 \times 1$ = 21·00	
	Wall, inner leaf = (4 + 0·5) 1·8 × 1 = 8·10	
	29·10 kN/m	Gk = 29·1

CURPE CONSULTANTS 46 Orburn Road Dunfield	Project: BRICKWORK & BLOCKWORK	Job ref
	Part of structure: Cavity wall.	Calc sheet no: C9 / 9, rev: 1
	Drawing ref / Calc by: R.A.S. / Date	Check by / Date

Ref BS5628	Calculations	Output
	Qk. Imposed $= \frac{6}{2} \times 1.5 \times 1.0 = 4.5\,kN/m$	$Q_k = 4.5$
	Design load on wall $n_w = 1.4(29.1) + 1.6(4.5)$	
	$n_w = 47.94\,kN/m$	$n_w = 47.94$
	Slenderness ratio and β	
CL.28.3.1.1	$h_{ef} = 0.75h$	
CL.28.4.1	$t_{ef} = \frac{2}{3}(102.5 + 100) = 135\,mm$	
	Slenderness ratio $= \frac{0.75 \times 4 \times 10^3}{135} = 22 < 27$	
CL.31	$e_x = \frac{t}{6} = 0.167t$	
	From Table 7, with S.R = 22 and $e_x = 0.167t$	
Table 7	$\beta = 0.473$ as before	$\beta = 0.473$
	Equate design load and resistance of wall	
	$47.94 \times 10^3 = \frac{0.473 \times 100 \times f_k \times 10^3}{3.5}$	
	f_k required $= 3.55\,N/mm^2$	f_k reqd $= 3.55$
	Select block size, strength and mortar	
	Using 440 × 215 × 100 solid blocks then	
	Shape factor $= \frac{215}{100} = 2.15$	
Table 2d	From Table 2(d) using mortar (IV), blocks with comp. strength $5\,N/mm^2$ gives $f_k = 4.4\,N/mm^2$, satisfactory	440×215×100 Solid blocks Comp. str. 5 $f_k = 4.4$ mortar (IV)

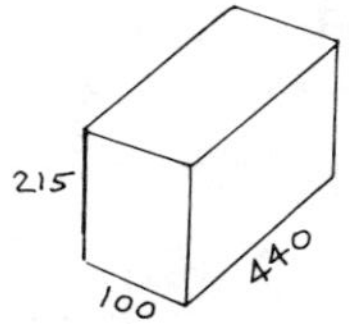

CURPE CONSULTANTS 46 Orburn Road Dunfield	Project: BRICKWORK & BLOCKWORK	Job ref
	Part of structure: Gravity retaining wall.	Calc sheet no: C9 / 10, rev: 1
	Drawing ref / Calc by: R.A.S. / Date	Check by / Date

Ref BS5628	Calculations	Output
	Ex. 5. Gravity Retaining Wall.	
	Wall 750 mm high. retains cohesionless soil of weight $17\,kN/m^3$ and $\phi = 30°$	
	Weight of brickwork taken as $20\,kN/m^3$	
	Coefficient of friction between conc/masonry and concrete/soil is 0.6	
	Weight of concrete foundation $24\,kN/m^3$	
	Determine minimum wall thickness	
	This will be based on the "middle third" rule for no tension in the brickwork.	
	Calculate for 1m length	

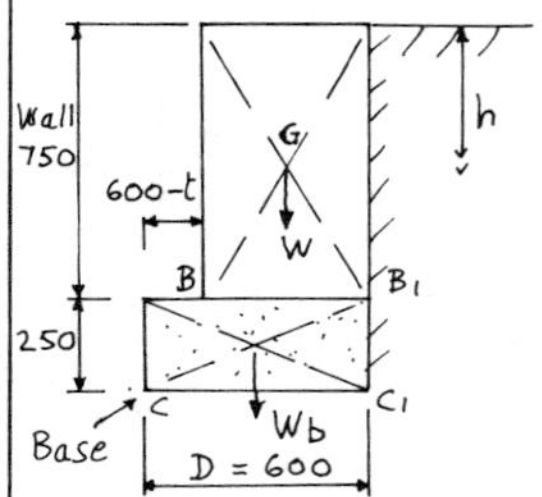

Section

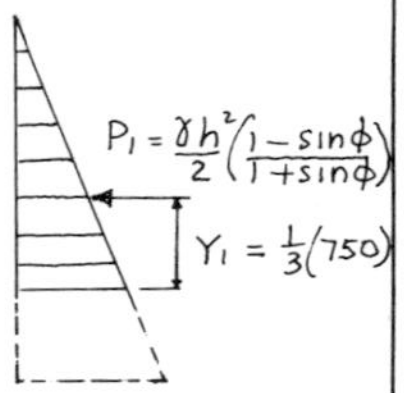

Horizontal pressure

Weight of wall $W = 20 \times 0.75 \times \frac{t}{10^3} \times 1 = 0.015t\ kN$

Horz force P_1 for depth $h = 0.75\,m$

$$P_1 = \frac{17 \times 0.75^2}{2}\left(\frac{1 - \sin 30}{1 + \sin 30}\right) = 1.593\,kN$$

G is the centre of gravity of the wall

CURPE CONSULTANTS 46 Orburn Road Dunfield	Project: BRICKWORK & BLOCKWORK	Job ref
	Part of structure: Gravity retaining wall.	Calc sheet no: C9 / 11 / rev
	Drawing ref / Calc by: R.A.S. / Date	Check by / Date

Ref	Calculations	Output
BS5628		

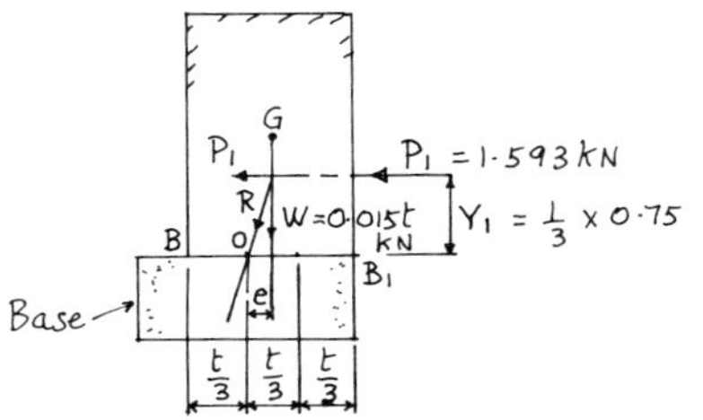

For no tension $e = \frac{t}{6}$ mm where e = "shift" of W.

Moments about O give $P_1 . Y_1 = W . e$.

$1.593 \times \frac{1}{3} \times 0.75 = 0.015t \times \frac{t}{6} \times \frac{1}{10^3}$

$t^2 = 159300 \quad \therefore t = 399$ mm (min)

Then design wall for 440 mm thick — *Output: Brick Wall 440 mm (215 + 10 + 215)*

Weight of wall $W = 20 \times 0.75 \times 0.44 \times 1 = 6.6$ kN

Weight of base $W_b = 24 \times 0.25 \times 0.6 \times 1 = 3.6$ kN

<u>Check for overturning of wall.</u>

Level B-B_1

Take moments about B

Overturning moment (O.T.M) $= P_1 . Y_1$

$= 1.593 \times \frac{0.75}{3} = 0.398$ kNm

Restoring moment (R.M) $= W . \frac{t}{2}$

$= 6.6 \times \frac{0.44}{2} = 1.452$ kN.m.

Factor of safety $= \frac{R.M}{O.T.M} = \frac{1.452}{0.398} = 3.64 > 1.5$ — *Output: Overturing wall sat.*

CURPE CONSULTANTS 46 Orburn Road Dunfield	Project: BRICKWORK & BLOCKWORK	Job ref
	Part of structure: Gravity retaining wall.	Calc sheet no: C9 / 12 / rev
	Drawing ref / Calc by: R.A.S. / Date	Check by / Date

Ref	Calculations	Output
BS5628		

Level C-C_1

Take moments about C.

O.T.M $= P_2 . Y_2$

$P_2 = \frac{17 (0.75 + 0.25)^2}{2} \left(\frac{1 - \sin 30}{1 + \sin 30} \right) = 2.833$ kN

$Y_2 = (0.75 + 0.25)\frac{1}{3} = \frac{1}{3}$

Then O.T.M $= 2.833 \times \frac{1}{3} = 0.944$ kNm

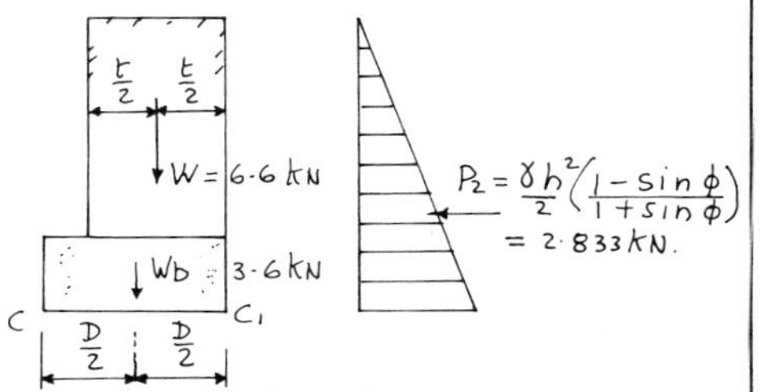

R.M. $= W\left(D - \frac{t}{2}\right) + \left(W_b \times \frac{D}{2}\right)$

$= 6.6\left(0.6 - \frac{0.44}{2}\right) + \left(3.6 \times \frac{0.6}{2}\right)$

$= 2.508 + 1.08 = 3.588$ kNm

F.O.S $= \frac{R.M.}{O.T.M} = \frac{3.588}{0.944} = 3.8 > 1.5$ — *Output: Overturning satisfactory*

<u>Check for sliding</u>

Level B-B_1

Force causing sliding $= P_1 = 1.593$ kN

Frictional resisting force $= \mu R$

$= 0.6 W = 0.6 \times 6.6 = 3.96$ kN

F.O.S $= \frac{3.96}{1.593} = 2.48 > 1.5$ — *Output: Sliding o.k wall/base*

Level C-C_1

Force causing sliding $= P_2 = 2.833$ kN.

CURPE CONSULTANTS 46 Orburn Road Dunfield	Project BRICKWORK & BLOCKWORK			Job ref	
	Part of structure Gravity retaining wall.			Calc sheet no C9 / 13	rev /
	Drawing ref	Calc by R. A. S	Date	Check by	Date

Ref	Calculations	Output
BS5628		
	Frictional resisting force $= \mu R$ $= 0.6\,(W + W_b) = 0.6(6.6 + 3.6) = 6.12$ kN F.O.S $= \frac{6.12}{2.833} = 2.16 > 1.5$	Sliding sat.
	Determine maximum stress in brickwork Max. moment $= P_1.Y_1 = 1.593 \times \frac{0.75}{3} = 0.398$ kNm Stresses $f = \frac{W}{A} + \frac{M}{Z}$ where $Z = \frac{BD^2}{6}$ $f = \frac{6.6 \times 10^3}{1 \times 10^3 \times 440} \pm \frac{0.398 \times 10^6 \times 6}{1 \times 10^3 \times 440^2}$ $f = 0.015 \pm 0.012 = 0.027$ and 0.003	
(CP121. PART I)	In accordance with CP121: Part I: 1973 Table 4.1. use special quality bricks with mortar designation (1) for earth retaining walls.	special quality bricks and mortar (1)
	Bearing pressures on soil Safe bearing pressure given as 175 kN/m² Resultant moment about centroidal axis of base $= P_2.Y_2 - W(\frac{D}{2} - \frac{t}{2})$ $= (2.833 \times \frac{1}{3}) - 6.6(\frac{0.6}{2} - \frac{0.44}{2})$ $= 0.944 - 0.528 = 0.416$ kN.m $f = \frac{(6.6 + 3.6)}{1 \times 0.6} \pm \frac{0.416 \times 6}{1 \times 6^2} = 17 \pm 6.93$ $f = 23.92$ kN/m² or 10.06 kN/m² Max. f < Safe bearing pressure $23.92 < 175$	Bearing press. sat.

10

BS 5268: structural use of timber: Part 2

The Code of Practice CP112, *The Structural Use of Timber,* has been replaced by BS 5268 as part of the British Standard Institution's integration work on Codes and Standards. The new Standard is intended to include a number of parts, the first two of which will cover general structural design. Part 2, *Permissible Stress Design,* is a revision of CP112: Part 2, which will be superseded.

From a purely structural view it is the grade stresses which are of prime importance, and these differ for each species and grade. To provide simplicity in design, softwoods and hardwoods having similar strength and stiffness properties have been grouped together in strength classes, which can be considered as being independent of species and grades. There are nine strength classes SC1 to SC9, with the first-mentioned the weakest. Examples will be based on timber of strength class SC3, this class covering a wide range of softwoods commonly used for structural work.

Stress grading

In this chapter, all timber used for structural work will be stress-graded in accordance with the requirements of BS 4978: 1973: *Timber Grades for Structural Use.* There are two principal ways of stress grading timber, these are by:

Visual stress grading
Machine stress grading

Two standard grades have been established for visual stress graded timber, these are designated:

GS General Structural Grade
SS Special Structural Grade

For machine stress grades there are two which are closely similar to the above visual grades; these are designated MGS and MSS. There are also two other machine grades identified by 'numbering' and designated M50 and M75.

Some samples of softwood species/grade combinations which satisfy the requirement for appropriate strength classes SC1, SC2, SC3 are as follows:

Strength class	*Species*	*Grade*
SC1	Western Red Cedar	GS
	European Spruce	GS
	Sitka Spruce	GS
SC2	Western Red Cedar	SS
	Douglas Fir (British)	GS
	European Spruce	M50/SS
	Sitka Spruce	M50/SS
SC3	Parana Pine	GS
	Redwood	GS/M50
	Whitewood	GS/M50
	Douglas Fir (British)	M50/SS
	Larch	GS
	European Spruce	M75
	Sitka Spruce	M75

Service exposure conditions

At various moisture contents, timber shrinks or swells and its strength properties increase or decrease. Permissible stresses should correspond to the moisture content that the particular member will attain in service. Some applications envisage the use of dry grade stresses with the material supplied in dry condition and adequately protected from the weather.

For service conditions where the material is either in contact with water or unprotected ground, or where the timber could attain an equilibrium moisture content above 18%, green or wet stresses should be used. For solid timber members more than 100 mm thick, it is normal to use wet stresses in design, irrespective of exposure conditions, because it is difficult and expensive to dry timber more than 75 mm thick artificially.

Grade stresses and moduli of elasticity for the dry exposure condition are given in Table 8 of BS 5268: Part 2 for the nine strength classes. Grade stress values for the wet exposure condition should be obtained by multiplying the dry stresses and moduli given in Table 8 by the modification factor K_2 from Table 14 of BS 5268: Part 2.

A modification factor K_1 is used to convert the geometrical properties of timber for the dry exposure conditions to wet exposure condition:

Wet exposure condition = geometrical properties for dry exposure condition $\times K_1$.

Values of K_1 and K_2 are listed as follows:

Geometrical property	K_1
Thickness, width, radius of gyration	1.02
Area	1.04
Section modulus	1.06
Second moment of area	1.08

Property	K_2
Bending parallel to grain	0.8
Tension parallel to grain	0.8
Compression parallel to grain	0.6
Compression perpendicular to grain	0.6
Shear parallel to grain	0.9
Mean and minimum modulus of elasticity	0.8

Duration of loading

The grade stresses given in BS 5268: Part 2 are applicable to long-term loading. Because timber can sustain a much greater load for a period of a few minutes than for a period of several years, grade stresses may be increased for other conditions of loading by the use of modification factors K_3, values of which are given below. These modification factors are applicable to all strength properties but not to moduli of elasticity.

Long-term loading	$K_3 = 1.00$
Medium-term loading	$K_3 = 1.25$
Short-term loading	$K_3 = 1.50$
Very short-term loading	$K_3 = 1.75$

Load-sharing systems

The grade stresses given in BS 5268: Part 2 are applicable to individual sections or pieces of timber. When four or more members can be considered to act together to support a common load, the appropriate grade stresses may be multiplied by a load-sharing modification factor K_8 which has a value of 1.1. The modification factor should be applied only to rafters, joists, trusses and wall studs spaced not further apart than 610 mm and with adequate provision for the lateral distribution of loads by means of purlins, binders, boarding, battens.

For the calculation of deflections, the mean modulus of elasticity should be used for load-sharing members, unless the imposed load is for an area intended for mechanical plant and equipment, storage, or floors subject to vibrations, in which case minimum modulus of elasticity is used. Special provisions are made for built-up beams, trimmer joists and lintels.

Flexural members

Permissible stresses should be taken as the product of the grade stress and the appropriate modification factors for service and loading. It is then compared with the applied stress in a member during structural design calculations.

Length and position of bearing

At any bearing on the side grain of timber, the permissible stress in compression perpendicular to the grain is dependent on the length and position of the bearing. Grade stresses for compression perpendicular to the grain apply to:

1) bearings of any length at the ends of a member,
2) bearings 150 mm or more in length at any position.

For bearings less than 150 mm long, located 75 mm or more from the end of a member, the grade stress should be multiplied by the modification factor K_4.

Length of bearing (mm)	10	15	25	40	50	75	100	$\geqslant 150$
K_4	1.74	1.67	1.53	1.33	1.20	1.14	1.10	1.00

Effective span

Generally taken as the distance between the centres of bearings.

Shear at notched ends

Notches are sometimes cut in members during construction to accommodate piping or ducting. Square cornered notches at the ends of flexural members cause severe stress concentrations which may be allowed for in accordance with Clause 14.4 of BS 5268: Part 2.

Form factor

Grade bending stresses apply to solid timber members of rectangular cross-section. For other shapes of cross-section the grade bending stress should be multiplied by the modification factor K_6, where:

K_6 = 1.18 for solid circular sections,
K_6 = 1.41 for solid square sections loaded on a diagonal.

Depth factor

The grade bending stresses given in Table 8 of BS 5268: Part 2 apply to material assigned to a strength class and having a depth h of 300 mm. For other depths of beams assigned to a strength class, the grade bending stress shall be multiplied by the depth modification factor K_7, where:

K_7 = 1.17 for solid beams having a depth $\leqslant$ 72 mm,

$K_7 = \left(\frac{300}{h}\right)^{0.11}$ for solid beams having a depth > 72 mm and < 300 mm,

$K_7 = 0.81\left(\frac{h^2 + 92300}{h^2 + 56800}\right)$ for solid beams having a depth > 300 mm.

Deflection and stiffness

Deflection should be restricted within appropriate limits having regard to possibility of damage to ceilings, finishings, etc. and to the functional needs. In addition to the deflection due to bending, shear deflection may be significant and should be taken into account. Shear deflection is given by

$$d = \frac{FM}{AG}$$

where F = 1.2 for rectangular sections,
M = bending moment at centre of span,
A = cross-sectional area,
G = modulus of rigidity, usually $E/16$.

For most general purposes this may be assumed to be found satisfactory if the deflection of the member when fully loaded does not exceed 0.003 of the span.

The deflection of solid timber members acting alone (no load sharing), should be calculated using the appropriate minimum modulus of elasticity. The deflections of load sharing systems should be calculated using the mean modulus of elasticity.

Lateral support

The depth-to-breadth ratios of solid rectangular section beams should be checked to ensure that there is no risk of buckling under loading. Maximum depth-to-breadth ratios for various degrees of lateral support are given in Table 17 of BS 5268: Part 2.

Timber columns

For short timber columns the mode of failure is by crushing of the material, with shear failures in the nature of compression creases at an angle of approximately 45° to the axis. For long, slender columns failure takes the form of buckling.

Practical compression members cannot support the full load suggested by the compressive strength parallel to the grain, and a modification factor for buckling has to be applied to enable the permissible axial load to be calculated. The load-carrying capacity of a timber column is based on the slenderness ratio, which is a function of the cross-section and its effective length. The end conditions govern the effective length chosen for design.

End restraints

The ends of the members may in general be restrained in position or in position and direction. These end conditions govern the chosen effective length, the slenderness ratio and consequently the permissible load which may be carried by the column.

Effective length

This can be derived from either the use of coefficients given in Table 19 of BS 5268: Part 2, or by considering the deflected form of the member as affected by any restraint and/or fixing moments, the effective length being the distance between adjacent points of zero bending between which the member is in single curvature.

Values of coefficients are for a column which is:

1) restrained at both ends in position and direction = 0.7
2) restrained at both ends in position and one end in direction = 0.85
3) restrained at both ends in position but not in direction = 1.0

The effective length is then obtained from:

$$L_E = \text{coefficient multiplied by } L$$

where L is the actual length.

Slenderness ratio

$$\text{Slenderness ratio} = \frac{L_E}{i}$$

where L_E is the effective length,
i is the radius of gyration.

This value should not exceed 180 for any compression member carrying dead and imposed loads, and 250 for wind forces only.

In many cases a column will have different effective length in relation to the two principal axes, thus both ratios will have to be calculated so that the critical one is properly used.

Members subject to axial compression

For compression members with slenderness ratio of less than 5, the permissible stress should be taken as the grade stress compression parallel to the grain, modified as appropriate for size, moisture content, duration of load and load sharing.

With slenderness ratio equal to or greater than 5, the permissible stress should be taken as the product of the grade stress compression parallel to the grain, modified as appropriate for size, moisture content, duration of load and load sharing, and the modification factor K_{12} given in Table 20 of BS 5268: Part 2, or calculated otherwise using a given equation.

The value of modulus of elasticity required for entry to Table 20 is the minimum (not the mean) for both compression members acting alone, and those in load-sharing systems, and the compression stress should be modified for duration of load.

Members subject to axial compression and bending

This occurs when members are subject to eccentric loading or lateral forces. A member restrained at both ends in position but not direction should be proportioned so that:

$$\frac{\sigma_{m,a\,||}}{\sigma_{m,adm\,||}\left(1 - \dfrac{1.5\sigma_{c,a\,||}}{\sigma_e}K_{12}\right)} + \frac{\sigma_{c,a\,||}}{\sigma_{c,adm\,||}} \leqslant 1.0$$

where $\sigma_{m,a}||$ = applied bending stress,
$\sigma_{m,adm}||$ = permissible bending stress,
$\sigma_{c,a}||$ = applied compression stress,
$\sigma_{c,adm}||$ = permissible compression stress (including K_{12}),
σ_e = Euler critical stress $\dfrac{\pi^2 E}{\left(\dfrac{L_E}{i}\right)^2}$.

For members in load-sharing systems, the permissible bending stress $\sigma_{m,adm}||$ and permissible compression stress $\sigma_{c,adm}||$, should be multiplied by the load-sharing stress modification factor K_8 which has a value of 1.1 for SC3 strength class.

The dimensions of compression members subject to bending should be such as to restrict deflection within satisfactory limits appropriate to the type of structure.

Bearing

The permissible bearing stress at the ends of a column is related to the grade stress parallel to the grain, with a slenderness ratio of zero. A load-sharing modification factor is used if appropriate. If the column bears on a timber sole plate or the load is applied through a timber beam or joist, then the permissible stress is related to the grade stress perpendicular to the grain, with appropriate duration of load and load-sharing modification factors applied.

Where wane is excluded, the full cross-section width may be used in calculating the applied stress. With wane present, the applied stress increases and a lower grade stress compression perpendicular to the grain is used in calculations.

Formwork

Formwork is a structure, normally temporary, used to contain poured concrete, moulding it to desired shapes and supporting it until it is capable of supporting itself. It consists of a face contact material, bearers, and a vertical supporting system, all of which are removed in a specified sequence after the concrete members have become self-supporting.

Timber is employed structurally as formwork for the construction of cast *in situ* slabs, beams and columns and designed in accordance with BS 5268: Part 2: 1984. It is recommended that timber of Strength Class SC3 should be the minimum quality adopted for use in formwork. The example given in this chapter is related to the design of timber soffit formwork supporting a simple *in situ* concrete floor slab.

The plywood decking and the softwood timber bearers should be capable of resisting the vertical loads arising from the self-weight of the formwork and the poured concrete. A loading of 1.5 kN/m^2 is allowed additionally to cater for the operatives placing the fresh concrete, and for the tools and plant used in the placing operations. In addition to strength it is also important that the deflection of individual members should be restricted to 1/270 of the span. It is found that this normally produces an acceptable flat concrete surface.

The vertical supports may be either timber props or adjustable steel props. Any defects such as eccentricity of loading from the bearers, or props erected out of plumb, should be avoided, as these may seriously reduce the load-carrying capacity of the props.

It is known that timber swells across the grain as it gets wetter, and under site conditions will usually be larger than the normal dry sizes. In the wet exposure condition, modification factors are employed to convert dry stresses to wet stresses, taking into account the wetness and the geometrical properties of the section size. This condition is relevant to the design of formwork and trench supports in excavations.

Timber support for shallow trench

As an example, calculations are provided for the design of timber boarding, walings and struts used for supporting the sides of a shallow trench. In the example, the formula for earth pressure is used for the sole purpose of obtaining the magnitudes of the applied forces and moments, necessary for the design of the timber components. In the case of deep excavations, appropriate earth pressure theories and formulae would have to be considered and applied, accounting also for the type of soils encountered in the surrounding area of the excavation.

Stud walling

As an example, calculations are provided for the vertical timber studs forming part of the frame of a load-bearing stud wall. Noggings are provided at mid-height and these reduce the effective length of the weaker axis of the stud to one half the overall length.

In this example, no allowance has been made for any strength afforded by the covering material, e.g. plasterboard both sides, in the design of the studs.

Table 20. Modification factor K_{12} for compression members

$E/\sigma_{c,\parallel}$	*Value of* K_{12}																			
	Values of slenderness ratio $\lambda (= L_e/i)$																			
	< 5	*5*	*10*	*20*	*30*	*40*	*50*	*60*	*70*	*80*	*90*	*100*	*120*	*140*	*160*	*180*	*200*	*220*	*240*	*250*
	Equivalent L_e/b *(for rectangular sections)*																			
	< 1.4	*1.4*	*2.9*	*5.8*	*8.7*	*11.6*	*14.5*	*17.3*	*20.2*	*23.1*	*26.0*	*28.9*	*34.7*	*40.5*	*46.2*	*52.0*	*57.8*	*63.6*	*69.4*	*72.3*
400	1.000	0.975	0.951	0.896	0.827	0.735	0.621	0.506	0.408	0.330	0.271	0.225	0.162	0.121	0.094	0.075	0.061	0.051	0.043	0.040
500	1.000	0.975	0.951	0.899	0.837	0.759	0.664	0.562	0.466	0.385	0.320	0.269	0.195	0.148	0.115	0.092	0.076	0.063	0.053	0.049
600	1.000	0.975	0.951	0.901	0.843	0.774	0.692	0.601	0.511	0.430	0.363	0.307	0.226	0.172	0.135	0.109	0.089	0.074	0.063	0.058
700	1.000	0.975	0.951	0.902	0.848	0.784	0.711	0.629	0.545	0.476	0.399	0.341	0.254	0.195	0.154	0.124	0.102	0.085	0.072	0.067
800	1.000	0.975	0.952	0.903	0.851	0.792	0.724	0.649	0.572	0.497	0.430	0.371	0.280	0.217	0.172	0.139	0.115	0.096	0.082	0.076
900	1.000	0.976	0.952	0.904	0.853	0.797	0.734	0.665	0.593	0.522	0.456	0.397	0.304	0.237	0.188	0.153	0.127	0.106	0.091	0.084
1000	1.000	0.976	0.952	0.904	0.855	0.801	0.742	0.677	0.609	0.542	0.478	0.420	0.325	0.255	0.204	0.167	0.138	0.116	0.099	0.092
1100	1.000	0.976	0.952	0.905	0.856	0.804	0.748	0.687	0.623	0.559	0.497	0.440	0.344	0.272	0.219	0.179	0.149	0.126	0.107	0.100
1200	1.000	0.976	0.952	0.905	0.857	0.807	0.753	0.695	0.634	0.573	0.513	0.457	0.362	0.288	0.233	0.192	0.160	0.135	0.116	0.107
1300	1.000	0.976	0.952	0.905	0.858	0.809	0.757	0.701	0.643	0.584	0.527	0.472	0.378	0.303	0.247	0.203	0.170	0.144	0.123	0.115
1400	1.000	0.976	0.952	0.906	0.859	0.811	0.760	0.707	0.651	0.595	0.539	0.486	0.392	0.317	0.259	0.214	0.180	0.153	0.131	0.122
1500	1.000	0.976	0.952	0.906	0.860	0.813	0.763	0.712	0.658	0.603	0.550	0.498	0.405	0.330	0.271	0.225	0.189	0.161	0.138	0.129
1600	1.000	0.976	0.952	0.906	0.861	0.814	0.766	0.716	0.664	0.611	0.559	0.508	0.417	0.342	0.282	0.235	0.198	0.169	0.145	0.135
1700	1.000	0.976	0.952	0.906	0.861	0.815	0.768	0.719	0.669	0.618	0.567	0.518	0.428	0.353	0.292	0.245	0.207	0.177	0.152	0.142
1800	1.000	0.976	0.952	0.906	0.862	0.816	0.770	0.722	0.673	0.624	0.574	0.526	0.438	0.363	0.302	0.254	0.215	0.184	0.159	0.148
1900	1.000	0.976	0.952	0.907	0.862	0.817	0.772	0.725	0.677	0.629	0.581	0.534	0.447	0.373	0.312	0.262	0.223	0.191	0.165	0.154
2000	1.000	0.976	0.952	0.907	0.863	0.818	0.773	0.728	0.681	0.634	0.587	0.541	0.455	0.382	0.320	0.271	0.230	0.198	0.172	0.160

Table 20 from BS 5268: Part 2

CURPE CONSULTANTS
46 Orburn Road
Dunfield

Contract	Structural Timber	Job ref
Part of structure	Floor Joists	Calc sheet No C10/1.
Drawing ref	Calculations by R.A.S. / Checked by	Date

Members BS5268 ref	CALCULATIONS	OUTPUT

Ex.1.

Floor Joists Strength Class SC3.

Span = 4 m. Spacing = 400 mm c/c

Loading. Imposed = $1 \cdot 5\ kN/m^2$

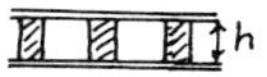

T & G. boarding = 0·11
Joists = 0·12
Plasterboard = 0·11
Skim = 0·10
Σ = $1 \cdot 94\ kN/m^2$

Long term loading is assumed.

T. 8 For SC3.
Grade stress bending parallel to grain = $5 \cdot 3\ N/mm^2$
" " shear " " " = 0·67 "
E mean value = $8800\ N/mm^2$

Modification factors

T. 15 K_3, duration of loading = 1·0

C. 13. K_8, load sharing system = 1·1

14.6 K_7, depth factor $= \left(\frac{300}{h}\right)^{0.11}$ For depth 225 = 1·032; For depth 200 = 1·046

Permissible stresses

Bending parallel to grain:
$5 \cdot 3 \times 1 \times 1 \cdot 1 \times 1 \cdot 032 = 6 \cdot 02\ N/mm^2$ (for 225)
$5 \cdot 3 \times 1 \times 1 \cdot 1 \times 1 \cdot 046 = 6 \cdot 10$ " (for 200)

Shear parallel to grain:
$0 \cdot 67 \times 1 \times 1 \cdot 1 = 0 \cdot 737\ N/mm^2$

U.D.L on 1 joist

$F = 1 \cdot 94 \times 0 \cdot 4 \times 4 = 3 \cdot 104\ kN.$

CURPE CONSULTANTS
46 Orburn Road
Dunfield

Contract	Structural Timber	Job ref
Part of structure	Floor Joists	Calc sheet No C10/2
Drawing ref	Calculations by R.A.S. / Checked by	Date

Members BS5268 ref	CALCULATIONS	OUTPUT

Bending $M = \frac{FL}{8} = \frac{3 \cdot 104 \times 4}{8} = 1 \cdot 552\ kNm$

Z reqd for depth 225 mm $= \frac{M}{\sigma_{m,adm,11}}$

$= \frac{1 \cdot 552 \times 10^6}{6 \cdot 02} = 257807\ mm^3$

Z reqd for depth 200 mm

$= \frac{1 \cdot 552 \times 10^6}{6 \cdot 10} = 254426\ mm^3$

C.14.7. Deflection Allowable = 0·003 × span

Actual $= \frac{5}{384}\frac{FL^3}{EI}$

Equating $0 \cdot 003 L = \frac{5 FL^3}{384 EI}$

Reqd $I = \frac{5 FL^2}{384\ E \times 0 \cdot 003}$

$= \frac{5 \times 3 \cdot 104 \times 10^3 \times 4^2 \times 10^6}{384 \times 8800 \times 0 \cdot 003}$

$= 24494949\ mm^4$

Section 200×40 provides:

$I = \frac{40 \times 200^3}{12} = 26666666\ mm^4$

$Z = \frac{40 \times 200^2}{6} = 266666\ mm^3$

Section 200×40 is thus adequate for bending and deflection with shear deflection small in this case and ignored.

Check section for shear

OUTPUT: 200×40 (SC3) adequate bending & deflection

CURPE CONSULTANTS
46 Orburn Road
Dunfield

Contract	Structural Timber	Job ref
Part of structure	Floor Joists.	Calc. sheet No C10/3
Drawing ref	Calculations by R.A.S. / Checked by	Date

Members ref	CALCULATIONS	OUTPUT

Shear

Maximum shear $V = \frac{F}{2} = \frac{3.104}{2} = 1.552$ kN.

Maximum shear stress at neutral axis

$= 1.5 \times$ average shear stress $= \frac{1.5\,V}{A}$

$= \frac{1.5 \times 1.552 \times 10^3}{40 \times 200} = 0.291\ \text{N/mm}^2$

This is less than permissible value of $0.737\ \text{N/mm}^2$.

Section adequate for shear.

C.14.2 **Bearing**

Assume bearing on wall plate

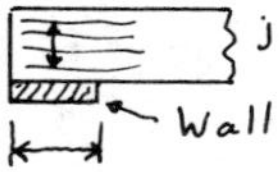

Bearing area on plan
= bearing length × joist thickness

Bearing stress $= \frac{\text{End reaction}}{\text{Bearing area}}$

$= \frac{1.552 \times 10^3}{100 \times 40} = 0.388\ \text{N/mm}^2$

This is lower than the grade stress of 1.7 compression perpendicular to the grain and without modification factors and is satisfactory

OUTPUT: USE 40×200 Strength Class SC3.

CURPE CONSULTANTS
46 Orburn Road
Dunfield

Contract	Structural Timber	Job ref
Part of structure	Timber Beam	Calc. sheet No C10/4
Drawing ref	Calculations by R.A.S. / Checked by	Date

Members ref BS5268	CALCULATIONS	OUTPUT

Ex. 2.

Timber Beam Strength Class SC3.
Section 300×75

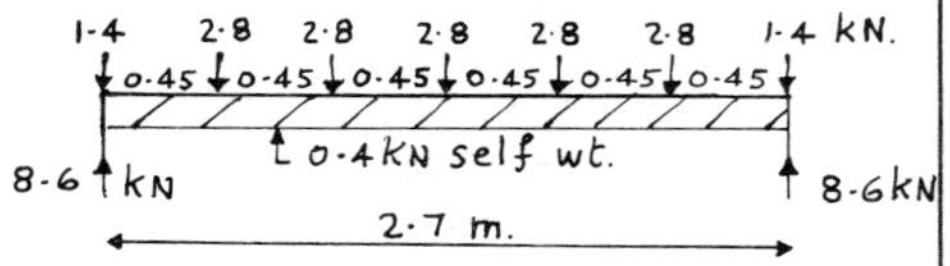

Loading - long term.

Check for bending, deflection, shear

Bending

Max. M at centre of span

$M = (8.6 - 1.4)1.35 - (2.8 \times 0.9) - (2.8 \times 0.45) - (0.2 \times \frac{2.7}{4})$

$= 9.72 - 2.52 - 1.26 - 0.135 = 5.805$ kNm

Permissible bending stress $\sigma_{m.adm}$

C.14.6 T.15. $=$ Grade stress bending $\times K_3 \times K_7$

$= 5.3 \times 1.0 \times 1.0 = 5.3\ \text{N/mm}^2$

Depth factor $K_7 = \left(\frac{300}{h}\right)^{0.11}$

T.15. $\therefore K_7 = \left(\frac{300}{300}\right)^{0.11} = 1.0$

$Z\ reqd = \frac{M}{\sigma_{m.adm}} = \frac{5.805 \times 10^6}{5.3}$

$= 1095283\ \text{mm}^3$

$Z\ given = \frac{bh^2}{6} = \frac{75 \times 300^2}{6} = 1125000\ \text{mm}^3$

Since 1095283 < 1125000. Adequate Bending

OUTPUT: 300×75 (SC3) Adequate bending

CURPE CONSULTANTS
46 Orburn Road
Dunfield

Contract	Structural Timber		Job ref
Part of structure	Timber Beam		Calc. sheet No C10/5
Drawing ref	Calculations by R.A.S.	Checked by	Date

Members ref	CALCULATIONS	OUTPUT
C.14.7.	<u>Deflection</u>	
	Since member is a principal member	
	i.e. acting alone, use min. E value	
	Allowable deflection = 0·003 × span.	
	$= 0{\cdot}003 \times 2{\cdot}7 \times 10^3 = 8{\cdot}1$ mm.	
	Actual deflection $= k.\frac{WL^3}{EI}$ for point loads	
	From Steel Designers Manual k = 0·0127	
	$\therefore$ deflection $= \frac{0{\cdot}0127 \times 16{\cdot}8 \times 10^3 \times 2{\cdot}7^3 \times 10^9 \times 12}{5800 \times 75 \times 300^3}$	
	$= 4{\cdot}29$ mm	
	Deflection for self weight $= \frac{5}{384}\frac{WL^3}{EI}$	
	$= \frac{5 \times 0{\cdot}4 \times 10^3 \times 2{\cdot}7^3 \times 10^9 \times 12}{384 \times 5800 \times 75 \times 300^3} = 0{\cdot}10$ mm	
	Total deflection = 4·29 + 0·10 = 4·39 mm	
	Since 4·39 < 8·1. Adequate for deflection	300×75 (SC3) Adequate for deflection
	<u>Note</u> Deflection due to shear is small	
	and has been disregarded here.	
	<u>Shear</u>	
	Max. applied shear force = 8·6 − 1·4 = 7·2 kN.	
	Max. shear stress $= \frac{1{\cdot}5 \times 7{\cdot}2 \times 10^3}{75 \times 300} = 0{\cdot}48$ N/mm^2	
	Permissible shear stress = Grade stress × K_3	
	$= 0{\cdot}67 \times 1{\cdot}0 = 0{\cdot}67$ N/mm^2	
	Since 0·48 < 0·67, Adequate for shear	300×75 (SC3) Adequate.

CURPE CONSULTANTS
46 Orburn Road
Dunfield

Contract	Structural Timber		Job ref
Part of structure	Timber Columm		Calc. sheet No C10/6
Drawing ref	Calculations by R.A.S.	Checked by	Date

Members ref BS 5268	CALCULATIONS	OUTPUT
	Ex. 3. <u>Timber Column</u> Section 100 × 100	
	Strength Class SC3	
	Effective length = 2·50 m.	
	Determine load carrying capacity:	
	(a) for long term loading	
	(b) for medium " "	
	<u>Radius of gyration</u> i	
	i min $= \sqrt{\frac{b^2}{12}} = \sqrt{\frac{d^2}{12}} = \sqrt{\frac{100^2}{12}} = 28{\cdot}86$	
	<u>Slenderness ratio</u>	
C.15.4	$\frac{L_E}{i} = \frac{2{\cdot}50 \times 10^3}{28{\cdot}86} = 86{\cdot}6 \not> 180$	
	<u>Modification factors</u>	
T. 15	$K_3 = 1{\cdot}0$ for long term	
	$= 1{\cdot}25$ " medium "	
T. 20	K_{12} see below	
	(a) For K_{12} use Table 20	
	$\frac{E\ min.}{\sigma_{c.//}} = \frac{5800}{6{\cdot}8 \times 1{\cdot}0} = 853$	
T. 20	$\therefore K_{12} = 0{\cdot}4666$	
	Permissible comp. stress	
	= grade stress comp. parallel to grain × K_3 × K_{12}	
	$= 6{\cdot}8 \times 1{\cdot}0 \times 0{\cdot}4666 = 3{\cdot}173$ N/mm^2	

CURPE CONSULTANTS
46 Orburn Road
Dunfield

Contract	Structural Timber	Job ref	
Part of structure	Timber Column.	Calc. sheet No.	C10/7
Drawing ref	Calculations by: R. A. S.	Checked by	Date

Members BS5268 ref	CALCULATIONS	OUTPUT
	$\therefore$ Capacity $= \frac{3.173 \times 100 \times 100}{10^3} = 31.73$ kN	Capacity 31.73 kN.
	(b) $\frac{E_{min}}{\sigma_{c.\parallel}} = \frac{5800}{6.8 \times 1.25} = 682$	
T. 20	$\therefore K_{12} = 0.4158$	
	Permissible comp. stress	
	= grade stress comp. parallel to grain $\times K_3 \times K_{12}$	
	$= 6.8 \times 1.25 \times 0.4158 = 3.534$ N/mm^2	
	$\therefore$ Capacity $= \frac{3.534 \times 100 \times 100}{10^3}$	
	$= 35.34$ kN.	Capacity 35.34 kN.
	Ex.4. Timber Column.	
	Check the adequacy of the column in the previous example if it is now subject to an axial load of 20kN and a bending moment of 200kN.mm from long term loading.	
	Permissible comp. stress $\sigma_{c.adm} = 3.173$ N/mm^2	
	Applied comp. stress $= \frac{20 \times 10^3}{100 \times 100} = 2.0$ N/mm^2	
	Applied bending stress $= \sigma_{m.a} = \frac{M}{Z}$	

CURPE CONSULTANTS
46 Orburn Road
Dunfield

Contract	Structural Timber	Job ref	
Part of structure	Timber Column	Calc. sheet No	C10/8
Drawing ref	Calculations by: R. A. S.	Checked by	Date

Members ref	CALCULATIONS	OUTPUT
	$\therefore \sigma_{m.a} = \frac{200 \times 10^3 \times 6}{100 \times 100^2} = 1.20$ N/mm^2	
	Permissible bending stress $\sigma_{m.adm}$.	
	= grade stress bending $\times K_3 \times K_7$	
	$= 5.30 \times 1.0 \times \left(\frac{300}{h}\right)^{0.11}$	
	$= 5.30 \times 1.0 \times \left(\frac{300}{100}\right)^{0.11}$	
	$= 5.30 \times 1.0 \times 1.128 = 5.978$ N/mm^2	
C. 15.6	Euler critical stress $\sigma_e = \frac{\pi^2 E}{\left(\frac{L_E}{i}\right)^2}$	
	$\therefore \sigma_e = \frac{\pi^2 \times 5800}{86.6^2} = 7.633$	
C. 15.6	Now $\frac{\sigma_{m.a}}{\sigma_{m.adm}\left(1 - \frac{1.5\sigma_{c.a}}{\sigma_e} \cdot K_{12}\right)} + \frac{\sigma_{c.a}}{\sigma_{c.adm}} \not> 1.0$	
	$= \frac{1.2}{5.978\left(1 - \frac{1.5 \times 2.0 \times 0.4658}{7.633}\right)} + \frac{2.0}{3.173}$	
	$= 0.2457 + 0.6303 = 0.876 < 1.0$	
	$\therefore$ Section adequate	100×100 (SC3) Section adequate.

CURPE CONSULTANTS
46 Orburn Road
Dunfield

Contract: Structural Timber | Job ref:
Part of structure: Stud Walling | Calc. sheet No: C10/9.
Drawing ref: | Calculations by: R.A.S. | Checked by: | Date:

Members BS5268 ref	CALCULATIONS	OUTPUT
	Ex. 5. Stud Walling Strength Class SC3	
	15 kN/m length (long term) loading	

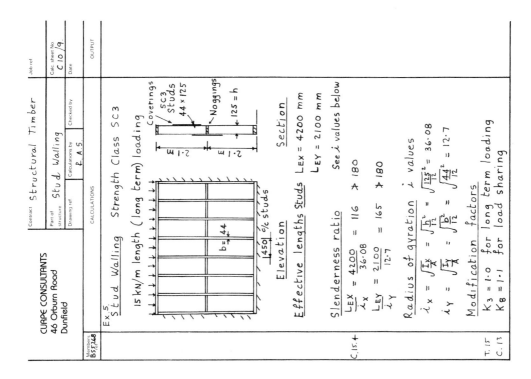

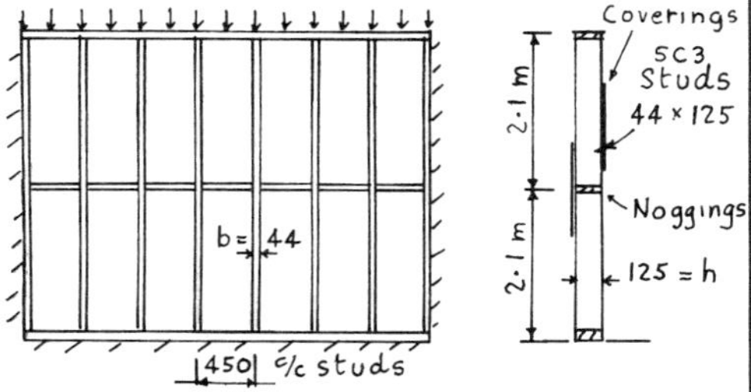

Elevation   Section

Members BS5268 ref	CALCULATIONS	OUTPUT
	Effective lengths Studs $L_{EX} = 4200$ mm	
	$L_{EY} = 2100$ mm	
	Slenderness ratio — See i values below	
C.15.4	$\frac{L_{EX}}{i_X} = \frac{4200}{36\cdot08} = 116 \not> 180$	
	$\frac{L_{EY}}{i_Y} = \frac{2100}{12\cdot7} = 165 \not> 180$	
	Radius of gyration i values	
	$i_X = \sqrt{\frac{I_X}{A}} = \sqrt{\frac{h^2}{12}} = \sqrt{\frac{125^2}{12}} = 36\cdot08$	
	$i_Y = \sqrt{\frac{I_Y}{A}} = \sqrt{\frac{b^2}{12}} = \sqrt{\frac{44^2}{12}} = 12\cdot7$	
	Modification factors	
T.15	$K_3 = 1\cdot0$ for long term loading	
C.13	$K_8 = 1\cdot1$ for load sharing	

CURPE CONSULTANTS
46 Orburn Road
Dunfield

Contract: Structural Timber | Job ref:
Part of structure: Stud Walling | Calc. sheet No: C10/10
Drawing ref: | Calculations by: R.A.S. | Checked by: | Date:

Members BS5268 ref	CALCULATIONS	OUTPUT
T.20	K_{12} from Table 20	
	$\frac{E_{min}}{\sigma_{c\parallel}} = \frac{5800}{6\cdot8 \times 1\cdot0} = 853 \quad \therefore K_{12} = 0\cdot1715$	
	Permissible compressive stress	
	= Grade stress comp. parallel to grain $\times K_3 \times K_8 \times K_{12}$	
	$= 6\cdot8 \times 1\cdot0 \times 1\cdot1 \times 0\cdot1715 = 1\cdot283$ N/mm²	
	Load to one stud $= 15 \times 0\cdot45 = 6\cdot75$ kN	
	Applied comp. stress $= \frac{load}{area} = \frac{6\cdot75 \times 10^3}{44 \times 125} = 1\cdot22$ N/mm²	
	Since $1\cdot22 < 1\cdot283$ Section adequate.	44 × 125 SC3 Adequate
	Noggings - Use same section as studs	
	Note - No allowance has been made for strength of wall coverings.	

CURPE CONSULTANTS 46 Orburn Road Dunfield	Contract: Structural Timber	Job ref.
	Part of structure: Timber support for shallow trench	Calc. sheet No. C10/11
	Drawing ref. / Calculations by R.A.S. / Checked by	Date

Members BS5268 ref	CALCULATIONS	OUTPUT

Ex.6. Timber support for Shallow Trench

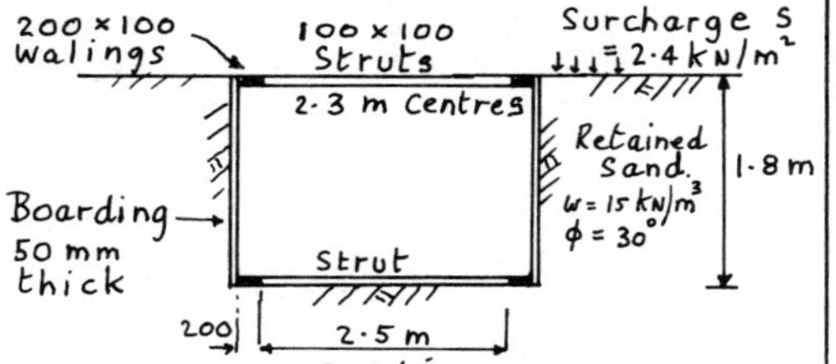

Section.

Horizontal forces per m length.

Sand:

$$P = \frac{wh^2}{2}\left(\frac{1-\sin\phi}{1+\sin\phi}\right) = \frac{15\times1.8^2}{2}\left(\frac{1-\sin 30}{1+\sin 30}\right)$$

$$= 24.3\left(\tfrac{1}{3}\right) = 8.1\ \text{kN}.$$

Surcharge:

$$P = sh\left(\frac{1-\sin\phi}{1+\sin\phi}\right) = 2.4\times1.8\times\tfrac{1}{3} = 1.44\ \text{kN}.$$

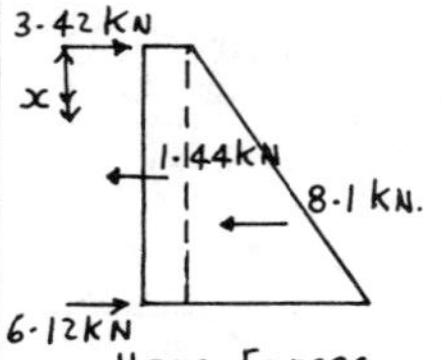

Hor$_z$ Forces

Reactions per m length

At top:

$$\frac{1.44}{2} + \frac{8.1}{3} = 3.42\ \text{kN}.$$

At bottom:

$$\frac{1.44}{2} + \left(\frac{2}{3}\times 8.1\right) = 6.12\ \text{kN}$$

Consider pressure due to sand i.e. (hor$_z$). At 1.8 m depth $= \frac{wh}{3}$

$$= 15\times1.8\times\tfrac{1}{3} = 9\ \text{kN/m}^2$$

CURPE CONSULTANTS 46 Orburn Road Dunfield	Contract: Structural Timber	Job ref
	Part of structure: Timber support for shallow trench	Calc. sheet No. C10/12
	Drawing ref. / Calculations by R.A.S. / Checked by	Date

Members BS5268 ref	CALCULATIONS	OUTPUT

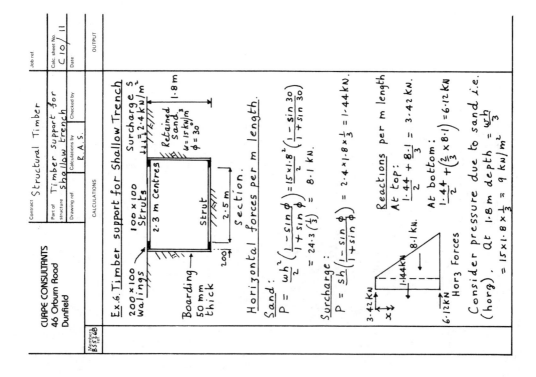

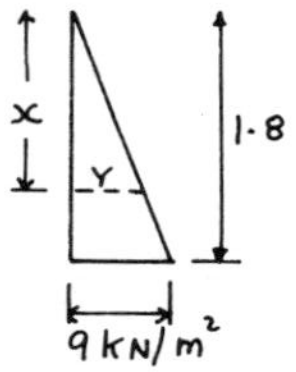

$9 : 1.8$

$Y : x$

$$\therefore Y = \frac{9x}{1.8} = 5x$$

Determine position of zero shear i.e. position of maximum bending moment for total loading applied to boarding.

At distance x from top

$$3.42 = \frac{1.44x}{1.8} + 5x.\frac{x}{2}$$

$$3.42 = 0.8x + 2.5x^2$$

$$2.5x^2 + 0.8x - 3.42 = 0$$

$$x^2 + 0.32x - 1.368 = 0 \quad \text{Quadratic eq.}$$

From which $x = 1.02$ m.

Then maximum bending moment

$$M = (3.42\times1.02) - (0.81\times0.51) - \left(2.60\times\frac{1.02}{3}\right)$$

$$M = 2.19\ \text{kNm}$$ — OUTPUT: M = 2.19 kNm

Strength Class SC3 timber

For wet exposure condition, grade stresses (dry) should be converted to wet stresses by using modification

T2. T14. factors K_1 and K_2

Design of boarding

$M = 2.19$ kNm for 1 m width.

CURPE CONSULTANTS 46 Orburn Road Dunfield	Project: Structural Timber	Job ref
	Part of structure: Timber support for shallow trench	Calc sheet no: C10/13 rev: 1
	Drawing ref: — Calc by: R.A.S. Date: —	Check by: — Date: —

Ref	Calculations	Output
	Wet exposure conditions apply	
	__Bending__	
	Permissible bending stress par. to grain	
	$= \text{grade stress} \times K_1 \times K_2 \times K_3 \times K_7$	
	$= 5.3 \times 1.06 \times 0.8 \times 1.25 \times 1.17 = 6.57\ \text{N/mm}^2$	
	Bending stress $= \dfrac{M}{Z} = \dfrac{2.19 \times 10^6 \times 6}{10^3 \times 50^2} = 5.25\ \text{N/mm}^2$	
	Since $5.25 < 6.57$ Satisfactory	
	__Shear__	
	Max. shear force $= 6.12$ kN/m width	
	Max. shear stress $= \dfrac{1.5 \times 6.12 \times 10^3}{10^3 \times 50} = 0.183\ \text{N/mm}^2$	
	Perm. shear stress parallel to grain	
	$= \text{grade stress} \times K_1 \times K_2 \times K_3$	
	$= 0.67 \times 1.04 \times 0.9 \times 1.25 = 0.78\ \text{N/mm}^2$	
	Since $0.183 < 0.78$ sat.	Use boarding 50mm thick
	__Design of bottom waling__ span 2·3m	
	Applied U.D.L per m length $= 6.12$ kN	
	Assume simply supported $M = \dfrac{wL^2}{8}$	
	$M = \dfrac{6.12 \times 2.3^2}{8} = 4.046$ kN m	
	Bending stress $= \dfrac{M}{Z} = \dfrac{4.046 \times 10^6 \times 6}{100 \times 200^2} = 6.07\ \text{N/mm}^2$	
	Perm. bending stress ∥ to grain	
	$= \text{grade stress} \times K_1 \times K_2 \times K_3 \times K_7$	$K_7 = 1.046$
	$= 5.3 \times 1.06 \times 0.8 \times 1.25 \times 1.046 = 5.87\ \text{N/mm}^2$	
	Since $6.07 > 5.87$ slightly overstressed	
	Reduce centres of struts to 2·2 m	
	giving bending stress $= 5.55 < 5.87$ sat.	Reduce centres of struts to 2·2 m
	__Shear__ Max. shear $= \dfrac{6.12 \times 2.2}{2} = 6.73$ kN	

CURPE CONSULTANTS 46 Orburn Road Dunfield	Project: Structural timber	Job ref
	Part of structure: Timber support for shallow trench	Calc sheet no: C10/14 rev: 1
	Drawing ref: — Calc by: RAS Date: —	Check by: — Date: —

Ref	Calculations	Output
	Max. shear stress $= 1.5 \times$ average	
	$= \dfrac{1.5 \times 6.73 \times 10^3}{200 \times 100} = 0.504\ \text{N/mm}^2$	Use walings 200×100 2·2 m span.
	Since $0.504 < 0.78$, shear sat.	
	__Design of bottom strut__ 2·2 m centres	
	Effective length $= 2.5$ m	
	Comp. force $= 6.12 \times 2.2 = 13.46$ kN	
	Radius of gyration $i = \sqrt{\dfrac{b^2}{12}} = \sqrt{\dfrac{100^2}{12}} = 28.86$	
	Slenderness ratio $= \dfrac{L_E}{i} = \dfrac{2.5 \times 10^3}{28.86} = 86 \not> 180$	
	Perm. comp. stress	
	$= \text{grade stress comp.} \parallel \text{to grain} \times K_1 \times K_2 \times K_3 \times K_{12}$	
	$= 6.8 \times 1.04 \times 0.6 \times 1.25 \times K_{12}$	
Table 20	$\dfrac{E\ \text{min}}{\sigma_{c.\parallel}} = \dfrac{5800}{6.8 \times 1.25} = 682 \quad \therefore K_{12} = 0.4293$	
	Perm. comp stress	
	$= 6.8 \times 1.04 \times 0.6 \times 1.25 \times 0.4293$	
	$= 2.27\ \text{N/mm}^2$	
	Applied comp. stress $= \dfrac{13.46 \times 10^3}{100 \times 100} = 1.34\ \text{N/mm}^2$	
	Since $1.34 < 2.27$ sat. as strut	Use 100×100 2·2 m centres
	__Check bearing__ for waling member	
	Bearing stress $= 1.34\ \text{N/mm}^2$	
	With no wane:	
	Grade stress comp ⊥ to grain $= 2.2\ \text{N/mm}^2$	
	Perm. stress $= 2.2 \times K_1 \times K_2 \times K_3 \times K_4$	

CURPE CONSULTANTS 46 Orburn Road Dunfield	Project: Structural Timber	Job ref
	Part of structure: Timber support for shallow trench	Calc sheet no: C10/15 rev: 1
	Drawing ref: — Calc by: RAS Date: —	Check by: — Date: —

Ref	Calculations	Output
	$= 2\cdot2 \times 1\cdot04 \times 0\cdot6 \times 1\cdot25 \times 1\cdot1 = 1\cdot88\,N/mm^2$	
	Since $1\cdot34 < 1\cdot88$ sat.	Bearing sat no wane
	With wane permitted:	
	Grade stress comp. ⊥ to grain $= 1\cdot7\,N/mm^2$	
	Perm. stress $= 1\cdot7 \times K_1 \times K_2 \times K_3 \times K_4$	
	$= 1\cdot7 \times 1\cdot04 \times 0\cdot6 \times 1\cdot25 \times 1\cdot1 = 1\cdot458\,N/mm^2$	
	Since $1\cdot34 < 1\cdot458$ sat.	Bearing sat wane permitted.
	Note. Strut centres altered from 2·3 m to 2·2 m to meet bending requirements for bottom waling. Span of bottom waling altered accordingly to 2·2 m.	

CURPE CONSULTANTS 46 Orburn Road Dunfield	Contract: Structural Timber	Job ref
	Part of structure: Forwork for conc. floor	Calc sheet No: C10/16
	Drawing ref: — Calculations by: R.A.S. Checked by: —	Date: —

Members ref BS5268	CALCULATIONS	OUTPUT

Ex.7. Formwork for insitu concrete floor

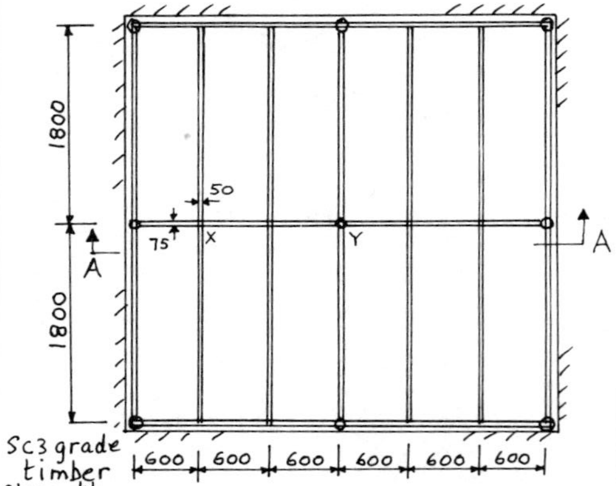

Plan

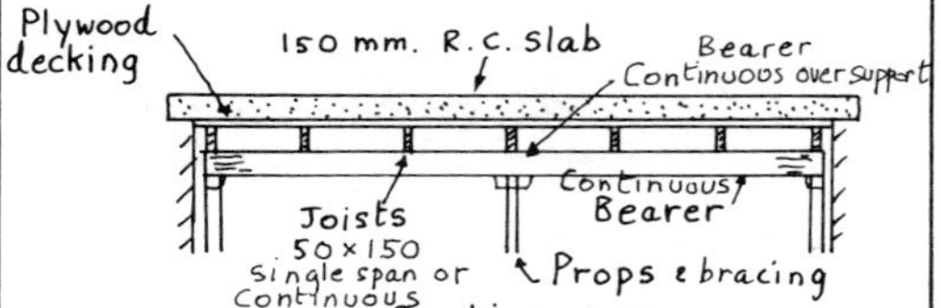

Section A-A

Design data

Loading.

	Imposed	1·50
	Plywood	0·14
	Slab	3·60
	Bearers & Joists	0·11
	Σ	= 5·35 kN/m^2

CURPE CONSULTANTS
46 Orburn Road
Dunfield

Project	Structural Timber	Job ref
Part of structure	Formwork for conc. floor	Calc sheet no C10/ 17/ rev
Drawing ref	Calc by R.A.S. Date	Check by Date

Ref	Calculations	Output

Three cases are considered for joists.

Case 1. Single span - simply supported

One span loaded

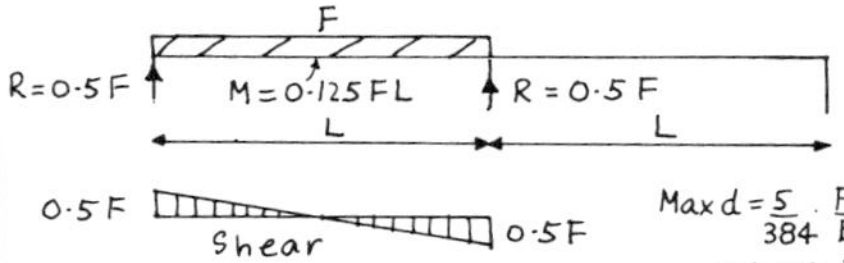

Max $d = \frac{5}{384} \cdot \frac{FL^3}{EI}$

$= 0{\cdot}013 \frac{FL^3}{EI}$

Case 2 Joists continuous over two spans

Both spans loaded

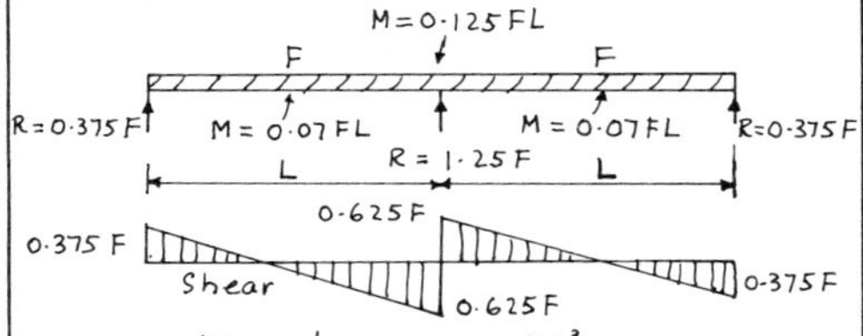

max. $d = 0{\cdot}00541 \frac{FL^3}{EI}$

Case 3. Joists continuous over two spans

One span loaded

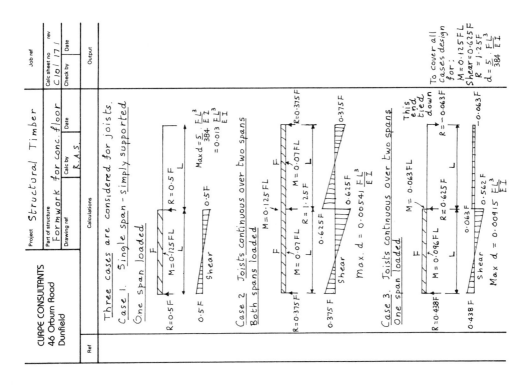

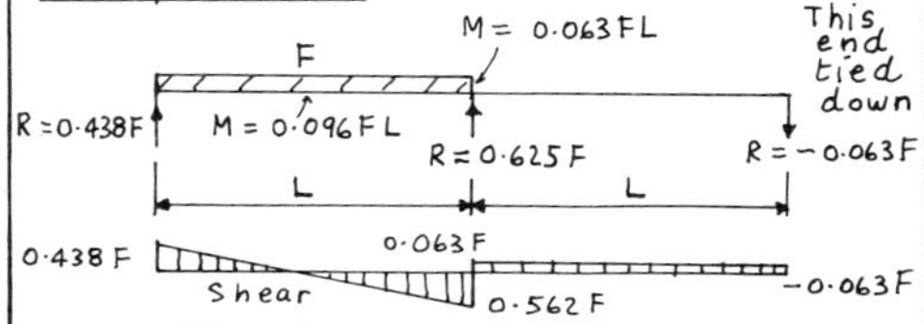

Max $d = 0{\cdot}00915 \frac{FL^3}{EI}$

Output: To cover all cases design for:
$M = 0{\cdot}125FL$
Shear $= 0{\cdot}625F$
$R = 1{\cdot}25F$
$d = \frac{5}{384} \cdot \frac{FL^3}{EI}$

CURPE CONSULTANTS
46 Orburn Road
Dunfield

Contract	Structural Timber	Job ref
Part of structure	Formwork for conc. floor	Calc sheet No C10/ 18
Drawing ref	Calculations by R.A.S. Checked by	Date

Members BS5268 ref	CALCULATIONS	OUTPUT

For wet exposure condition

T.2. T.14. T.15.

Bending stress = grade stress $\times K_1 \times K_2 \times K_3$

$= 5{\cdot}3 \times 1{\cdot}06 \times 0{\cdot}8 \times 1{\cdot}4$

$= 6{\cdot}29\ N/mm^2$

Output: $K_3 = 1{\cdot}4$ for formwork

Shear stress = grade stress $\times K_1 \times K_2 \times K_3$

$= 0{\cdot}67 \times 1{\cdot}04 \times 0{\cdot}9 \times 1{\cdot}4$

$= 0{\cdot}87\ N/mm^2$

Mean E value = mean E $\times K_1 \times K_2$

$= 8800 \times 1{\cdot}08 \times 0{\cdot}8$

$= 7603\ N/mm^2$

Minimum E value = min. E $\times K_1 \times K_2$

$= 5800 \times 1{\cdot}08 \times 0{\cdot}8$

$= 5011\ N/mm^2$

Design of joists

Load F to one joist $= 5{\cdot}35 \times 0{\cdot}6 \times 1{\cdot}8 = 5{\cdot}778$ kN

$M = \frac{FL}{8} = \frac{5{\cdot}778 \times 1{\cdot}8}{8} = 1{\cdot}30$ kNm.

Z reqd $= \frac{M}{\text{Permissible stress}}$

C.13. C.14.6.

Permissible stress = grade stress $\times K_8 \times K_7$

$= 6{\cdot}29 \times 1{\cdot}1 \times 1{\cdot}079$

$= 7{\cdot}465\ N/mm^2$

Output: $K_7 = 1{\cdot}079$ for 150 mm depth.

Z reqd $= \frac{1{\cdot}30 \times 10^6}{7{\cdot}465} = 174146\ mm^3$

If $b = 50$ mm then $\frac{50h^2}{6} = 174146$

CURPE CONSULTANTS
46 Orburn Road
Dunfield

Project: Structural Timber			Job ref	
Part of structure: Formwork for conc. floor			Calc sheet no: C10/ 19	rev: 1
Drawing ref	Calc by: R.A.S.	Date	Check by	Date

Ref	Calculations	Output
	Minimum $h = \sqrt{\frac{174146 \times 6}{50}} = 144.56$ mm	50×150 section bending sat.
a.14.7.	Deflection Allowable $= \frac{L}{270}$	
	Actual $= \frac{5}{384}\frac{FL^3}{EI}$	
	Equating $\frac{L}{270} = \frac{5FL^3}{384EI}$	
	I reqd $= \frac{5FL^2 \times 270}{384E}$	
	$= \frac{5 \times 5.778 \times 10^3 \times 1.8^2 \times 10^6 \times 270}{384 \times 7603}$	I reqd = 8656455 mm^4
	If $b = 50$ then $\frac{bh^3}{12} = 8656455$	
	$h^3 = \frac{8656455 \times 12}{50} = 2077549$	
	Minimum $h = 127.6$ mm	
	50 × 150 section sat	50×150 section deflection sat.
	Check for shear	
	Max shear force $= 0.625F$	
	$= 0.625 \times 5.778$	
	$= 3.61$ kN.	
	Max shear stress $= 1.5 \times$ average	
	$= \frac{1.5 \times 3.61 \times 10^3}{50 \times 150} = 0.722\ N/mm^2$	
	Permissible stress = grade stress × K_8	
	$= 0.87 \times 1.1 = 0.957\ N/mm^2$	
	Since $0.722 < 0.957$ sat.	
	Joists sat. for bending, shear and deflection.	50×150 section shear sat. Use 50×150 strength class SC3.

CURPE CONSULTANTS
46 Orburn Road
Dunfield

Project: Structural Timber			Job ref	
Part of structure: Formwork for conc. floor.			Calc sheet no: C10/ 20	rev: 1
Drawing ref	Calc by: R.A.S.	Date	Check by	Date

Ref	Calculations	Output
	Design of Bearers	
	Bearers are designed as continuous over two spans. Two cases are considered.	
	Case I Both spans loaded	

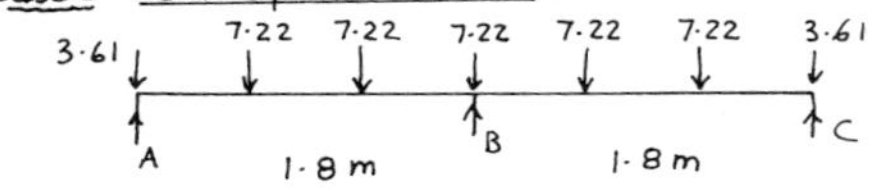

Ref	Calculations	Output
	Point load $= 1.25F = 1.25 \times 5.778 = 7.22$ kN	
	Max M occurs at B $= -0.334PL$	
	$= -0.334 \times 7.22 \times 1.8 = 4.34$ kN m	$M = 4.34$ kN m
	$K_7 = 1.0$ assumed only	
	Z reqd $= \frac{4.34 \times 10^6}{6.29 \times 1.0} = 689984\ mm^3$	
	If $b = 75$ mm then $\frac{75h^2}{6} = 689984$	b assumed as 75 mm.
	$h^2 = \frac{689984 \times 6}{75} = 55198$	
	$\therefore h = 235$ mm	min. $h = 235$ mm. for bending
	Reactions:	
	$R_A = (0.33 \times 7.22 \times 2) + 3.61 = 8.38$ kN	
	$R_B = (1.33 \times 7.22 \times 2) + 7.22 = 26.42$ kN	
	$R_C = 8.38$ kN	
	Max shear $= 9.62$ kN at B.	
	Deflection $= 0.0152\frac{PL^3}{EI}$	
	$= \frac{0.0152 \times 7.22 \times 10^3 \times 1.8^3 \times 10^9}{5011 \times I}$	
	This is not critical case	

CURPE CONSULTANTS
46 Orburn Road
Dunfield

Project: Structural Timber | Job ref
Part of structure: Formwork for conc. floor | Calc sheet no: C10/21/1 | rev
Drawing ref | Calc by: R.A.S. | Date | Check by | Date

Ref	Calculations	Output
	Case 2. One span loaded.	
	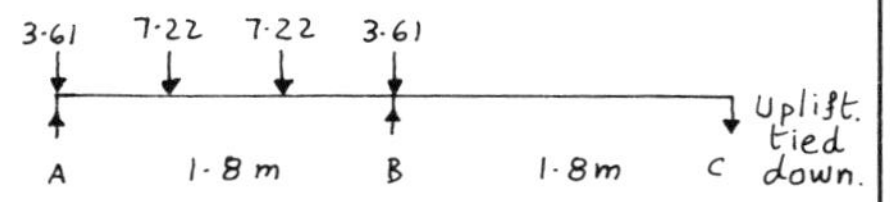	
	$M_B = -0.167\,PL$	
	$= -0.167 \times 7.22 \times 1.8 = 2.17$ kNm	
	Since $2.17 < 4.34$ not critical.	Max. design M = 4.34 kNm
	Reactions:	
	$R_A = (0.417 \times 7.22 \times 2) + 3.61 = \mathbf{9.62}$ kN.	
	$R_B = (0.67 \times 7.22 \times 2) + 3.61 = 13.28$ kN	
	$R_C = (0.083 \times 7.22 \times 2) = 1.19$ kN uplift.	
	Max. shear $= 8.42$ kN.	
	Since $8.42 < 9.62$ not critical.	Max. design shear force = 9.62 kN.
	Deflection $= 0.025 \frac{PL^3}{EI}$	
	Allowable $\frac{L}{270}$	
	Equating I reqd $= \frac{0.025 PL^2 \times 270}{E}$	
	$= \frac{0.025 \times 7.22 \times 10^3 \times 1.8^2 \times 10^6 \times 270}{5011}$	
	$= 31510955 \text{ mm}^4$	
	If $b = 75$ mm then $\frac{75h^3}{12} = 31510955$	
	Minimum $h = 171.47$ mm	min. h = 172 for deflection
	Try 75 × 250 section for bending.	
	$Z = \frac{bh^2}{6} = \frac{75 \times 250^2}{6} = 781250 \text{ mm}^3$	
	Since $781250 > 689984$ bending sat.	Bending sat.

CURPE CONSULTANTS
46 Orburn Road
Dunfield

Project: Structural Timber | Job ref
Part of structure: Formwork for conc. floor | Calc sheet no: C10/22/1 | rev
Drawing ref | Calc by: R.A.S. | Date | Check by | Date

Ref	Calculations	Output
	Check section for shear	
	Max. shear force $= 9.62$ kN	
	Max. shear stress $= \frac{1.5 \times 9.62 \times 10^3}{75 \times 250} = 0.77 \text{ N/mm}^2$	
	Since $0.77 < 0.87$ shear sat.	shear sat.
	Deflection	
	$I = \frac{bh^3}{12} = \frac{75 \times 250^3}{12} = 97656249 \text{ mm}^4$	
	Since $31510955 < 97656249$	75 × 250 Deflection sat.
	Check bearing stress at point X on plan	
	Bearing stress $= \frac{7.22 \times 10^3}{50 \times 75} = 1.92 \text{ N/mm}^2$	
	Grade stress perp. to grain (no wane) $= 2.2 \text{ N/mm}^2$	
	Permissible stress for wet exposure condition = grade stress $\times K_1 \times K_2 \times K_3 \times K_4 \times 1.2$	1.2 for formwk
	$= 2.2 \times 1.04 \times 0.6 \times 1.4 \times 1.14 \times 1.2 = 2.63 \text{ N/mm}^2$	
	Since $1.92 < 2.63$ bearing sat.	Bearing sat at X.
	Check bearing stress at point Y on plan	
	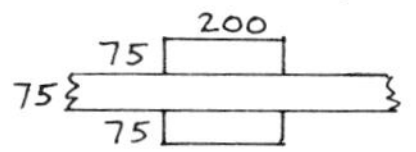 Timber pieces adequately connected to bearer.	
	Area provided $= 200 \times 225 = 45000 \text{ mm}^2$	
	Permissible stress $= 2.2 \times 1.04 \times 0.6 \times 1.4 \times 1.0 \times 1.2$ $= 2.32 \text{ N/mm}^2$	
	Bearing stress $= \frac{26.42 \times 10^3}{45000} = 0.59 \text{ N/mm}^2$	
	Minimum length reqd if no timber pieces are provided $= \frac{26.42 \times 10^3}{2.32 \times 75}$	
	$= 152$ mm.	Provide min bearing length 152 mm

11

Introduction to reinforced concrete design

British Standards

It is of considerable benefit for engineers to have a guide to good design practice, which is the issue of experience gained from many years of practice combined with the results of research. When all engineers use the same guide then each can follow, understand and check the others' calculations. This guide to good design practice is called a Code of Practice or a British Standard.

The first attempt to produce such a document for reinforced concrete was in 1934 and through a process of development, refinement and evolution, CP110 *The Structural Use of Concrete* was published in 1972. This was revised and published as a British Standard, BS 8110: 1985. As was its predecessor, this publication is based on the limit state philosophy of design and in fact, referring to CP110, BS 8110 states 'there are no major changes in principle' but 'the text has largely been rewritten'. Some of the background is given in the foreword of BS 8110 and is recommended reading for all students of structural design.

Clause 2.1.1 of BS 8110 states 'the aim of design is the achievement of an acceptable probability that structures being designed will perform satisfactorily during their intended life', i.e. they will not reach a limit state, or become unfit for use. This can happen, as indicated in 2.2.1, 'by collapse, overturning, buckling (ultimate limit states), deformation, cracking, vibration, etc. (serviceability limit states) and that the structure will . . . be durable'. In order to ensure safety against reaching a limit state it is important that factors of safety be employed.

Factors of safety

Due to the fact that many accidental causes of failure can occur in a structure, it is necessary to introduce factors of safety. In the elastic method of design these were all on the material stresses, while in the load factor method the factors of safety were on the loads. In limit state design however, there are partial factors of safety on both the materials and the loads. These combine in a design to ensure a global factor of safety on the structure or element.

Partial factors of safety for materials

Obviously, if a large number of tests were done on either the concrete or the reinforcement the results obtained would follow the normal frequency distribution curve shown in Fig. 11.1.

BS 8110 refers not to the mean strength f_m, but to the characteristic strength which it defines in 2.4.2.1 as 'the value . . . below which 5% of all possible test results would be expected to fall'. This characteristic strength is based on samples which are produced under ideal conditions, and there is a difference between these and those actually experienced on site. In the case of concrete it may be segregated, not compacted,

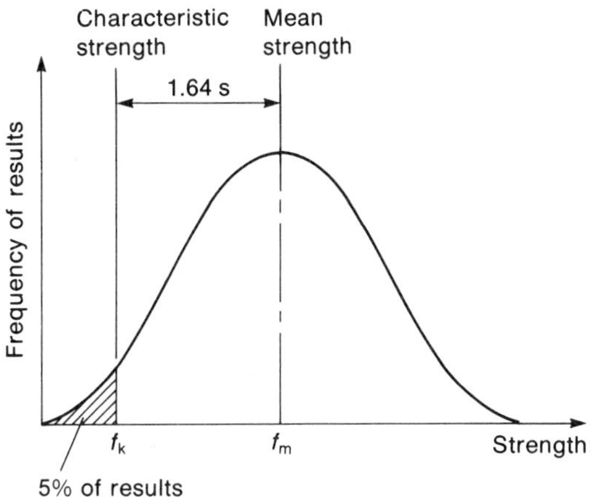

Fig. 11.1 *Normal frequency distribution curve*

badly cured, etc., while reinforcement can be misshapen or corroded.

To account for these effects, partial factors of safety, γ_m, are introduced and the design strengths are obtained by dividing the characteristic strengths by the appropriate γ_m, that is:

$$\text{Design strength} = \frac{f_k}{\gamma_m}$$

While more detailed information can be obtained in BS 8110: Part 2: 1985, Table 11.1 gives the γ_m values for the ultimate limit state.

Table 11.1.

Values of γ_m for the ultimate limit state	
Reinforcement	1.15
Concrete in flexure or axial load	1.50
Shear strength without shear reinforcement	1.25
Bond strength	1.4
Others (e.g. bearing stress)	$\geqslant$ 1.5

Partial factors of safety for loads

It is not possible to have a statistical analysis of loads and their loading patterns as was done for the materials. However guidance for the assessment of the characteristic loads is obtained from BS 648: 1964: *Schedule of Weights of Building Materials,* BS 6399: Part 1: 1984 *Code of Practice for Dead and Imposed Loads,* CP3: Chapter V: Part 2: 1972: *Wind Loads,* and CP 2004 for nominal earth loads.

The partial factors of safety for loads, γ_f, are necessary for a number of reasons:

1) errors in calculation,
2) inaccurate construction,
3) unanticipated increases in load,
4) unforeseen stress distribution,
5) the importance of the limit state being considered.

The design load is obtained by multiplying the characteristic load by the partial factor of safety, that is:

$$\text{Design load} = F_k \gamma_f$$

The values of γ_f depend on the load combination and limit state being considered and the design loads should be arranged so as to produce the most severe stresses. This is illustrated in relation to cantilevered elements in Chapter 13.

The design process

Clause 2.2.1 of BS 8110 states: 'The usual approach is to design on the most critical limit state and then to check that the remaining limit states will not be reached.' This is usually done by designing for the ultimate limit state (bending and shear) and subsequently checking the serviceability limit states, e.g. deflection and cracking. Others may be checked as deemed necessary by the engineer.

There are two ways in BS 8110 in which sections can be designed, viz. by design charts and by design formulae. In the examples of reinforced concrete design illustrated in this book, the designs have been done using the design charts contained in Part 3 of BS 8110. However, each student should have a grasp of the design formulae. As has been stated in the Preface, it is not the intention to produce a theoretical book, thus there is not a great emphasis placed on the theory and the formulae, but the following is outlined in the hope that it may form a bridge between this book and others of a more theoretical nature.

Table 11.2. Partial factor of safety at various load combinations

Load combination	*Load type*					
	Dead		*Imposed*		*Earth and water pressure*	*Wind*
	Adverse	*Beneficial*	*Adverse*	*Beneficial*		
1) Dead and Imposed (and earth and pressure)	1.4	1.0	1.6	0	1.4	
2) Dead and Wind (and earth and water pressure)	1.4	1.0	—	—	1.4	1.4
3) Dead and Wind and Imposed (and earth and water pressure)	1.2	1.2	1.2	1.2	1.2	1.2

Design formulae

Reinforced concrete design and theory is based on the following assumptions:

1) Plane sections before bending remain plane after bending,
2) the concrete in tension has cracked and all the tensile stresses are taken by the reinforcement,
3) there is good bond between the reinforcement and the concrete; therefore the strain in the reinforcement is given by the theoretical strain in the adjacent concrete,
4) the stresses in all bars are equal; therefore the resultant tensile force acts at the centroid of the reinforcement.

BS 8110 gives the stress–strain curves to be used in design (see Clause 3.4.4.1). It is necessary to consider singly and doubly reinforced beams, both flanged and rectangular.

Singly reinforced beams

Consider the rectangular beam shown in Fig. 11.2 together with its associated stress distributions. When the beam in Fig. 11.2(a) is subjected to pure bending, it has been determined by experiment that the stress distribution at failure is as shown in Fig. 11.2(b). When the partial safety factors are introduced the stress distribution equates to Fig. 11.2(c). The simplified stress distribution adopted by BS 8110 is shown in Fig. 11.2(d), and this is used in the derivation of the design formula.

The ultimate moment of resistance M_u is the result of the action of the couple formed by the tensile and compressive forces i.e.

$$M_u = \text{tensile force} \times \text{lever arm } (Z)$$
$$= \text{compressive force} \times \text{lever arm } (Z)$$

where $Z = (d - 0.45x)$, i.e. $x = \dfrac{(d - Z)}{0.45}$

From the compressive force

$$M_u = 0.45 f_{cu} b 0.9x(d - 0.45x) \qquad [1]$$

To ensure that an over-reinforced beam is not designed (since it can fail explosively without warning) and that an under-reinforced beam is designed (which gives visible warning of impending failure as the reinforcement yields), BS 8110 limits the value of x to $\dfrac{d}{2}$.

Substituting this value in eqn. [1] above,

$$M_u = 0.45 f_{cu} b 0.9 \frac{d}{2}\left(d - 0.45\frac{d}{2}\right)$$

i.e. $$M_u = 0.156 f_{cu} b d^2 \qquad [2]$$

$$\therefore \quad \frac{M_u}{f_{cu} b d^2} = 0.156 \qquad [3]$$

This value of 0.156 is called, in BS 8110, K' and obviously if K is calculated on the basis of:

$$K = \frac{M}{f_{cu} b d^2} \qquad [4]$$

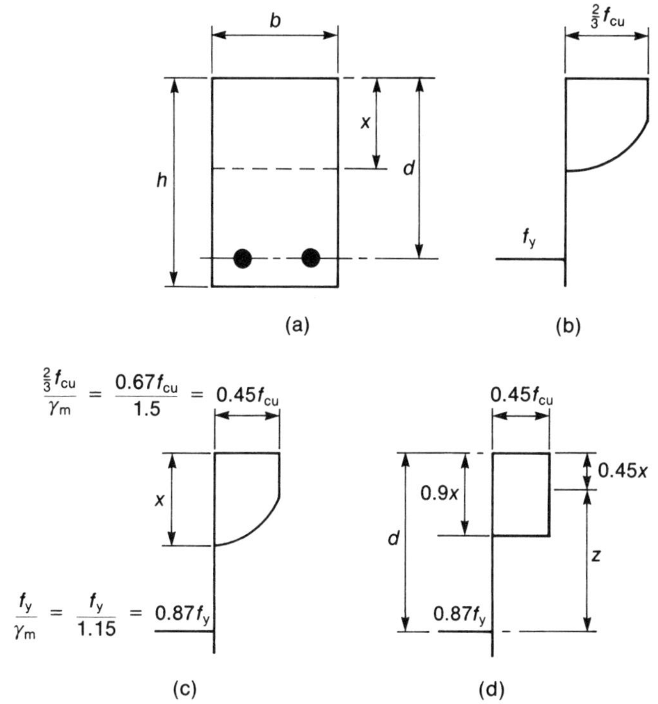

Fig. 11.2 *Singly reinforced beam and associated stress distributions*

and is found to be less than 0.156, then M_u has not been reached. Hence $0.45f_{cu}$ has not been exceeded and thus the beam does not require compressive reinforcement.

The area of reinforcement required is found from a consideration of the tensile force:

$$M = 0.87 f_y A_s Z$$

$$\therefore \quad A_s = \frac{M}{0.87 f_y Z} \qquad [5]$$

Doubly reinforced beams

The moment of resistance of a doubly reinforced beam can be considered as consisting of two component parts. As the load is applied the section acts as a singly reinforced beam until its ultimate moment of resistance is reached. This means that the concrete is overstressed and must be reinforced. To keep horizontal forces equal, tensile reinforcement must be added until the compressive and tensile forces balance. This is illustrated in Fig. 11.3; in Fig. 11.3(a) the singly rein-

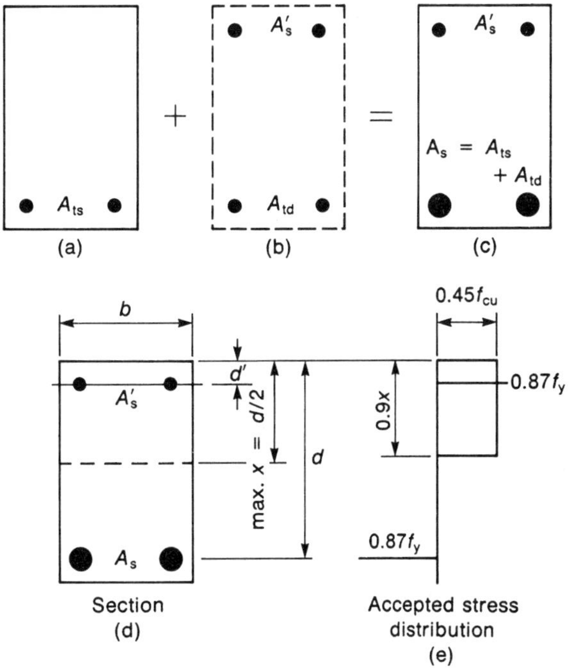

Fig. 11.3 *Doubly reinforced beam and associated stress distribution*

forced beam is shown with A_{ts} denoting the area of tension reinforcement in the singly reinforced beam. The ultimate moment of resistance of such a beam has been shown to be

$$M_{us} = K'f_{cu}bd^2 \quad \text{[6], as [2] above.}$$

In Fig. 11.3(b) the compression reinforcement is added A'_s and the additional tensile reinforcement required to balance it $= A_{td}$.

Fig. 11.3(c) shows (a) and (b) combined to give the doubly reinforced section. Figs. 11.3(d) and 11.3(e) show a typical section and the accepted stress distribution. The limiting compressive stress in the reinforcement is given in BS 8110, Fig. 2.2, as $\frac{f_y}{\gamma_m}$, i.e. $0.87f_y$ as it is in tension.

The moment of resistance of the section in Fig. 11.3(b) can be found in two ways:

1) take moments about the compressive reinforcement:

$$M_{ud} = A_{td}\,0.87f_y(d - d') \quad [7]$$

2) take moments about the tensile reinforcement:

$$M_{ud} = A'_s\,0.87f_y(d - d') \quad [8]$$

Combining eqns. [6] and [8],

$$M_u = A'_s\,0.87f_y(d - d') + K'f_{cu}bd^2$$

but $M_u = Kf_{cu}bd^2$

$$\therefore \quad A'_s = \frac{Kf_{cu}bd^2 - K'f_{cu}bd^2}{0.87f_y(d - d')}$$

$$\text{i.e.} \quad A'_s = \frac{(K - K')f_{cu}bd^2}{0.87f_y(d - d')} \quad [9]$$

It is clear from a comparison of equations [7] and [8] that

$$A_{td} = A'_s \quad [10]$$

Combining eqns. [4], [5] and [7],

$A_s = A_{ts}$ (eqn. [5]) $+ A_{td}$ (eqn. [7]) or A'_s (eqn. [10])

when the singly reinforced section is at its compressive stress limit $K = K'$

$$\therefore \quad A_s = \frac{K'f_{cu}bd^2}{0.87f_yZ} + A'_s \quad [11]$$

Thus equations [9] and [11] can be used to calculate the main reinforcement in a doubly reinforced beam by taking the following steps.

1) Calculate the ultimate design bending moment.
2) Calculate K (equation [4]).
3) Calculate A'_s (equation [9]).
4) Calculate A_s (equation [11]).

Flanged beams

It is common that a reinforced concrete floor is poured monolithically with the beam downstands forming a complete structural section as indicated in Fig. 11.4. The actual width to be considered acting in conjunction with the downstand is called the effective flange width and is determined in accordance with Clause 3.4.1.5 of BS 8110. This is to account for the difference between the theoretical and actual stress distributions as illustrated in Fig. 11.5. There are two cases which must be considered:

1) when the neutral axis lies within the flange,
2) when the neutral axis lies below the flange.

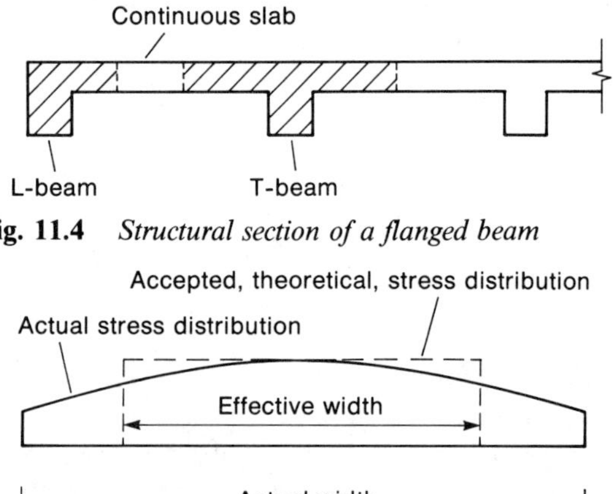

Fig. 11.4 *Structural section of a flanged beam*

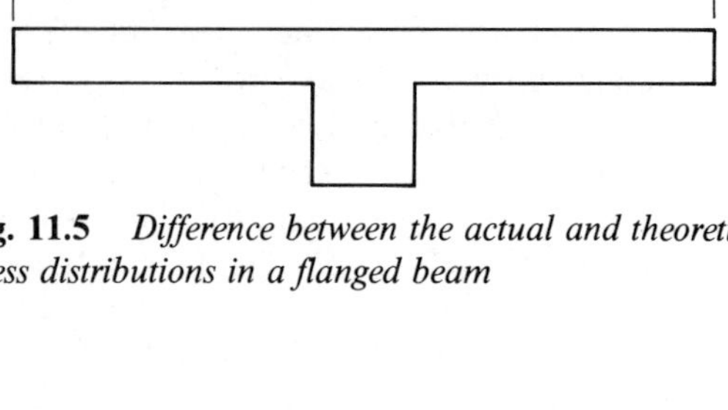

Fig. 11.5 *Difference between the actual and theoretical stress distributions in a flanged beam*

1) Neutral axis within the flange
In this case the stress distribution is exactly as shown in Fig. 11.2(d), since the concrete is assumed to have cracked in tension below the neutral axis. Thus the calculations are as for a singly reinforced beam, except that in the middle of a span the breadth of the section is the effective flange width, while at the support of a continuous member where the bending moment is reversed it is exactly the same as a rectangular beam, either singly or doubly reinforced as the moment to be resisted demands (see Fig. 11.6).

To verify the position of the neutral axis, the equation used previously still applies, i.e.

$$X = \frac{d - Z}{0.45}$$

2) Neutral axis below the flange
Where the neutral axis falls in the rib, the effective section includes a small rectangle of concrete shown shaded in Fig. 11.7. BS 8110 gives the

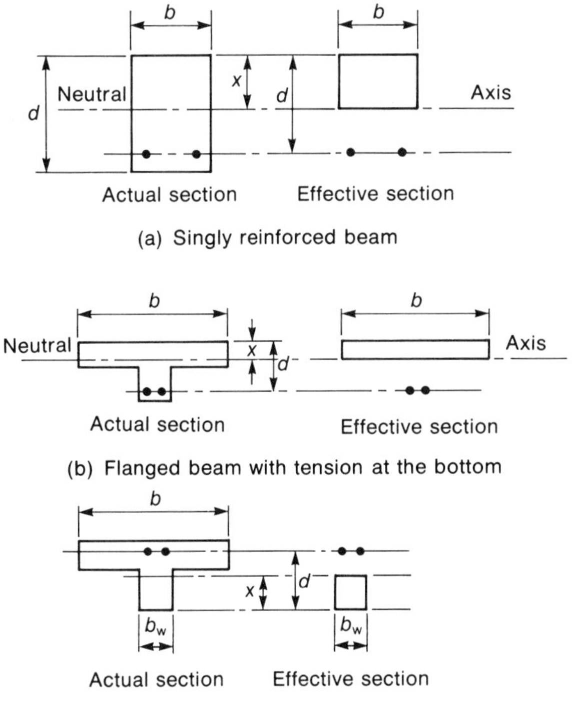

(a) Singly reinforced beam

(b) Flanged beam with tension at the bottom

(c) Flanged beam with tension at the top

Fig. 11.6 *T-beams which are equivalent to rectangular beams*

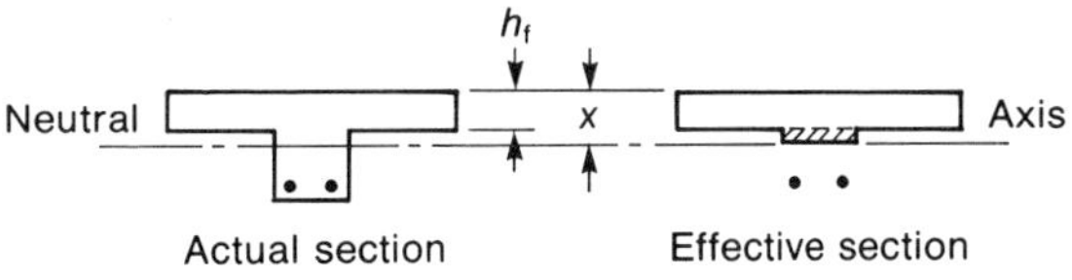

Fig. 11.7 *Neutral axis below the flange*

following formula for calculating the area of tensile reinforcement:

$$A_s = \frac{M + 0.1 f_{cu} b_w d(0.45d - h_f)}{0.87 f_y (d - 0.5h_f)}$$

This accounts for the shaded rectangle.

However, a simple method of design is achieved if two assumptions are made:

1) the shaded rectangle and its compressive force is neglected,
2) the flange thickness is less than $0.9x$, i.e. the depth of the rectangular stress block. If $h_f > 0.9x$ the design will be conservative.

Take moments about the tensile reinforcement:

$$M_u = 0.45 f_{cu} b h_f (d - 0.5h_f) \qquad [12]$$

This gives the ultimate moment capacity and if this value exceeds the applied bending moment then the area of reinforcement can be calculated as shown below using equation [13]. In the rare event of the applied bending moment exceeding the ultimate moment capacity the concrete section should be increased.

Take moments about the centre of the flange:

$$M = 0.87 f_y A_s (d - 0.5h_f)$$

$$\therefore \quad A_s = \frac{M}{0.87 f_y (d - 0.5h_f)} \qquad [13]$$

When the applied moment and the section sizes are known, the equation will yield the area of tension reinforcement.

Design charts for beams in bending
The above formulae have been derived using the simplified stress block shown in Fig. 11.2(d). However the design charts given in Part 3 of BS 8110 are based on the rectangular parabolic stress distribution shown in Fig. 11.2(c). This gives

slightly more economical results as well as being more accurate and these charts are utilized in the remainder of the reinforced concrete designs illustrated in this book.

Shear in beams

Another ultimate limit state is that of shear. The analysis of shear in reinforced concrete beams is complex and has been the subject of research for many years. However, it is agreed that in a beam without shear reinforcement, the shear resistance is composed of three major parts, namely:

1) the uncracked concrete in the compression zone (20–40%),
2) the interlock of the aggregate across the crack zone (35–50%),
3) the dowel action of the longitudinal reinforcement (15–25%).

The first and last of these can, to some extent, be quantified but obviously the second cannot. Thus Table 3.9 of BS 8110 which gives values of the permissible shear stress, v_c is based on the effective depth and the percentage of anchored, tensile reinforcement.

The design or actual shear stress is calculated from

$$v = \frac{V}{b_v d} \qquad [14]$$

where V = the shear force due to ultimate loads,
b_v = the breadth of the section which, for a flanged beam, should be taken as the rib width,
d = the effective depth.

Where v exceeds v_c, shear reinforcement must be provided to account for the difference $(v - v_c)$. However, v must never exceed the maximum of $0.8\sqrt{f_{cu}}$ or 5 N/mm^2, even with shear reinforcement provided.

The resistance to shear is normally provided in one of the following two ways:

1) Vertical shear reinforcement, i.e. links or binders,
2) inclined shear reinforcement, i.e. bent-up or inclined bars plus vertical shear reinforcement.

The total shear resistance is equal to the shear resistance of the plain concrete section and the shear resistance of the links plus the shear resistance of the inclined bars.

Shear resistance of links

When a reinforced concrete beam is broken in the laboratory under the action of shear it is difficult to determine precisely the angle of the shear crack. If it is assumed that the angle is 45° then the theoretical shear resistance calculated, as shown in Fig. 11.8, compares well with the experimental value. The shear resistance due to vertical reinforcement is provided by the number of bars crossing the potential crack, i.e. (d/s_v).

Now, resolving the forces vertically, the shear resistance provided by the vertical links:

$$= \text{Area} \times \text{stress}$$

$$= \frac{d}{s_v} A_{sv} \frac{f_{yv}}{\gamma_m}$$

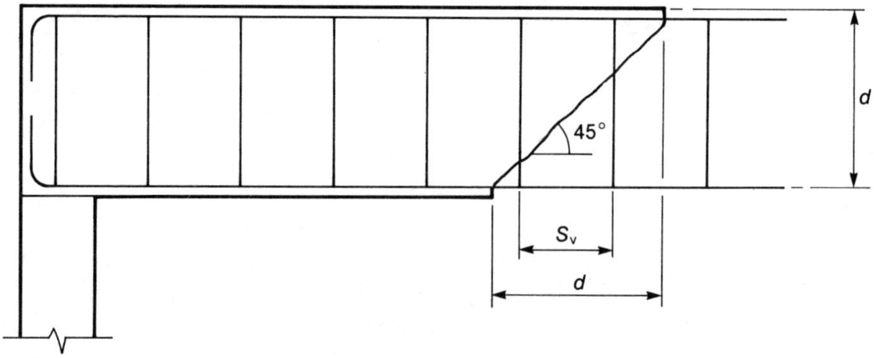

Fig. 11.8 *Theoretical shear resistance of links*

And with $\gamma_m = 1.15$ for reinforcement

$$= \frac{d}{s_v} A_{sv} 0.87 f_{yv}$$

For a beam, therefore, the total shear resistance will be provided by the addition of the shear resistance of the beam itself and that of the links, i.e.:

$$V = c \text{ for concrete} + V \text{ for links}$$

$$= v_c bd + \frac{d}{s_v} A_{sv} 0.87 f_{yv}$$

But

$$v = \frac{V}{bd} \qquad [15]$$

$$\therefore \quad \frac{V}{bd} = v = v_c + \frac{A_{sv}}{s_v b} 0.87 f_{yv}$$

$$\therefore \quad \frac{A_{sv}}{s_v} = \frac{(v - v_c)}{0.87 f_{yv}} \qquad [16]$$

Equation [16] corresponds to Table 3.8 of BS 8110.

Notwithstanding the links calculated from equation [16], it must also be established that in a doubly reinforced beam, sufficient links have been provided to stop the compression reinforcement buckling (see 3.12.7.1 of BS 8110 and Chapter 18). It should be noted that for the tension reinforcement to be considered in a shear calculation, it must extend beyond the section a distance at least equal to the effective depth, except at an end support where all the reinforcement can be used provided the code requirements with regard to the detailing are met.

Generally throughout this book, the links will be in mild steel having an $f_{yv} = 250\ \text{N/mm}^2$. Using mild steel ($f_y = 250\ \text{N/mm}^2$) allows the main reinforcement to be placed more accurately since the radius to which mild steel can be bent is two diameters as opposed to three diameters for high yield reinforcement.

Shear resistance of inclined or bent-up bars

It is not permissible, according to BS 8110, to have all the shear resistance provided by bent-up bars. Clause 3.4.5.6 states:

'The design shear resistance of a system of bent-up bars may be calculated by assuming that the bent-up bars form the tension members of one or more single systems of trusses in which the concrete forms the compression members.... At least 50% of the shear resistance provided by the steel should be in the form of links.'

This truss analogy is shown in Fig. 11.9.

The pitch $\quad s = 2(d - d') \cot \theta \qquad [17]$

Resolving vertically on the section X–X:

The shear resistance due to the inclined bars

$$= \frac{f_y}{\gamma_m} A \sin \theta$$

$$= 0.87 f_y A \sin \theta \qquad [18]$$

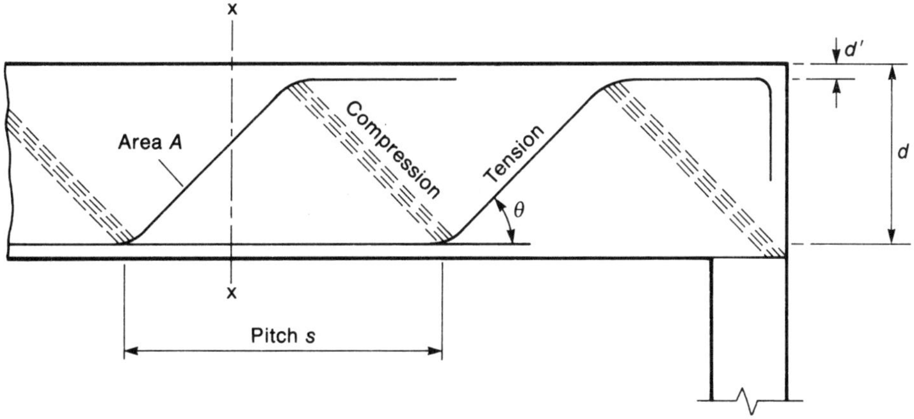

Fig. 11.9 *Truss analogy in inclined or bent-up bars*

When the pitch is less than that in equation [17], the shear resistance is increased in inverse proportion (i.e. if the pitch is halved the shear resistance is doubled). The arrangement shown in Fig. 11.9 is called a single system, but if the spacing is halved it is called a double system. In a double system:

$$\text{Shear resistance} = 2 \times 0.87 f_y A \sin\theta$$
$$= 1.74 f_y A \sin\theta \qquad [19]$$

A common choice for θ is 45° and f_y is equal to 460 N/mm^2, thus simplifying equations [18] and [19] to:

$$\text{Shear resistance for a single system} = 0.26A \text{ kN}$$

and

$$\text{Shear resistance for a single system} = 0.28A \text{ kN}$$

where A is the area of the reinforcement in mm^2.

The above formulae are for rectangular beams. However, the following points should be noted:

1) For flanged beams, only the rib is considered to resist the shear force.
2) In a doubly reinforced beam the anchored tension reinforcement only is considered when calculating $\frac{100A_s}{b_v d}$ for Table 3.9, BS 8110.

Having dealt with the requirements for the ultimate limit states of bending and shear it remains to outline the requirements for the serviceability limit states of deflection and cracking. It is not intended to deal with these in the same detail as bending and shear but simply to state the requirements of BS 8110 which will be illustrated in subsequent chapters.

Deflection

The method for checking deflection is as follows:

1) Choose the basic (span/effective depth) ratio from Table 3.10 if the span is less than 10 m, or refer to Clause 3.4.6.4 if the span is greater than 10 m and the engineer wishes to restrict the deflection to avoid damage to the finishes or partitions.
2) This value is modified by a factor from Table 3.11, depending on the amount of tension reinforcement, and a factor from Table 3.12 depending on the amount of compression reinforcement.
3) The modified ratio gives the permissible $1/d$ ratio; when this is compared with the actual $1/d$ ratio it can be seen whether or not deflection is satisfactory.

Cracking

The method of controlling flexural cracking in a reinforced concrete beam is by ensuring that the spacing of the reinforcement in tension does not exceed stated maximum values. These values are given in Clause 3.12.11.2 and Table 3.30 of BS 8110.

Occasionally there may be an advantage in calculating the actual crack width under service loads. A method for the calculation is given in section three of BS 8110: Part 2: 1985.

Solid slabs

The design of reinforced concrete solid slabs is basically the same as that outlined above for beams with respect to bending deflection and cracking. There are some differences in the rules for cracking but, as for beams, the requirements are based on bar spacing. However, one difference does occur when a slab is subjected to shear, from point loads. These produce punching shear which should be checked in accordance with Clause 3.7.7. This is illustrated in Chapter 17 when dealing with pad foundations.

12

Simply supported beams and slabs: worked examples

CURPE CONSULTANTS 46 Orburn Road Dunfield	Project	Job ref
		Sheet no 12/1
	Sub-section S. S. BEAM - TENSION	Made by B.C.
		Date
	Structural Summary Sheet	Checked by
		Date

A SIMPLY SUPPORTED RECTANGULAR REINFORCED CONCRETE BEAM, REINFORCED IN TENSION ONLY

A reinforced concrete beam is required to span 6.50 m between the centres of supporting brick piers which are 300 mm wide. The beam carries a 1.75 m height of brickwork which consists of 2No. 100 mm skins with a 50 mm cavity between them. The beam is to be designed for mild conditions of exposure and is to satisfy BS8110 with the following material strengths:

f_{cu} (the characteristic strength of the concrete) $= 40\ N/mm^2$

f_y (the characteristic strength of the reinforcement)

$= 460\ N/mm^2$ for the main reinforcement and

$= 250\ N/mm^2$ for the links

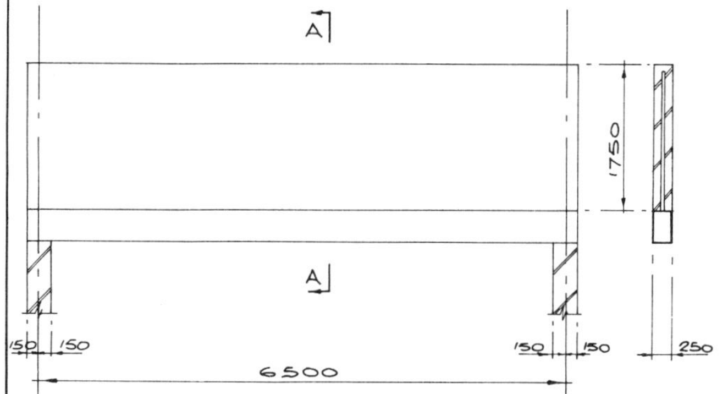

ELEVATION OF THE BEAM　　　A-A

CURPE CONSULTANTS 46 Orburn Road Dunfield	Project		Job ref
	Part of structure: S.S. BEAM - TENSION		Calc sheet no 12/2 rev /
	Drawing ref	Calc by B.C. Date	Check by Date

Ref	Calculations	Output
	<u>To DETERMINE THE BEAM DIMENSIONS</u>	
table 3.10	basic $\frac{\text{span}}{\text{eff. depth}} = 20$	
	$\therefore$ eff. depth $= \frac{6500}{20} = 325\text{mm}$	
table 3.4	cover, for mild conditions of exposure, $= 20\text{mm}$	
	Assuming 20mm ϕ main bars and 8mm links, the overall depth	
	'h' $= 325 + 20 + 8 + 20/2$	
	$= 363\text{mm}$	
	say 'h' = 400mm	h = 400mm
	$\therefore$ eff. depth 'd' $= 400 - (20 + 8 + 20/2)$	
	$= 362\text{mm}$	d = 362mm
	Let the breadth be determined by the brickwork dimension ie 250mm.	b = 250mm
3.4.1.6.	<u>CHECK THE SLENDERNESS LIMITS</u>	
	$60 b_c = 60 \times 250 \times 10^{-3} = 15\text{m}$	
	$\frac{250 b_c^2}{d} = \frac{250^3 \times 10^{-3}}{362} = 43.16\text{m}$	
	both greater than 6.20m	Slenderness Limits O.K.

CURPE CONSULTANTS 46 Orburn Road Dunfield	Project		Job ref
	Part of structure: S.S. BEAM - TENSION		Calc sheet no 12/3 rev /
	Drawing ref	Calc by B.C. Date	Check by Date

Ref	Calculations	Output
3.4.1.2	<u>THE EFFECTIVE SPAN</u>	
	The above calculations were based on a span of 6.50 m. Now that the effective depth is known this must be confirmed	
	i.e. the span is the smaller of	
	(i) 6.50m	
	or (ii) $6.20 + 0.362 = 6.562\text{ m}$	$l = 6.50\text{m}$
	<u>LOADING - for a 1m length of beam</u>	
BS 648	Brickwork weighs $55\text{ kg/m}^2/25\text{ mm}$ thickness	
	$\therefore$ load from b'w'k $= \frac{1.75 \times 200 \times 55}{25} = 770\text{ kg/m}$	
	Self weight $= 0.25 \times 0.400 \times 2400 = 240$	
	Total $g_k = 1010$	
	$\therefore g_k = 1010 \times 9.81 \times 10^{-3} = 9.91\text{ kN/m}$	
	There is no imposed load $\therefore q_k = 0$	
table 2.1	Design Load $F = 1.4G_k + 1.6Q_k$	
	$= (1.4 \times 9.91 + 1.6 \times 0)\,6.50$	
	$= 90.18\text{ kN}$	F = 90.18kN
	<u>BENDING MOMENT</u>	
	$M_{max} = \frac{Fl}{8} = \frac{90.18 \times 6.50}{8}$	
	$= 73.27\text{ kN.m}$	M = 73.27kN.m.

CURPE CONSULTANTS 46 Orburn Road Dunfield	Project			Job ref	
	Part of structure: S.S. BEAM- TENSION			Calc sheet no: 12 / 4	rev: /
	Drawing ref	Calc by: B.C.	Date	Check by	Date

Ref	Calculations	Output
	MAIN REINFORCEMENT	
	$\frac{M}{bd^2} = \frac{73.27 \times 10^6}{250 \times 362^2}$	
	$= 2.24 \ N/mm^2$	
chart 2	$\frac{100A_s}{bd} = 0.6$	
	$A_s = \frac{0.6 \times 250 \times 362}{100}$	
	$= 543 \ mm^2$	
	3T16 bars = $603mm^2$	Main reinf. 3T16 bars
	When the effective depth was calculated the main bars were assumed to be 20mm ϕ. The fact that 16mm ϕ bars are being used changes the effective depth. The small difference of 2mm would be ignored in practice. The following calculation justifies this decision.	
	New 'd' = 400 - (20 + 8 + 16/2)	
	= 364 mm	
	$\frac{M}{bd^2} = \frac{73.27 \times 10^6}{250 \times 364^2}$	
	$= 2.21 \ N/mm^2$	
chart 2	$\frac{100A_s}{bd} = 0.59$	

CURPE CONSULTANTS 46 Orburn Road Dunfield	Project			Job ref	
	Part of structure: S.S. BEAM - TENSION			Calc sheet no: 12 / 5	rev: /
	Drawing ref	Calc by: B.C.	Date	Check by	Date

Ref	Calculations	Output
	$\therefore A_s = \frac{0.59 \times 250 \times 364}{100}$	
	$= 536.9 \ mm^2$	
	ie 3T16 bars still required.	
	Since 16mm ϕ bars are to be used, the effective depth corresponding to these ie. 364 mm, will be used in the remainder of this design.	
3.4.5	CHECK SHEAR - Worst at the support	
	Shear force 'V' = F/2	
	= 90.18/2	
	= 45.09 kN	
3.4.5.2.	Design shear stress	
	$v = \frac{V}{b_v d} = \frac{45.09 \times 10^3}{250 \times 364}$	
	$= 0.50 \ N/mm^2$	$v = 0.50 N/mm^2$
	This should not exceed either $5 N/mm^2$ or $0.8\sqrt{f_{cu}} = 0.8\sqrt{40} = 5.06 \ N/mm^2$. The value of $0.5 \ N/mm^2$ is acceptable.	
Fig 3.53	For a simply supported beam only 50% of the mid span reinforcement is carried to the support and anchored. In this case there are three bars at mid span and two will be carried through to the support.	

CURPE CONSULTANTS 46 Orburn Road Dunfield	Project	Job ref
	Part of structure: S. S. BEAM – TENSION	Calc sheet no: 12 / 6 rev: 1
	Drawing ref; Calc by: BC; Date	Check by; Date

Ref	Calculations	Output
	∴ At the support $\frac{100A_s}{bd}$	
	$= \frac{100 \times 402}{250 \times 364}$	
	$= 0.44$	
table 3.9	$v_c = 0.49\ N/mm^2$ which may be enhanced by $(f_{cu}/25)^{1/3}$ i.e $(40/25)^{1/3} = 1.16$	
	$\therefore v_c = 0.49 \times 1.16 = 0.57\ N/mm^2$	$v_c = 0.57\ N/mm^2$
table 3.8	$0.5 v_c = 0.285 < v$ (= 0.5 page 12.5)	
	$v_c + 0.4 = 0.97 > v$	
	$\therefore 0.5 v_c < v < v_c + 0.4$	
	∴ Minimum links are required throughout the whole length of the beam.	
	$A_{sv} \geq 0.4 b_v s_v / 0.87 f_{yv}$	
	$\therefore s_v \leq A_{sv}\, 0.87 f_{yv} / 0.4 b_v$	
	with 8mm ϕ links $A_{sv} = 101 mm^2$	
	$\therefore s_v \leq \frac{101 \times 0.87 \times 250}{0.4 \times 250}$	
	$\leq 219.68 mm$	
3.4.5.5.	max $s_v = 0.75d = 0.75 \times 364$	
	$= 273 mm$	
	∴ USE 8 mm links at 200 mm c/c	R8 @ 200 c/c

CURPE CONSULTANTS 46 Orburn Road Dunfield	Project	Job ref
	Part of structure: S.S. BEAM – TENSION	Calc sheet no: 12 / 7 rev: 1
	Drawing ref; Calc by: B.C.; Date	Check by; Date

Ref	Calculations	Output
3.4.6	CHECK DEFLECTION	
table 3.10	basic $l/d = 20$	
Eqn 8	service stress f_s	
	$= \frac{5 f_y A_{s\,req}}{8 A_{s\,prov}} \times \frac{1}{\beta_b}$	
	$= \frac{5}{8} \times \frac{460 \times 543}{603} \times 1$	
	$= 258.89\ N/mm^2$	
	At mid span $\frac{M}{bd^2} = 2.24\ N/mm^2$	
Eqn 7	Modification factor for tension reinf.	
	$= 0.55 + \frac{477 - 258.89}{120(0.9 + 2.24)}$	
	$= 1.13$	
	∴ Permissible $l/d = 20 \times 1.13$	
	$= 22.6$	
	Actual $l/d = \frac{6500}{364}$	
	$= 17.86$	
	∴ SATISFACTORY	Deflection O.K.

CURPE CONSULTANTS
46 Orburn Road
Dunfield

Project				Job ref	
Part of structure S.S. BEAM – TENSION				Calc sheet no 12/8	rev /
Drawing ref	Calc by B. C.	Date		Check by	Date

Ref	Calculations	Output
3.4.7	CHECK CRACKING	
3.12.11.2.	Crack control is provided by the distance between bars in tension	
Table 3.30	Maximum clear distance between bars = 160mm	
	At mid span, where worst tensile cracking would be expected, the actual clear distance	

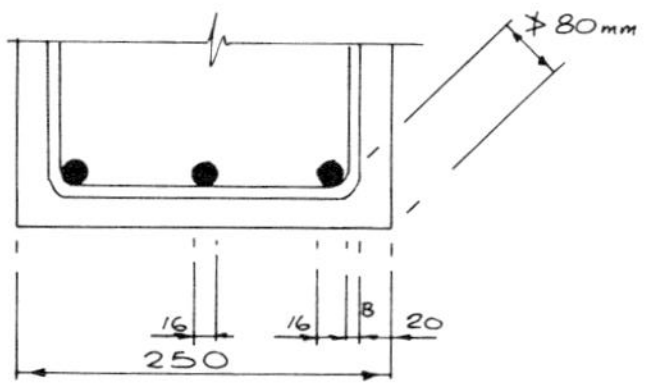

$$= \frac{250 - 2(20 + 8) - 3 \times 16}{2}$$

$= 73\ mm < 160mm$

3.12.11.2.5 The corner distance should not be greater than $160/2 = 80mm$.

Actual distance $= \sqrt{28^2 + 28^2}$

$= 39.60mm < 80mm$

∴ SATISFACTORY — **Output:** Cracking O.K.

CURPE CONSULTANTS
46 Orburn Road
Dunfield

Project	Job ref
Sub-section S. S. BEAM – TENSION & COMPRESSION	Sheet no 12/9
	Made by BC
	Date
Structural Summary Sheet	Checked by
	Date

A SIMPLY SUPPORTED RECTANGULAR REINFORCED CONCRETE BEAM, REINFORCED IN TENSION AND COMPRESSION

A reinforced concrete floor slab is introduced on top of the brickwork in the previous simply supported beam design

Section A-A on page 12/1 - 139 now becomes:-

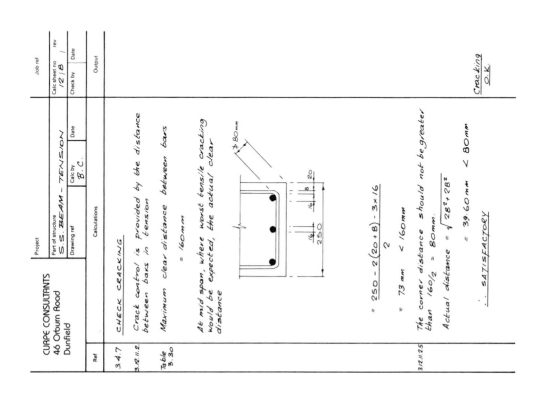

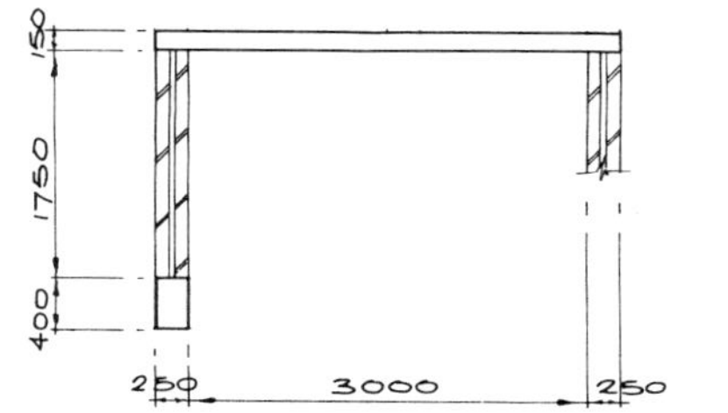

Assume : (i) there is a 50mm screed on top of the slab,

(ii) the imposed load on the slab = $5{\cdot}00\ kN/m^2$

CURPE CONSULTANTS 46 Orburn Road Dunfield	Project	Job ref
	Part of structure: S.S. BEAM - TENSION & COMPRESSION	Calc sheet no: 12/10, rev: 1
	Drawing ref; Calc by: B.C; Date	Check by; Date

Ref	Calculations	Output
	ADDITIONAL LOADING	
	Dead:	
	Self weight of the slab $= 2400 \times 0.150$ $= 360$ kg/m²	
BS648	Screed weighs 30 kg/m²/13 mm thickness	
	∴ 50 mm thickness $= \frac{30 \times 50}{13} = 115.38$	
	Total $= 475.38$	
	In force units $= 475.38 \times 9.81 \times 10^{-3}$ $= 4.66\ kN/m^2$	
	On a 1 m span of beam the additional dead load $= 4.66 \times (3/2 + 0.25)$ $= 8.16\ kN/m$	
	Imposed:	
	On a 1 m span of the beam $= 5.0(3/2 + 0.25)$ $= 8.75\ kN/m$	
Table 2.1	DESIGN LOAD	
	$F = 1.4\,G_K + 1.6\,Q_K$	
	$= [1.4(8.16 + 9.91) + (1.6 \times 8.75)]\,6.5$	
	$= 255.44\ kN$	F = 255.44 kN

CURPE CONSULTANTS 46 Orburn Road Dunfield	Project	Job ref
	Part of structure: S.S. BEAM TENSION & COMPRESSION	Calc sheet no: 12/11, rev: 1
	Drawing ref; Calc by: B.C.; Date	Check by; Date

Ref	Calculations	Output
	BENDING MOMENT	
	$M_{max.} = Fl/8 = \frac{255.44 \times 6.5}{8}$	
	$= 207.55\ kN.m$	M = 207.55 kN.m
	MAIN REINFORCEMENT	
	$\frac{M}{bd^2} = \frac{207.55 \times 10^6}{250 \times 362^2}$	
	$= 6.34\ N/mm^2$	
	This value cannot be read from chart 2, therefore the beam must be doubly reinforced. To chose the appropriate design chart d'/d must be known.	
	$d' = 20 + 8 + 20/2$	
	$= 38\ mm$	
	Note: Since the loading is increased it is likely that at least 20 mm ϕ bars will be required. Both d' and d have been calculated on this basis. Thus $d = 362\ mm$.	
	$d'/d = 38/362 = 0.10$	
	With $f_{cu} = 40\ N/mm^2$ and $f_y = 460\ N/mm^2$	
	USE CHART No. 12	
	When using the doubly reinforced beam charts, for simply supported beams the value of x/d should be less than 0.5 to guard against the beam being over-reinforced.	

CURPE CONSULTANTS
46 Orburn Road
Dunfield

Project			Job ref	
Part of structure S.S. BEAM TENSION & COMPRESSION			Calc sheet no 12/12	rev 1
Drawing ref	Calc by B.C.	Date	Check by	Date

Ref	Calculations	Output
chart No. 12	Assume $\frac{100A_s'}{bd} = 0.8$	
	$\therefore A_s' = \frac{0.8 \times 250 \times 362}{100}$	
	$= 724\ mm^2$	
	4T16 bars gives $A_s' = 804\ mm^2$	Top reinf.
	Actual $\frac{100A_s'}{bd} = \frac{100 \times 804}{250 \times 362}$	
	$= 0.89$	
	$\frac{100A_s}{bd} = 1.77$	
	$A_s = \frac{1.77 \times 250 \times 362}{100}$	
	$= 1601.85\ mm^2$	
	2T25mm bars + 2T20mm bars $= 1610\ mm^2$	Btm. reinf.
3.4.5	CHECK SHEAR - Worst at the support	
	Shear Force $V = F/2$	
	$= 255.44/2$	
	$= 127.72\ kN$	
3.4.5.2	Design shear stress	
	$v = \frac{V}{b_v d} = \frac{127.72 \times 10^3}{250 \times 362}$	
	$= 1.41\ N/mm^2$	$v = 1.41\ N/mm^2$

CURPE CONSULTANTS
46 Orburn Road
Dunfield

Project			Job ref	
Part of structure S.S. BEAM TENSION & COMPRESSION			Calc sheet no 12/13	rev 1
Drawing ref	Calc by B.C.	Date	Check by	Date

Ref	Calculations	Output
	This does not exceed $5\ N/mm^2$ or $0.8\sqrt{f_{cu}}$ and is therefore acceptable.	
Fig 3.53	Assume 2T25 bars are carried to the support.	
	$\therefore \frac{100A_s}{bd} = \frac{100 \times 982}{250 \times 362}$	
	$= 1.09$	
table 3.9	$v_c = 0.665\ N/mm^2$ - may be enhanced by $(f_{cu}/25)^{1/3}$ ie $(40/25)^{1/3} = 1.16$	
	ie $v_c = 0.665 \times 1.16 = 0.77\ N/mm^2$	$v_c = 0.77\ N/mm^2$
table 3.8	$0.5\,v_c = 0.385\ N/mm^2 < v$ of $1.41\ N/mm^2$	
	$v_c + 0.4 = 1.17\ N/mm^2 < v$ of $1.41\ N/mm^2$	
	$\therefore (v_c + 0.4) < v < 0.8\sqrt{f_{cu}}$	
	ie $A_{sv} \geq b_v s_v (v - v_c) / 0.87 f_{yv}$	
	$s_v \leq \frac{A_{sv}\ 0.87\ f_{yv}}{b(v - v_c)}$	
	$\leq \frac{101 \times 0.87 \times 250}{250(1.41 - 0.77)}$	
	$\leq 137.30\ mm$	
3.4.5.5	max $s_v = 0.75d = 0.75 \times 362$	
	$= 271.5\ mm$	

CURPE CONSULTANTS 46 Orburn Road Dunfield	Project		Job ref	
	Part of structure S.S. BEAM TENSION & COMPRESSION		Calc sheet no 12/14	rev 1
	Drawing ref	Calc by B.C. / Date	Check by	Date

Ref	Calculations	Output
3.12.7.1.	To contain compression reinforcement	
	minimum diameter of link = $\frac{1}{4} \times 16 = 4mm$ maximum spacing of link = $12 \times 16 = 192mm$	
	∴ USE R8mm links at 125 mm c/c	R8 @ 125 c/c
3.4.6.	CHECK DEFLECTION	
table 3.10	basic $l/d = 20$	
Eq^n 8	service stress f_s	
	$= \frac{5 \times 460}{8} \times \frac{1601.85}{1610} \times 1$	
	$= 286.04 \; N/mm^2$	
	$M/bd^2 = 6.34 \; N/mm^2$	
Eq^n 7	Modification factor for tension reinf.	
	$= 0.55 + \frac{477 - 286.04}{120(0.9 + 6.34)}$	
	$= 0.77$	
table 3.12	Modification factor for compression reinf.	
	$\frac{100 A_s' \text{ prov}}{bd} = \frac{100 \times 804}{250 \times 362}$	
	$= 0.89$	
	Modification factor $= 1.23$	
	∴ Permissible $l/d = 20 \times 0.77 \times 1.23$	
	$= 18.94$	

CURPE CONSULTANTS 46 Orburn Road Dunfield	Project		Job ref	
	Part of structure S.S. BEAM TENSION & COMPRESSION		Calc sheet no 12/15	rev 1
	Drawing ref	Calc by B.C. / Date	Check by	Date

Ref	Calculations	Output
	Actual $l/d = 6500/362$	
	$= 17.96 < 18.94$	
	∴ SATISFACTORY	Deflection O.K.
3.4.7	CHECK CRACKING	
	The rules are as in previous example.	
	Actual clear distance	
	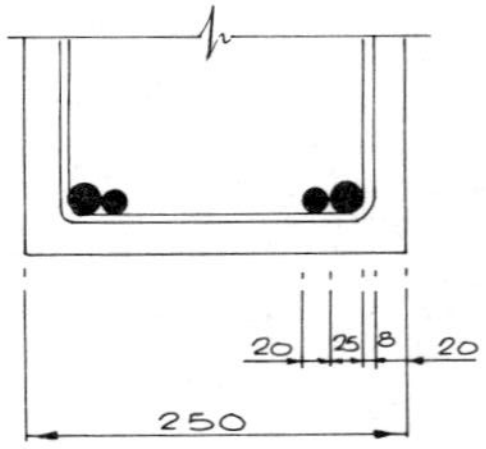	
	$= 250 - 2(20 + 8 + 25 + 20) = 104mm < 160 mm$	
	Corner distance = 39.60mm, as previously	
	∴ SATISFACTORY	Cracking O.K.

CURPE CONSULTANTS
46 Orburn Road
Dunfield

Project	Job ref
Sub-section: S.S. FLANGED BEAM	Sheet no 12/16
	Made by BC
	Date
Structural Summary Sheet	Checked by
	Date

A SIMPLY SUPPORTED FLANGED BEAM

The reinforced concrete floor slab introduced in the doubly reinforced concrete beam previously designed is now positioned so as to come into the side of the beam.

The section through the beam now becomes:

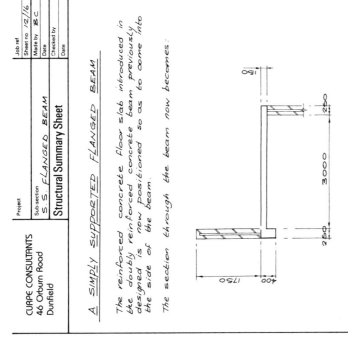

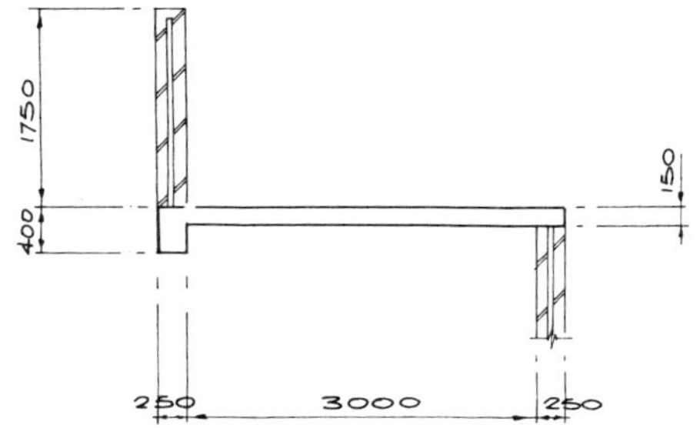

CURPE CONSULTANTS
46 Orburn Road
Dunfield

Project			Job ref	
Part of structure: S.S. FLANGED BEAM			Calc sheet no 12/17	rev 1
Drawing ref	Calc by B.C.	Date	Check by	Date

Ref	Calculations	Output
	The loading and the maximum bending moment will be as calculated in the previous beam design. - see pages 12/10 and 12/11	
	i.e. $F = 255.44$ kN	
	$M = 207.55$ kN.m	
	Due to the position of the slab, there will be a part of the slab which will act in conjunction with the beam to form a flanged beam.	
3.4.1.5	EFFECTIVE FLANGE WIDTH	
	lesser of (i) $250 + \frac{6500}{10} = 900$ mm	
	or (ii) $\frac{3000}{2} + 250 = 1750$ mm	
	$\therefore b = 900$ mm	$b = 900$ mm
	MAIN REINFORCEMENT	
	$\frac{M}{bd^2} = \frac{207.55 \times 10^6}{900 \times 362^2}$	
	$= 1.76$ N/mm²	
chart 2	$\frac{100A_s}{bd} = 0.46$	
	$A_s = \frac{0.46 \times 900 \times 362}{100}$	
	$= 1498.68$ mm²	
	2T25 mm bars + 2T20 mm bars = 1610 mm²	2T25+2T20

CURPE CONSULTANTS 46 Orburn Road Dunfield	Project	Job ref
	Part of structure: S.S. FLANGED BEAM	Calc sheet no: 12/18 rev: 1
	Drawing ref / Calc by: B.C. / Date	Check by / Date

Ref	Calculations	Output

It is important, in a flanged beam, to check the position of the neutral axis. If the neutral axis lies within the flange the concrete in compression is of rectangular cross-section. Thus the use of the charts, which are for rectangular beams only, is justified.

There are two ways in which the neutral axis position can be checked.

a) Using design charts

Using a doubly reinforced beam chart, say No. 12, it is clear that the position read for $\frac{100A_s}{bd}$ lies to the left of

the line $\frac{x}{d} = 0.3$

ie $\frac{x}{d} < 0.3$

$x < 0.3 \times 362$ mm

$x < 108.6$ mm

∴ N.A. lies within the flange

3.4.4.4 b) Using formulae

$K' = 0.156$ since redistribution = 0%

$K = \frac{M}{bd^2 f_{cu}} = \frac{207.54 \times 10^6}{900 \times 362^2 \times 40}$

$= 0.044$

$K < K'$ ∴ no compression reinf. required - this confirms the above calculation for main reinforcement.

CURPE CONSULTANTS 46 Orburn Road Dunfield	Project	Job ref
	Part of structure: S.S. FLANGED BEAM	Calc sheet no: 12/19 rev: 1
	Drawing ref / Calc by: B.C. / Date	Check by / Date

Ref	Calculations	Output

$Z = d\left\{0.5 + \sqrt{\left(0.25 - \frac{K}{0.9}\right)}\right\}$

$= 362\left\{0.5 + \sqrt{\left(0.25 - \frac{0.044}{0.9}\right)}\right\}$

$= 343.34$ mm

or $Z = 0.95d = 0.95 \times 362$

$= 343.90$mm

$Z = 343.34$ mm

$x = \frac{(d - z)}{0.45} = \frac{362 - 343.34}{0.45}$

$= 41.47$mm

∴ N.A. lies within the flange

Whatever method is adopted it is clear that the charts may be used.

3.4.5 CHECK SHEAR

For shear 'b' is the rib width, thus the shear calculation is the same as that indicated in the previous beam design, with the exception of the links for compression reinforcement.

3.4.6 CHECK DEFLECTION

$\frac{b_w}{b} = \frac{250}{900} = 0.28$

table 3.10 basic $l/d = 16$

CURPE CONSULTANTS
46 Orburn Road
Dunfield

Project			Job ref	
Part of structure: S.S. FLANGED BEAM			Calc sheet no: 12/20	rev: 1
Drawing ref	Calc by: B.C.	Date	Check by	Date

Ref	Calculations	Output
Eqn 8	Service stress f_s	
	$= \frac{5 \times 460}{8} \times \frac{1498.68}{1610} \times 1$	
	$= 267.62\ N/mm^2$	
	$M/bd^2 = 1.76\ N/mm^2$	
Eqn 7	Modification factor for tension reinf.	
	$= 0.55 + \frac{477 - 267.62}{120(0.9 + 1.76)}$	
	$= 1.21$	
	Permissible $l/d = 16 \times 1.21$	
	$= 19.36$	
	Actual $l/d = \frac{6500}{362}$	
	$= 17.96 < 19.36$	
	∴ SATISFACTORY	Deflection O.K
3.4.7	CHECK CRACKING	
	The arrangement of tension reinforcement is similar to the doubly reinforced beam previously designed. The check for cracking is therefore similar and is satisfactory.	

CURPE CONSULTANTS
46 Orburn Road
Dunfield

Project	Job ref
Sub-section: S.S. SLAB	Sheet no: 12/21
	Made by: B C
	Date
Structural Summary Sheet	Checked by
	Date

A SIMPLY SUPPORTED REINFORCED CONCRETE SLAB.

A reinforced concrete slab is required to span between supporting brickwalls as shown below. The slab is to be designed to satisfy BS8110 with loads and material strengths as follows:

f_{cu} (the characteristic strength of the concrete) $= 35\ N/mm^2$

f_y (the characteristic strength of the reinforcement)

$= 460\ N/mm^2$

Imposed loading on the slab $= 2.50\ kN/m^2$

Finishes and partitions on the slab $= 1.75\ kN/m^2$

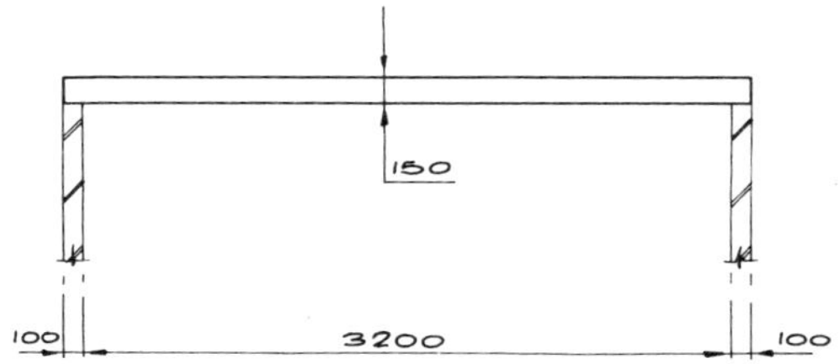

SECTION THROUGH THE SLAB

CURPE CONSULTANTS 46 Orburn Road Dunfield	Project	Job ref
	Part of structure: S.S. SLAB	Calc sheet no: 12/22 rev: 1
	Drawing ref: / Calc by: B.C. / Date:	Check by: / Date:

Ref	Calculations	Output
	EFFECTIVE DEPTH	
table 3.4.	For mild conditions of exposure, the cover $= 20mm$	
	Assuming 12mm ϕ main bars	
	$d = 150 - (20 + 12/2)$	
	$= 124mm$	$d = 124mm$
3.4.1.2	EFFECTIVE SPAN	
	smaller of (i) $3200 + 100 = 3300$	
	or (ii) $3200 + 124 = 3324$	
	$\therefore$ span $= 3300$	$l = 3300mm$
	LOADING	
	Dead: Finishes & partitions $= 1.75$ kN/m^2	
	Self wt. $= 0.15 \times 2400 \times 9.81 \times 10^{-3} = 3.53$	
	$\therefore$ total $g_k = 5.28$	
	Imposed: $q_k = 2.50 kN/m^2$	
table 2.1	Design Load $F = 1.4G_k + 1.6Q_k$	
	$= (1.4 \times 5.28 + 1.6 \times 2.50)3.30$	
	$= 37.59 kN/m$ strip	$F = 37.59kN$
	BENDING MOMENT	
	$M_{max} = \frac{Fl}{8} = \frac{37.59 \times 3.3}{8}$	
	$= 15.51 kN.m$	$M = 15.51 kN.m$

CURPE CONSULTANTS 46 Orburn Road Dunfield	Project	Job ref
	Part of structure: S.S. SLAB	Calc sheet no: 12/23 rev: 1
	Drawing ref: / Calc by: B.C. / Date:	Check by: / Date:

Ref	Calculations	Output
	MAIN REINFORCEMENT	
	$\frac{M}{bd^2} = \frac{15.51 \times 10^6}{10^3 \times 124^2}$	
	$= 1.01 N/mm^2$	
chart 2	$\frac{100A_s}{bd} = 0.27$	
	$A_s = \frac{0.27 \times 10^3 \times 124}{10^2}$	
	$= 334.8 mm^2/m$	
	T12 mm ϕ bars at 300 mm c/c ($377 mm^2/m$)	Main Reinf T12 @ 300 c/c
table 3.27	CHECK MINIMUM AREA OF REINFORCEMENT	
	Minimum percentage, in both directions,	
	$100A_s/A_c = 0.13$	
	$\therefore A_s = 0.13 \times 10^3 \times 150 \times 10^{-2}$	
	$= 195 mm^2/m$	
	T10 mm ϕ bars at 350 mm c/c ($225 mm^2/m$)	Distribution T10 @ 350 c/c
3.5.7	CHECK DEFLECTION	
table 3.10	basic $l/d = 20$	
Eq^n 8	service stress f_s	
	$= \frac{5 \times 460 \times 334.8}{8 \times 377} \times 1$	
	$= 255.32 N/mm^2$	

CURPE CONSULTANTS 46 Orburn Road Dunfield	Project	Job ref
	Part of structure: S. S. SLAB	Calc sheet no: 12/24 rev: 1
	Drawing ref: — Calc by: B.C. Date: —	Check by: — Date: —

Ref	Calculations	Output
Eqⁿ 7	Modification factor for tension reinf.	
	$= 0.55 + \frac{477 - 255.32}{120(0.9 + 1.01)}$	
	$= 1.52$	
	Permissible $l/d = 20 \times 1.52$ $= 30.34$	
	Actual $l/d = \frac{3300}{124}$ $= 26.61 < 30.34$	
	∴ SATISFACTORY	Deflection O.K.
3.5.5	CHECK SHEAR - worst at support	
	N.B. Under normal conditions of loading shear will seldom prove to be a problem in a solid slab. However the following illustrates the typical calculations required if deemed necessary	
	In the case where heavy point loads are applied to the slab, punching shear should be checked. This will be illustrated in the design of a pad foundation – page 177 - 185	
	Shear Force $= \frac{F}{2} = \frac{37.59}{2}$ $= 18.80$ kN	

CURPE CONSULTANTS 46 Orburn Road Dunfield	Project	Job ref
	Part of structure: S. S. SLAB	Calc sheet no: 12/25 rev: 1
	Drawing ref: — Calc by: B.C. Date: —	Check by: — Date: —

Ref	Calculations	Output
eqⁿ 21	shear stress $= \frac{V}{bd} = \frac{18.80 \times 10^3}{10^3 \quad 124}$	
	$= 0.15\ N/mm^2$	
table 3.9	design concrete shear stress is obviously very much larger than $0.15\ N/mm^2$	
table 3.17	No shear reinforcement is required	shear O.K.
3.5.8	CHECK CRACKING	
3.12.11.27	clear spacing between bars should not exceed $3d = 3 \times 124 = 372$mm or 750mm	
	Also under (a) (2) no further check is required.	
	∴ SATISFACTORY	Cracking O.K

13

Cantilever beams and slabs: worked examples

CURPE CONSULTANTS 46 Orburn Road Dunfield	Project	Job ref
		Sheet no 13/1
	Sub-section CANTILEVER BEAMS & SLABS	Made by BC
		Date
	Structural Summary Sheet	Checked by
		Date

CANTILEVER BEAMS AND SLABS

The reinforced concrete floor shown in plan below is required to carry the following loads and to satisfy BS 8110. The conditions of exposure are mild and the material strengths are as indicated.

Loading on the slab:

i) imposed = 5 $^{kN}/_{m^2}$

ii) finishes and partitions = 4·50 $^{kN}/_{m^2}$.

Material strengths:

f_{cu} (the characteristic strength of the concrete) = 35 $^{N}/_{mm^2}$

f_y (the characteristic strength of the reinforcement)

= 460 $^{N}/_{mm^2}$ for the main reinforcement and

= 250 $^{N}/_{mm^2}$ for the links.

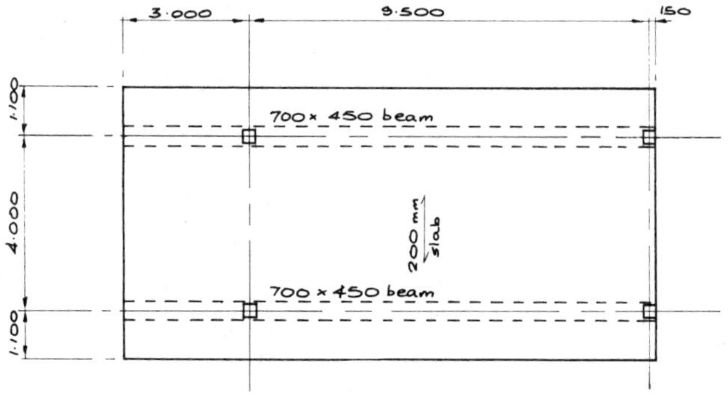

PLAN OF FLOOR

CURPE CONSULTANTS 46 Orburn Road Dunfield	Project			Job ref
	Part of structure CANTILEVER BEAMS & SLABS			Calc sheet no 13/2 rev /
	Drawing ref	Calc by B.C.	Date	Check by / Date

Ref	Calculations	Output

SLAB DESIGN

LOADING

Dead: Finishes & partitions $= 4.50\ kN/m^2$

Self wt $= \frac{0.2 \times 2400 \times 9.81}{10^3} = 4.71$

Total $g_k = 9.21\ kN/m^2$

Imposed: $q_k = 5.00\ kN/m^2$

2.4.3.1. LOADING COMBINATIONS

2.4.3.1.1. Loads must be combined in such a way as to give the worst stresses.

table 2.1. Max. load $= 1.4 g_k + 1.6 q_k$

$= 1.4 \times 9.21 + 1.6 \times 5.0$

$= 20.89\ kN/m^2$

Min. load $= 1.0 g_k$

$= 9.21\ kN/m^2$

When a 1m strip of slab is considered the above loads become kN/m.

Combination 1. - Max. support & min. span moments

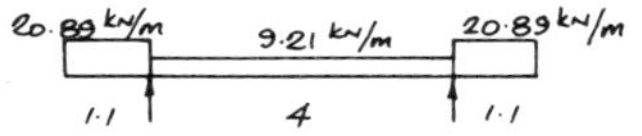

Max support $M = \frac{20.89 \times 1.1^2}{2} = 12.64\ kN.m$ — Output: Max suppt. M = 12.64 kN.m

CURPE CONSULTANTS 46 Orburn Road Dunfield	Project			Job ref
	Part of structure CANTILEVER BEAMS & SLABS			Calc sheet no 13/3 rev /
	Drawing ref	Calc by BC	Date	Check by / Date

Ref	Calculations	Output

Min. span $M = \frac{9.21 \times 4^2}{8} - 12.64$

$= 5.78\ kN.m$

Combination 2. - Max. span & min support moments

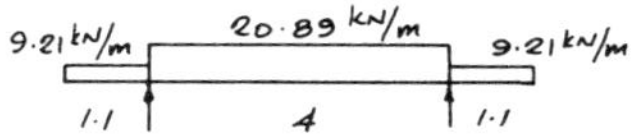

Max. span $M = \frac{20.89 \times 4^2}{8} - \frac{9.21 \times 1.1^2}{2}$

$= 36.21\ kN.m$ — Output: Max. span M = 36.21 kN.m

Min support $M = \frac{9.21 \times 1.1^2}{2}$

$= 5.57\ kN.m$

Combination 3 - Max shear force

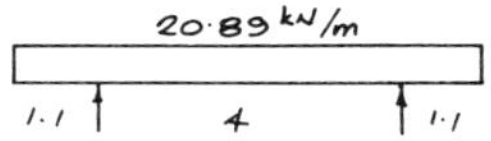

$V_{max} = 20.89 \times 2$

$= 41.78\ kN.$ — Output: V = 41.78 kN

EFFECTIVE DEPTH

table 3.4 For mild conditions of exposure, the cover $= 20mm.$

CURPE CONSULTANTS 46 Orburn Road Dunfield	Project	Job ref
	Part of structure: CANTILEVER BEAMS & SLABS	Calc sheet no: 13/4 rev: 1
	Drawing ref: — Calc by: B.C. Date: —	Check by: — Date: —

Ref	Calculations	Output
	Assuming 12mm ϕ main bars	
	$d = 200 - (20 + 12/2)$	
	$= 174$ mm.	$d = 174$ mm
	MAIN REINFORCEMENT	
	1. Mid span.	
	$\dfrac{M}{bd^2} = \dfrac{36.21 \times 10^6}{10^3 \times 174^2}$	
	$= 1.20\ N/mm^2$	
chart 2	$\dfrac{100A_s}{bd} = 0.31$	
	$A_s = \dfrac{0.31 \times 10^3 \times 174}{10^2}$	
	$= 539.4\ mm^2/m$	
	T12 mm ϕ bars at 200 mm c/c (566 mm^2/m)	Mid span Main Reinf T12 @ 200 c/c
	2. Support	
	$\dfrac{M}{bd^2} = \dfrac{12.64 \times 10^6}{10^3 \times 174^2}$	
	$= 0.42\ N/mm^2$	
chart 2	$\dfrac{100A_s}{bd} = 0.12$	
table 3.27	min. amount of reinf. $\dfrac{100A_s}{A_c} = 0.13$	

CURPE CONSULTANTS 46 Orburn Road Dunfield	Project	Job ref
	Part of structure: CANTILEVER BEAMS & SLABS	Calc sheet no: 13/5 rev: 1
	Drawing ref: — Calc by: B.C. Date: —	Check by: — Date: —

Fef	Calculations	Output
	$A_s = \dfrac{0.13 \times 10^3 \times 200}{10^2}$	
	$= 260\ mm^2/m$	
	T10 mm ϕ bars at 300 mm c/c (262 mm^2/m)	Support Main Reinf T10 @ 300 c/c
	Since this is the minimum amount it will also be used for the secondary reinf.	
3.5.7	CHECK DEFLECTION	
	There are two critical positions for deflection :-	
	a) Mid span	
table 3.10	basic $l/d = 26$	
Eq^n 8	service stress f_s	
	$= \dfrac{5 \times 460}{8} \times \dfrac{539.4}{566} \times 1$	
	$= 273.99\ N/mm^2$	
Eq^n 7	Modification factor for tension reinf.	
	$= 0.55 + \dfrac{477 - 273.99}{120(0.9 + 1.20)}$	
	$= 1.36$	
	Permissible $l/d = 26 \times 1.36$	
	$= 35.36$	
	Actual $l/d = 4000/174 = 22.99 < 35.36$	
	$\therefore$ SATISFACTORY	

CURPE CONSULTANTS
46 Orburn Road
Dunfield

Project: | Part of structure: CANTILEVER BEAMS & SLABS | Drawing ref: | Calc by: B.C. | Date: | Job ref: | Calc sheet no: 13/6 rev: 1 | Check by: | Date:

Ref	Calculations	Output
	b) At support	
table 3.10	basic $l/d = 7$	
Eq^n 8	service stress f_s	
	$= \frac{5 \times 460 \times 260}{8 \times 262} \times 1$	
	$= 285.31\ N/mm^2$	
Eq^n 7	Modification factor for tension reinf.	
	$= 0.55 + \frac{477 - 285.31}{120(0.9 + 0.42)}$	
	$= 1.80$	
	Permissible $l/d = 7 \times 1.76$	
	$= 12.32$	
	Actual $l/d = \frac{1100}{174}$	
	$= 6.32 < 12.32$	
	∴ SATISFACTORY	Deflection O.K
3.5.5	CHECK SHEAR	
	- see note on page 12/24	
3.5.8	CHECK CRACKING	
	- as page 12/25	

CURPE CONSULTANTS
46 Orburn Road
Dunfield

Project: | Part of structure: CANTILEVER BEAMS & SLABS | Drawing ref: | Calc by: B.C. | Date: | Job ref: | Calc sheet no: 13/7 rev: 1 | Check by: | Date:

Ref	Calculations	Output
	BEAM DESIGN	
	LOADING - Each beam carries 3.1 m width of slab.	
	Imposed loading on beam	
	$= 5.0 \times 3.1 = 15.5\ kN/m$	$q_k = 15.5\ kN/m$
	Dead loading from slab to beam	
	$= 9.21 \times 3.1 = 28.54\ kN/m$	
	Self wt of beam downstand	
	$= 2400 \times 9.81 \times 10^{-3} \times 0.45 \times 0.5$	
	$= 5.30\ kN/m$	
	Total $g_k = 33.84\ kN/m$	$g_k = 33.84\ kN/m$
2.4.3.1	LOADING COMBINATIONS	
	As with the slab design the loads must be combined to give the worst stresses	
	Combination 1 - Max. span & min. support moments	

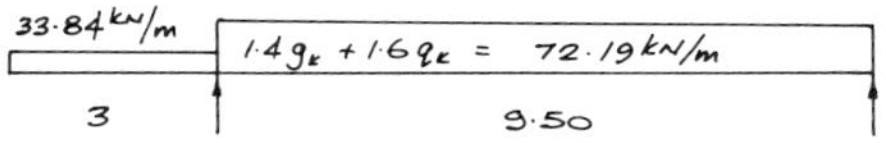

Ref	Calculations	Output
	Min. support $M = \frac{33.84 \times 3^2}{2}$	
	$= 152.28\ kN.m$	
	Max. span $M = \frac{72.19 \times 9.5^2}{8} - \frac{152.28}{2}$	

CURPE CONSULTANTS
46 Orburn Road
Dunfield

Project		Job ref
Part of structure: CANTILEVER BEAMS & SLABS		Calc sheet no 13/8 rev 1
Drawing ref	Calc by B.C. Date	Check by Date

Ref	Calculations	Output

= 738·25 kN.m — **Output:** Max. span M = 738·25 kN.m

Combination 2 – Max. support & min. span moments

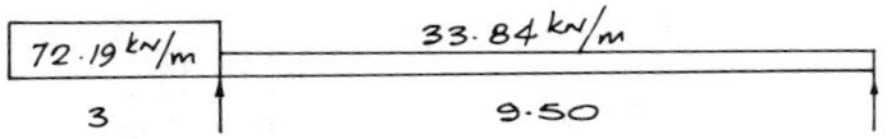

Max. support $M = \frac{72\cdot19 \times 3^2}{2}$

$= 324\cdot86$ kN.m — **Output:** Max. suppt M = 324·86 kN.m

Min span $M = \frac{33\cdot84 \times 9\cdot5^2}{8} - \frac{324\cdot86}{2}$

$= 219\cdot33$ kN.m

Combination 3 – Max. shear force

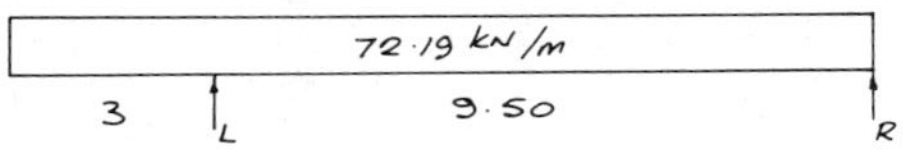

M/R : $9\cdot50L = \frac{72\cdot19 \times 12\cdot5^2}{2}$

$L = 593\cdot67$ kN

$R = 308\cdot71$ kN

Shear forces are
i) $3 \times 72\cdot19 = 216\cdot57$ kN
ii) $593\cdot67 - 216\cdot57 = 377\cdot1$ kN
iii) 308·71 kN

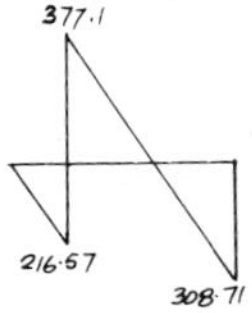

SF DIA

CURPE CONSULTANTS
46 Orburn Road
Dunfield

Project		Job ref
Part of structure: CANTILEVER BEAMS & SLABS		Calc sheet no 13/9 rev 1
Drawing ref	Calc by B.C. Date	Check by Date

Ref	Calculations	Output

∴ Max. shear force = 377·1 kN — **Output:** V = 377·1 kN

EFFECTIVE DEPTH

(table 3.4) cover = 20mm

(3.3.1.2) cover should not be less than the size of the main bar.
Assuming 32mm bars

$d = 700 - (32 + 32/2)$

$= 652$ mm — **Output:** d = 652 mm

(3.4.1.5) EFFECTIVE FLANGE WIDTH

Actual flange width

$= 1.1 + 4/2 = 3.1$m

$l_z \simeq 0\cdot85 \times 9\cdot50 = 8\cdot075$m

then flange width $= 450 + \frac{8075}{5}$

$= 2065$ mm.

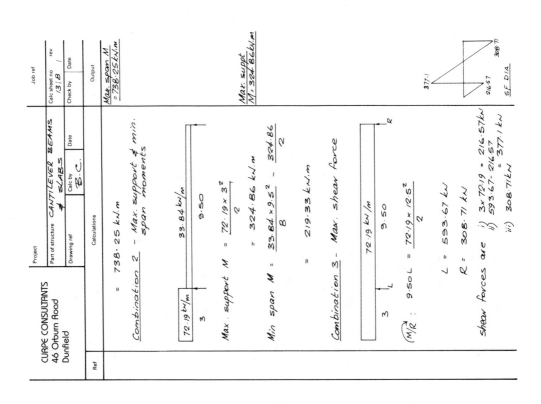

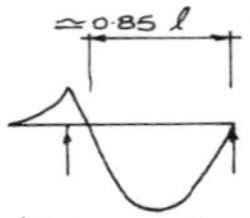

APPROX B.M. DIA.

b is lesser i.e. 2·065m — **Output:** b = 2·065m

MAIN REINFORCEMENT

a) At the support – a rectangular section

$\frac{M}{bd^2} = \frac{324\cdot86 \times 10^6}{450 \times 652^2}$

$= 1\cdot70$ N/mm²

(chart 2) $\frac{100 A_s}{bd} = 0\cdot45$

CURPE CONSULTANTS
46 Orburn Road
Dunfield

Project		Job ref
Part of structure CANTILEVER BEAMS & SLABS		Calc sheet no 13/10 rev /
Drawing ref	Calc by B.C	Date	Check by	Date

Ref	Calculations	Output
	$A_s = \frac{0.45 \times 450 \times 652}{100}$	
	$= 1320.3\ mm^2$	
	3T25 mm ϕ bars – 1470 mm²	At Suppt. 3T25
	b) At mid span – a flanged section	
	$\frac{M}{bd^2} = \frac{738.25 \times 10^6}{2065 \times 652^2}$	
	$= 0.84$	
chart 2	$\frac{100 A_s}{bd} = 0.23$	
	$A_s = \frac{0.23 \times 652 \times 2065}{100}$	
	$= 3096.67\ mm^2$	
table 3.27	Minimum amount of reinforcement	
	$\frac{b_w}{b} = \frac{450}{2065} = 0.22 < 0.4$	
	$\frac{100 A_s}{b_w h} = 0.18$	
	$A_s = \frac{0.18 \times 450 \times 700}{100}$	
	$= 567\ mm^2$	
	4T32 mm ϕ bars – 3220 mm²	At mid span 4T32

CURPE CONSULTANTS
46 Orburn Road
Dunfield

Project		Job ref
Part of structure CANTILEVER BEAMS & SLABS		Calc sheet no 13/11 rev /
Drawing ref	Calc by B.C.	Date	Check by	Date

Ref	Calculations	Output
	CHECK POSITION OF NEUTRAL AXIS	
	This has been explained on page 12/18. The method used here is the design formulae.	
	$K' = 0.156$ since redistribution = 0%	
	$K = 0.84/35 = 0.024$	
	$K < K'$ ∴ no compression reinf required – this confirms the above calculation for main reinforcement.	
	$Z = d\left\{0.5 + \sqrt{\left(0.25 - \frac{K}{0.9}\right)}\right\}$	
	$= 652\left\{0.5 + \sqrt{\left(0.25 - \frac{0.024}{0.9}\right)}\right\}$	
	$= 634.12\ mm$	
	or $Z = 0.95d = 0.95 \times 652 = 619.4\ mm$	
	$Z = 619.4\ mm$	
	$x = \frac{d - Z}{0.45} = \frac{652 - 619.4}{0.45}$	
	$= 72.44\ mm$	
	ie $< 150mm$	
	∴ N.A. lies within the flange	

CURPE CONSULTANTS
46 Orburn Road
Dunfield

Project			Job ref	
Part of structure CANTILEVER BEAMS & SLABS			Calc sheet no 13/12	rev 1
Drawing ref	Calc by B.C.	Date	Check by	Date

Ref	Calculations	Output
3.4.5	CHECK SHEAR	
	Max shear force $V = 377.1\,kN$	
3.4.5.2	Design shear stress	
eq^n 3	$v = \frac{V}{b_v d} = \frac{377.1 \times 10^3}{450 \times 652}$	
	$= 1.29\ N/mm^2$	$v = 1.29\ N/mm^2$
	$< 5\ N/mm^2$ & $0.8\sqrt{f_{cu}}$ ie acceptable	
	At the support $\frac{100 A_s}{bd}$	
	$= \frac{100 \times 1470}{450 \times 652}$	
	$= 0.50$	
table 3.9	$v_c = 0.50\ N/mm^2$ – may be enhanced by $(f_{cu}/25)^{1/3}$ ie $(35/25)^{1/3} = 1.119$	
	ie $v_c = 0.56\ N/mm^2$	$v_c = 0.56\ N/mm^2$
	$v_c + 0.4 = 0.96\ N/mm^2 < v$	
table 3.8	$v_c + 0.4 < v < 0.8\sqrt{f_{cu}}$	
	$A_{sv} \geq \frac{b_v s_v (v - v_c)}{0.87 f_{yv}}$	
	with 10mm links $A_{sv} = 157\,mm^2$	
	$\therefore s_v \leq \frac{157 \times 0.87 \times 250}{450 \times (1.29 - 0.56)}$	

CURPE CONSULTANTS
46 Orburn Road
Dunfield

Project			Job ref	
Part of structure CANTILEVER BEAMS & SLABS			Calc sheet no 13/13	rev 1
Drawing ref	Calc by B.C.	Date	Check by	Date

Ref	Calculations	Output
	$s_v \leq 103.95\,mm$	
3.4.5.5	max $s_v = 0.75d$	
	$= 0.75 \times 652$	
	$= 489\ mm$	
	$\therefore$ USE R10 links at 100 mm c/c	links R10 @ 100 c/c
3.4.6.	CHECK DEFLECTION	
	As was illustrated in the slab design the deflection must be checked at mid span and at the support. The principles shown previously apply except that the basic span/effective depth ratios from table 3.10 are different values since	
	$\frac{b_w}{b} < 0.3$	
3.4.7	CHECK CRACKING.	
3.12.11.2	Crack control is provided by the distance between the bars in tension	
table 3.30	Maximum clear distance between bars	
	$= 160\,mm$	
	Maximum corner distance $= 80\,mm$	
	At mid span where there are 4T32 bars the actual clear distance	
	$= [450 - (2 \times 32 + 4 \times 32)]\ 1/3$	
	$= 86\,mm$	
	At the support where there are 3T25 bars	

CURPE CONSULTANTS
46 Orburn Road
Dunfield

Project			Job ref	
Part of structure CANTILEVER BEAMS & SLABS			Calc sheet no 13/14	rev 1
Drawing ref	Calc by BC	Date	Check by	Date

Ref	Calculations	Output
	the actual clear distance	
	$= [450 - (2 \times 32 + 3 \times 25)]\,½$	
	$= 155.5$ mm	
	ie both < 160 mm.	
	Also, by inspection the corner distance will be satisfactory	
	∴ SATISFACTORY	Cracking O.K.

14

Reinforced concrete staircases

With the exception of special cases, such as spiral or scissor stairs, the normal types of staircase may be divided into two broad groups, namely those spanning transversely and those spanning longitudinally.

1) Transverse spans

There are three main forms of transverse span staircase as shown in Fig. 14.1:

1) Stairs spanning between supporting walls at each side (Fig. 14.1(a)).
2) Stairs cantilevering from a support at one side only (Fig. 14.1(b)).
3) Stairs cantilevering across a central spine beam (Fig. 14.1(c)).

2) Longitudinal spans

Stairs which span longitudinally span between supports at the top and bottom of the flights and are completely unsupported at the sides. There are again three main forms of construction, as shown in Fig. 14.2:

1) Beams or walls supporting the staircase at the inside edges of the landings (Fig. 14.2(a)).
2) Beams or walls supporting the staircase at the outside edges of the landings (Fig. 14.2(b)).
3) The flight spans on to the landings, which in turn span at right angles to the flight on to supporting walls or beams (Fig. 14.2(c)).

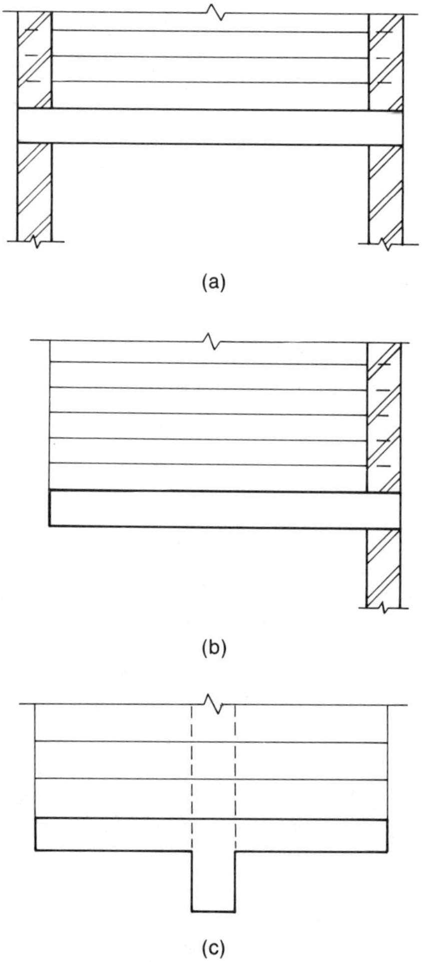

Fig. 14.1

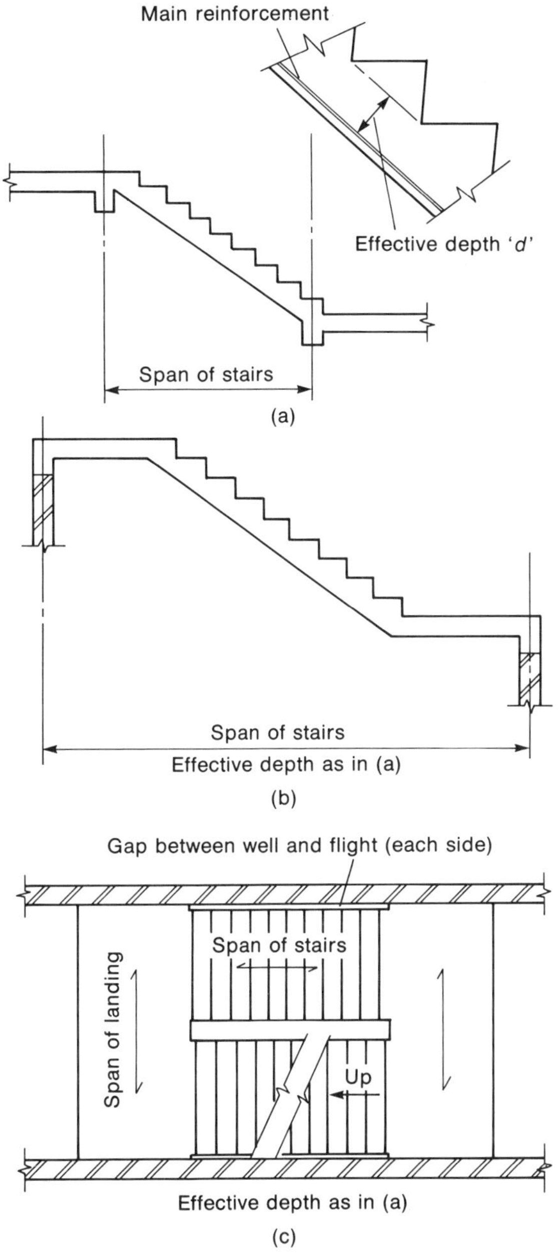

Fig. 14.2

Definitions

Most of the terminology used in staircase design is illustrated in Fig. 14.3. However, the term 'span' requires definition in the case of stair 2(c). In this the landings are supporting members and BS 8110 Clause 3.10.1.3 reads: 'When the staircase is built monolithically at its ends into structural members spanning at right angles to its span, the effective span should be

$$l_a + 0.5(l_{b1} + l_{b2})$$

where l_a is the clear horizontal distance between the supporting members,
l_{b1} is the breadth of the supporting member at one end or 1.8 m, whichever is the smaller,
l_{b2} is the breadth of the supporting member at the other end or 1.8 m, whichever is the smaller.'

The use of this clause is illustrated on page 14/4. The staircase designed below is of the type 2(c), illustrated in Fig. 14.2(c).

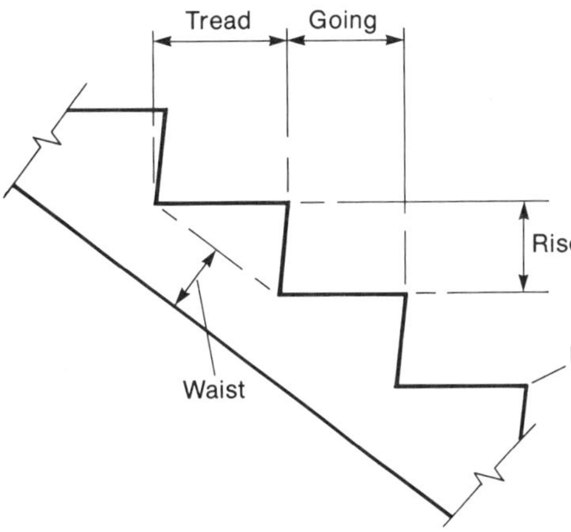

Fig. 14.3

CURPE CONSULTANTS 46 Orburn Road Dunfield	Project	Job ref	
		Sheet no	14/1
	Sub-section STAIRCASE	Made by	BC
		Date	
	Structural Summary Sheet	Checked by	
		Date	

A REINFORCED CONCRETE STAIRCASE

A reinforced concrete staircase is shown below. It is to be designed to satisfy BS 8110 and the following data:

i) the imposed load is 3 kN/m²;

ii) the finishes are - 13 mm plaster on the underside of the flight and landings, 25 mm terrazzo tiles on 13 mm screed on the treads and the landings;

iii) f_{cu} (the characteristic strength of the concrete) = 35 N/mm²;

iv) f_y (the characteristic strength of the reinforcement) = 460 N/mm²;

v) the conditions of exposure are mild.

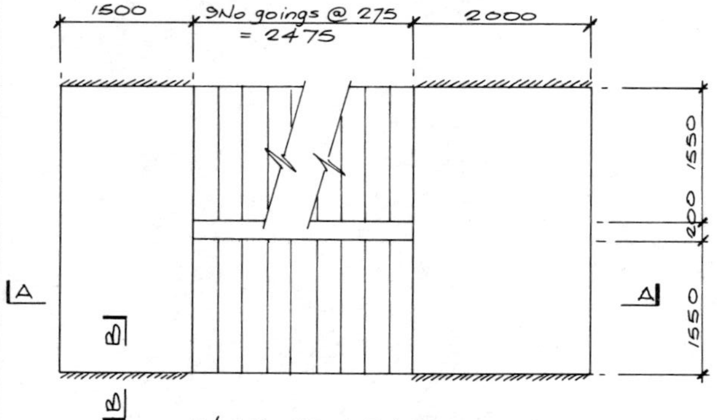

PLAN OF STAIRCASE

Supported edges shown thus - see B-B

CURPE CONSULTANTS 46 Orburn Road Dunfield	Project	Job ref	
		Sheet no	14/2
	Sub-section STAIRCASE	Made by	BC
		Date	
	Structural Summary Sheet	Checked by	
		Date	

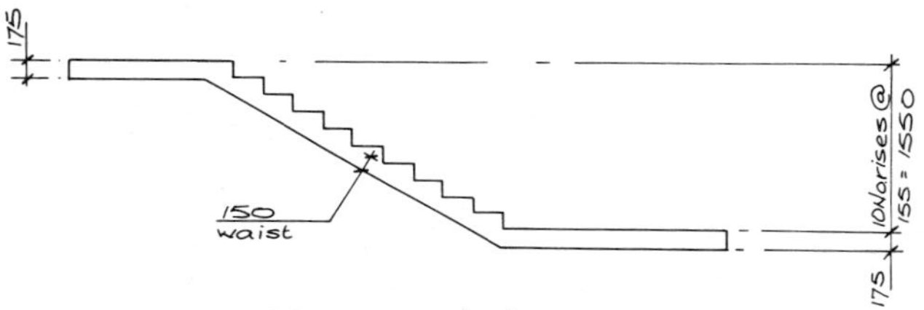

SECTION A-A

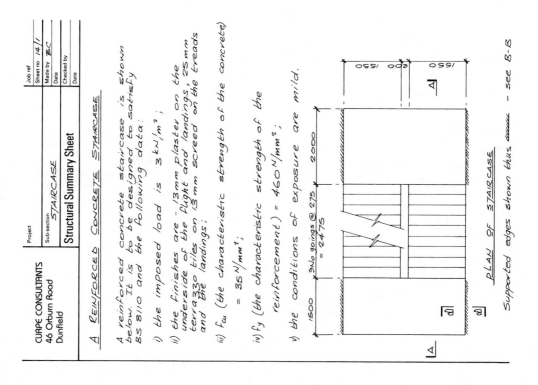

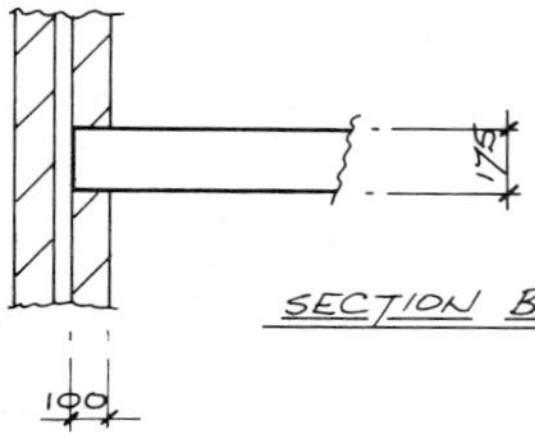

SECTION B-B

CURPE CONSULTANTS 46 Orburn Road Dunfield	Project	Job ref
	Part of structure: STAIRCASE	Calc sheet no: 14/3 rev: /
	Drawing ref · Calc by: B.C. · Date	Check by · Date

Ref	Calculations	Output

LOADING

To find the weight of the waist on plan, the weight on the slope must be increased by the ratio:

$$\frac{\sqrt{(R^2 + G^2)}}{G}$$

where R and G are the Rise and the Going respectively.

In this case $= \frac{\sqrt{(155^2 + 275^2)}}{275}$

$= 1.15$

Also the weight of the steps must be taken into account. A simple method of doing this is to regard the steps as a solid slab of thickness equal to half the rise, and no increase is necessary to obtain the weight on plan

FLIGHT LOADING ON PLAN

BS 648 — Dead load. (kN/m^2)

Waist	$= 0.15 \times 23.6 \times 1.15$	$= 4.07$	
steps	$= 0.155 \times 0.5 \times 23.6$	$= 1.83$	
Terrazzo	$= 54 \times 9.81 \times 10^{-3}$	$= 0.53$	
Screed	$= 30 \times 9.81 \times 10^{-3}$	$= 0.29$	
Plaster	$= 22 \times 9.81 \times 10^{-3} \times 1.15$	$= 0.25$	
	total g_k	$= 6.97$	$g_k = 6.97\ kN/m^2$

CURPE CONSULTANTS 46 Orburn Road Dunfield	Project	Job ref
	Part of structure: STAIRCASE	Calc sheet no: 14/4 rev: /
	Drawing ref · Calc by: B.C. · Date	Check by · Date

Ref	Calculations	Output

Imposed load

$q_k = 3.00\ kN/m^2$ — Output: $q_k = 3.00\ kN/m^2$

table 2.1 — Design load $F = 1.4 G_k + 1.6 Q_k$

or $n = 1.4 g_k + 1.6 q_k$

$= 1.4 \times 6.97 + 1.6 \times 3$

$= 14.56\ kN/m^2$ — Output: $n = 14.56\ kN/m^2$

3.10.1.3 — THE EFFECTIVE SPAN

eq^n 47 — span $= l_a + 0.5(l_{b,1} + l_{b,2})$

$= 2.475 + 0.5(1.5 + 1.8\ max)$

$= 4.125\ m$ — Output: $l = 4.125\ m$

DESIGN OF THE FLIGHT

The structural system for the flight can be equated to:

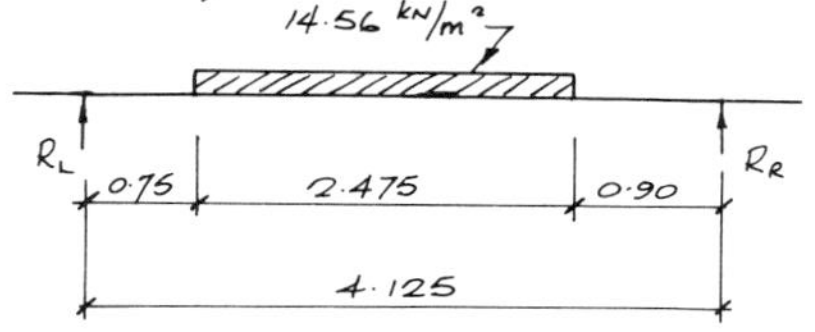

For a 1m width of flight

$$R_L = \frac{14.56 \times 2.475\ (2.475/2 + 0.9)}{4.125}$$

$= 18.67\ kN$

CURPE CONSULTANTS
46 Orburn Road
Dunfield

Project			Job ref	
Part of structure: STAIRCASE			Calc sheet no: 14/5	rev: 1
Drawing ref	Calc by: B.C.	Date	Check by	Date

Ref	Calculations	Output
	$\therefore R_R = 14.56 \times 2.475 - 18.67$	
	$= 17.36\ kN$	
	The position of the maximum bending moment is at the position of zero shear force.	
	ie from $R_L = \dfrac{18.67}{14.56} + 0.75$	
	$= 2.03\ m$	
	<u>Maximum Bending Moment</u>	
	$= 18.67 \times 2.03 - \left(\dfrac{2.03 - 0.75}{2}\right)^2 \times 14.56$	
	$= 25.97\ kN.m$	max. B.M = 25.97kN.m
	<u>EFFECTIVE DEPTH</u>	
table 3.4	cover = 20mm	
	Assuming 12mm ϕ bars	
	$d = 150 - (20 + 12/2)$	
	$= 124mm$	d = 124 mm
	<u>MAIN REINFORCEMENT</u>	
	$\dfrac{M}{bd^2} = \dfrac{25.97 \times 10^6}{10^3 \times 124^2}$	
	$= 1.69\ N/mm^2$	
chart 2	$\dfrac{100A_s}{bd} = 0.45$	

CURPE CONSULTANTS
46 Orburn Road
Dunfield

Project			Job ref	
Part of structure: STAIRCASE			Calc sheet no: 14/6	rev: 1
Drawing ref	Calc by: B.C.	Date	Check by	Date

Ref	Calculations	Output
	$A_s = \dfrac{0.45 \times 10^3 \times 124}{10^2}$	
	$= 558\ mm^2/m$	
	<u>T12 mm ϕ bars at 200 mm c/c (566 mm^2/m)</u>	Main Reinf T12@ 200 c/c
3.10.2	<u>CHECK DEFLECTION</u>	
table 3.10	basic $l/d = 26$ since it is effectively continuous over the supports and top reinforcement will be provided.	
Eq^{n}8	service stress f_s	
	$= \dfrac{5 \times 460 \times 558}{8 \quad\quad 566} \times 1$	
	$= 283.44\ N/mm^2$	
Eq^{n}7	Modification factor for tension reinf.	
	$= 0.55 + \dfrac{477 - 283.44}{120(0.9 + 1.69)}$	
	$= 1.17$	
3.10.2.2	percentage area of flight	
	$= \dfrac{2.475}{4.125} \times 100$	
	$= 60\%$	
	$\therefore$ an additional factor of 15% may be included	
	Permissible $l/d = 26 \times 1.17 \times 1.15$	
	$= 34.98$	

CURPE CONSULTANTS 46 Orburn Road Dunfield	Project		Job ref	
	Part of structure: STAIRCASE		Calc sheet no: 14/7	rev: 1
	Drawing ref	Calc by: B. C. / Date	Check by	Date

Ref	Calculations	Output
	Actual $l/d = \frac{4125}{124}$	
	$= 33.27 < 34.98$	
	ie SATISFACTORY	Deflection O.K
	SECONDARY REINFORCEMENT	
table 3.27	$\frac{100 A_s}{A_c} = 0.13$	
	$A_s = \frac{0.13 \times 10^3 \times 150}{10^2}$	
	$= 195\ mm^2/m$	
	T10mm ϕ bars at 400mm c/c $(196.5\ mm^2/m)$	Distribution T10@400 c/c
	DESIGN OF THE LANDINGS	
	It is assumed that the load from the stairs is carried uniformly over the whole width of the landings.	
	LOADING OF LANDINGS	
BS648	Dead loading — kN/m^2	
	self weight $= 0.175 \times 23.6 = 4.13$	
	terrazzo finish $= 54 \times 9.81 \times 10^{-3} = 0.53$	
	screed $= 30 \times 9.81 \times 10^{-3} = 0.29$	
	plaster $= 22 \times 9.81 \times 10^{-3} = 0.22$	
	total $g_k = 5.17$	$g_k = 5.17\ kN/m^2$

CURPE CONSULTANTS 46 Orburn Road Dunfield	Project		Job ref	
	Part of structure: STAIRCASE		Calc sheet no: 14/8	rev: 1
	Drawing ref	Calc by: B. C. / Date	Check by	Date

Ref	Calculations	Output
	The load from the flight has to be included.	
	Design load for 1.50m landing per 1m of span:	
	$= (1.4 \times 5.17 + 1.6 \times 3.00)1.5 + 18.67$ (ie R_L)	
	$= 36.73\ kN/m$	n for 1.5 landing $= 36.73\ kN/m$
	Design load for 2.0m landing per 1m of span	
	$= (1.4 \times 5.17 + 1.6 \times 3.00)2.0 + 17.36$ (ie R_R)	
	$= 41.44\ kN/m$	n for 2.0 landing $= 41.44\ kN/m$
	EFFECTIVE DEPTH	
table 3.4	cover = 20mm	
	Assuming 12mm ϕ bars	
	$d = 175 - (20 + 12/2)$	
	$= 149\ mm$	$d = 149\ mm$
3.4.1.2	EFFECTIVE SPAN	
	smaller of (i) $(1.55 + 0.2 + 1.55) + 0.1 = 3.40m$	
	or (ii) $(1.55 + 0.2 + 1.55) + 0.149 = 3.449m$	
	$\therefore l = 3.40m$	$l = 3.40m$

CURPE CONSULTANTS 46 Orburn Road Dunfield	Project			Job ref	
	Part of structure STAIRCASE			Calc sheet no 14 / 9	rev 1
	Drawing ref	Calc by B.C.	Date	Check by	Date

Ref	Calculations	Output
	BENDING MOMENT FOR 1·50 LANDING	
	$M_{max} = \frac{36.73 \times 3.40^2}{8}$	
	$= 53.07$ kN.m.	
	$\frac{M}{bd^2} = \frac{53.07 \times 10^6}{1500 \times 149^2}$	
	$= 1.59$ N/mm^2	
	$\frac{100A_s}{bd} = 0.41$	
	$A_s = \frac{0.41 \times 1500 \times 149}{100}$	
	$= 916$ mm^2	
	9T12 mm ϕ bars – 1020 mm^2	Main reinf 9T12 ie T12 @ 175 c/c
	The checks for deflection and cracking can be carried out as previously illustrated but it should be noted that the basic span/eff. depth ratio for the landings = 20.	
	The 2.0m wide landing is designed in a similar manner.	

15

Continuous beams and slabs – one way: worked examples

CURPE CONSULTANTS 46 Orburn Road Dunfield	Project	Job ref
		Sheet no 15/1
	Sub-section CONTINUOUS BEAMS & SLABS – ONE WAY	Made by BC.
		Date
	Structural Summary Sheet	Checked by
		Date

CONTINUOUS BEAMS AND SLABS - ONE WAY

The reinforced concrete floor shown in plan below is required to carry the following loads and to satisfy BS 8110. Details are as follows.

i) Imposed loading on the slab = 3 kN/m^2

ii) Finishes and partitions = 2.50 kN/m^2

iii) f_{cu} (the characteristic strength of the concrete) = 35 N/mm^2

iv) f_y (the characteristic strength of the reinforcement)

= 460 N/mm^2 for the main reinforcement and

= 250 N/mm^2 for the links.

v) The conditions of exposure are mild

vi) There is to be a 2 hour fire resistance

vii) The floor is one in a four storey building

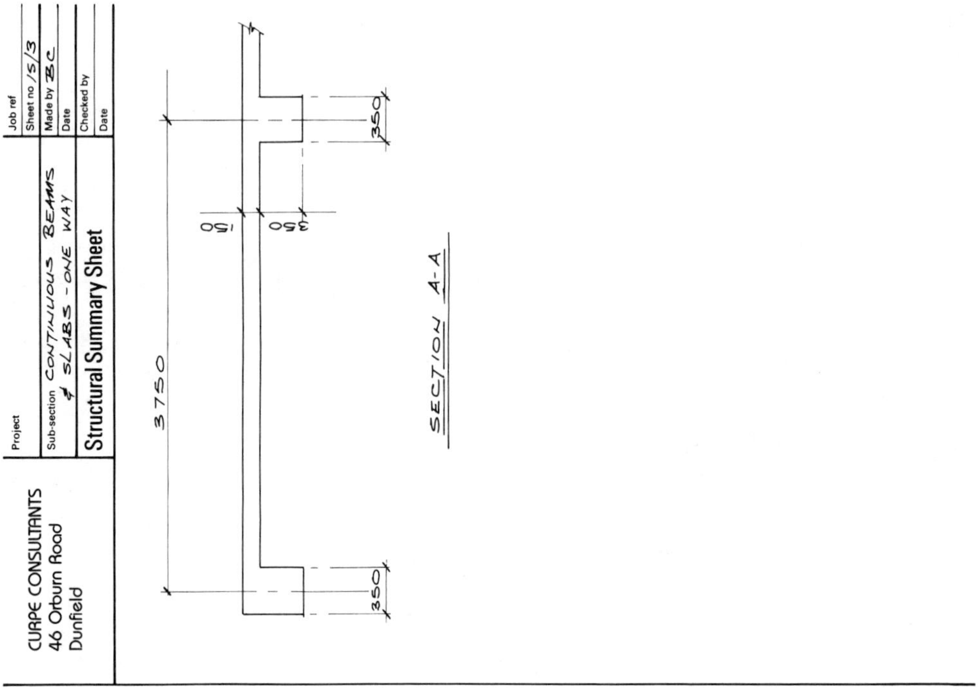
CURPE CONSULTANTS
46 Orburn Road
Dunfield
Project
Sub-section CONTINUOUS BEAMS & SLABS - ONE WAY
Structural Summary Sheet
Job ref
Sheet no 15/3
Made by BC
Date
Checked by
Date
3750
150
350
350
350
SECTION A-A

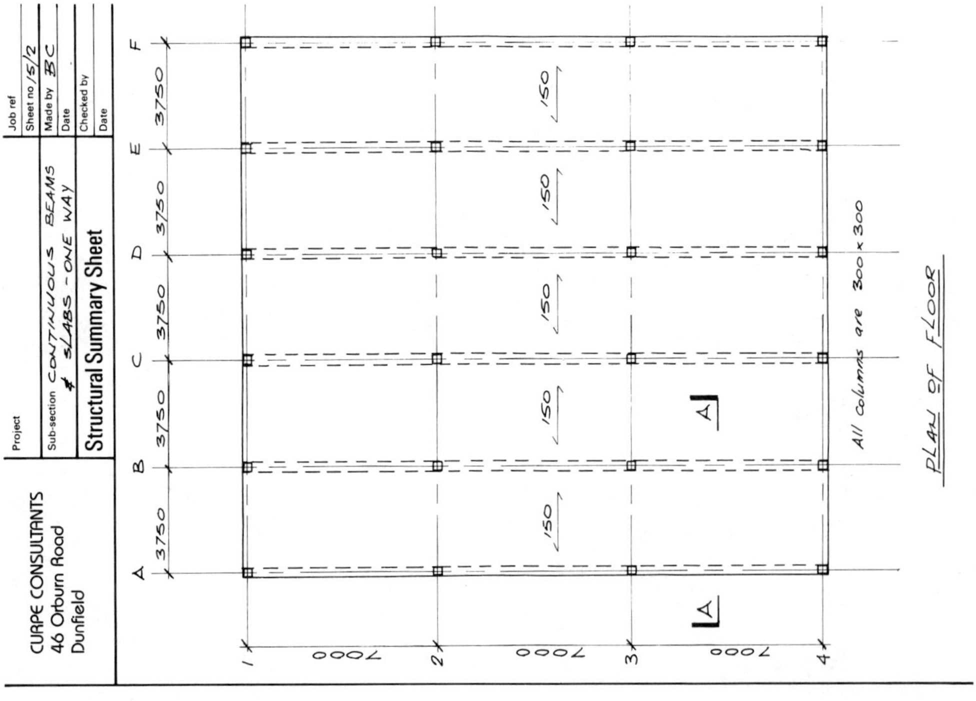
CURPE CONSULTANTS
46 Orburn Road
Dunfield
Project
Sub-section CONTINUOUS BEAMS & SLABS - ONE WAY
Structural Summary Sheet
Job ref
Sheet no 15/2
Made by BC
Date
Checked by
Date
A
B
C
D
E
F
3750
3750
3750
3750
3750
150
150
150
150
150
1
2
3
4
7000
7000
7000
A
A
All columns are 300 x 300
PLAN OF FLOOR

CURPE CONSULTANTS
46 Orburn Road
Dunfield

Project		Job ref
Part of structure: CONTINUOUS BEAMS & SLABS - ONE WAY		Calc sheet no 15/4, rev /
Drawing ref	Calc by B.C. / Date	Check by / Date

Ref	Calculations	Output
	SLAB DESIGN	
	LOADING	
	Dead : Finishes & partitions = $2.50\ kN/m^2$	
	Self wt. = $\frac{0.15 \times 2400 \times 9.81}{10^3}$ = 3.53	
	Total g_k = $6.03\ kN/m^2$	
	Imposed : q_k = $3.00\ kN/m^2$	
	Total Design load 'F' for 1m width	
	= $[(1.4 \times 6.03) + (1.6 \times 3)]\,3.75$	
	= 49.66 kN/m width.	F = 49.66kN
3.5.2.3 3.5.2.4 table 3.13	ULTIMATE BENDING MOMENTS	
	At outer support $M = 0$	
	Near middle of end span $M = 0.086F\ell$	
	$= 0.086 \times 49.66 \times 3.75$	
	= 16.02 kN.m	
	At 1st int. suppt. $M = -0.086F\ell$	
	= − 16.02 kN.m	
	Mid. of int. spans $M = 0.063F\ell$	
	= 11.73 kN.m	
	Int. suppt. $M = -0.063F\ell$	
	= − 11.73 kN.m	

CURPE CONSULTANTS
46 Orburn Road
Dunfield

Project		Job ref
Part of structure: CONTINUOUS BEAMS & SLABS - ONE WAY		Calc sheet no 15/5, rev /
Drawing ref	Calc by B.C. / Date	Check by / Date

Ref	Calculations	Output
	EFFECTIVE DEPTH	
table 3.4.	For mild conditions of exposure, the	
	cover = 20mm	
table 3.5	For 2 hour fire resistance, the cover	
	= 25mm	cover = 25mm
	Assuming 12mm dia. main bars	
	$d = 150 - (25 + 12/2)$	
	= 119 mm	d = 119mm
	MAIN REINFORCEMENT - b = 1.000m	

Position	$\frac{M}{bd^2}$ N/mm^2	$\frac{100A_s}{bd}$	A_s mm^2/m	Bars
Near middle of end span	1.13	0.3	357	T12 @ 300 (377)
1st int. suppt.	1.13	0.3	357	T12 @ 300 (377)
Mid. int. spans	0.83	0.22	261.8	T10 @ 300 (262)
Int. suppt	0.83	0.22	261.8	T10 @ 300 (262)

Ref	Calculations	Output
3.5.7.	CHECK DEFLECTION	
	The worst position for deflection is at the middle of the end span	
table 3.10	basic ℓ/d = 26	

CURPE CONSULTANTS 46 Orburn Road Dunfield	Project			Job ref	
	Part of structure CONTINUOUS BEAMS & SLABS - ONE WAY			Calc sheet no 15/6	rev 1
	Drawing ref	Calc by B.C.	Date	Check by	Date

Ref	Calculations	Output
Eqⁿ 8	service stress f_s	
	$= \frac{5 \times 460}{8} \times \frac{357}{377} \times 1$	
	$= 272.25\ N/mm^2$	
Eqⁿ 7	Modification factor for tension reinf.	
	$= 0.55 + \frac{477 - 272.25}{120(0.9 + 1.13)}$	
	$= 1.39$	
	Permissible $l/d = 26 \times 1.39$	
	$= 36.14$	
	Actual $l/d = 3750/119$	
	$= 31.51$	
	∴ SATISFACTORY	Deflection O.K.
3.5.5	CHECK SHEAR	
	The maximum shear will occur at the 1st interior support	
table 3.13	$V = 0.6F = 0.6 \times 49.66$	
	$= 29.80\ kN$	
eqⁿ 21	$v = \frac{V}{bd} = \frac{29.80 \times 10^3}{10^3 \times 119}$	
	$= 0.25\ N/mm^2$	

CURPE CONSULTANTS 46 Orburn Road Dunfield	Project			Job ref	
	Part of structure CONTINUOUS BEAMS & SLABS - ONE WAY			Calc sheet no 15/7	rev 1
	Drawing ref	Calc by B.C.	Date	Check by	Date

Ref	Calculations	Output
table 3.9	Permissible shear stress is greater than $0.25\ N/mm^2$	
table 3.17	No shear reinforcement is required	Shear O.K
3.5.8	CHECK CRACKING	
	- as page 12/25	
	CHECK MINIMUM AREA OF REINF.	
	This amount will provide the secondary or distribution reinf.	
table 3.27	$\frac{100 A_s}{A_c} = 0.13$	
	for 1m width	
	$A_s = \frac{0.13 \times 10^3 \times 150}{10^2}$	
	$= 195\ mm^2/m$	
	T10 @ 400 mm c/c - $196.5 mm^2/m$	Dist. reinf. T10 @ 400
	This is supplied by the designed main reinf. in all cases.	
3.12.3	TIES	
3.12.3.4	INTERNAL TIES	
3.12.3.4.2	Force to be resisted	
	$= \frac{(g_k + q_k)}{7.5} \cdot \frac{l_r}{5} \cdot F_t$	
	$F_t = (20 + 4n_o) = 20 + 4 \times 4$	
	$= 36\ kN$	

CURPE CONSULTANTS 46 Orburn Road Dunfield	Project	Job ref
	Part of structure: CONTINUOUS BEAMS & SLABS - ONE WAY	Calc sheet no: 15/8 rev: 1
	Drawing ref: — Calc by: B.C. Date: —	Check by: — Date: —

Ref	Calculations	Output
	OR $F_t = 60 > 36$	
	$\therefore F_t = 36$ kN	
3.12.3.4.1	l_r has two values	
	i) In the direction of the span	
	$l_r = 3.750$ m	
	ii) In the transverse direction	
	$l_r = 7.000$ m	
	a) Force in the direction of the span	
	$= \frac{(6.03 + 3)}{7.5} \cdot \frac{3.75}{5} \cdot 36$	
	$= 32.51$ kN $< F_t$	
	$\therefore$ Design for 36 kN	
	Area of reinf. required	
	$= \frac{36 \times 10^3}{460}$	
	$= 78{\cdot}26\ \text{mm}^2/\text{m}$	
	b) Force in the transverse direction	
	$= \frac{(6.03 + 3)}{7.5} \cdot \frac{7}{5} \cdot 36$	
	$= 60.69$ kN $> F_t$	
	$\therefore$ Design for 60.69 kN	

CURPE CONSULTANTS 46 Orburn Road Dunfield	Project	Job ref
	Part of structure: CONTINUOUS BEAMS & SLABS - ONE WAY	Calc sheet no: 15/9 rev: 1
	Drawing ref: — Calc by: B.C. Date: —	Check by: — Date: —

Ref	Calculations	Output
	Area of reinf. required	
	$= \frac{60.69 \times 10^3}{460}$	
	$= 131.92\ \text{mm}^2/\text{m}$	
3.12.3.2	both areas are satisfied by the main reinforcement and the distribution reinforcement designed previously. i.e. the tie reinforcement is NOT additional.	
3.12.3.5	PERIPHERAL TIE	
	$1.0 F_t = 36$ kN	
	Area of reinforcement required to resist this force	
	$= \frac{36 \times 10^3}{460}$	
	$= 78.26\ \text{mm}^2/\text{m}$	
	This is satisfied by the designed reinforcement.	
	TIES SATISFACTORY	Ties O.K.

CURPE CONSULTANTS 46 Orburn Road Dunfield	Project	Job ref
	Part of structure CONTINUOUS BEAMS & SLABS - ONE WAY	Calc sheet no 15/10 rev /
	Drawing ref; Calc by B.C.; Date	Check by; Date

Ref	Calculations	Output
	BEAM DESIGN	
	There are two different beams in the floor. Beams on grid lines A and F are similar, but handed, while beams on grid lines B to E inclusive are similar.	
	1. Beam on grid line A.	
	LOADING - beam carries 3.75/2 = 1.875m width of slab.	
	Imposed loading on beam	
	$= 3.00 \times 1.875 = 5.625$ kN/m	$q_k = 5.625$ kN/m
	Dead loading from slab to beam	
	$= 6.03 \times 1.875 = 11.31$ kN/m	
	Self wt. of beam downstand	
	$= 2400 \times 9.81 \times 10^{-3} \times 0.35 \times 0.35$	
	$= 2.88$ kN/m	
	Total $g_k = 14.19$ kN/m	$g_k = 14.19$ kN/m
	DESIGN LOAD $F = 1.4G_k + 1.6Q_k$	
	$= (1.4 \times 14.19 + 1.6 \times 5.625)7$	
	$= 202.10$ kN	$F = 202.10$kN

CURPE CONSULTANTS 46 Orburn Road Dunfield	Project	Job ref
	Part of structure CONTINUOUS BEAMS & SLABS - ONE WAY	Calc sheet no 15/11 rev /
	Drawing ref; Calc by B.C.; Date	Check by; Date

Ref	Calculations	Output
3.4.3 table 3.6.	ULTIMATE BENDING MOMENTS & SHEAR FORCES	

Position	B. M (kN. m)	S.F. (kN)
At outer suppt	0	$0.45F = 90.95$
Near mid. end span	$0.09Fl = 127.32$	—
At 1st int. suppt.	$-0.11Fl = -155.62$	$0.6F = 121.26$
Mid. int. spans	$0.07Fl = 99.03$	—

Ref	Calculations	Output
	EFFECTIVE DEPTH	
table 3.4	For mild conditions of exposure, the cover = 20mm	
table 3.5	For 2 hour fire resistance, the cover = 30mm	cover = 30mm
	Assuming 25 mm dia. main bars and 10mm links	
	$d = 500 - (30 + 10 + 25/2)$	
	$= 447.5$ mm	$d = 447.5$mm

CURPE CONSULTANTS
46 Orburn Road
Dunfield

Project: | Job ref:
Part of structure: CONTINUOUS BEAMS & SLABS – ONE WAY | Calc sheet no 15/12 rev 1
Drawing ref: | Calc by BC | Date: | Check by: | Date:

Ref	Calculations	Output

3.4.1.5 EFFECTIVE FLANGE WIDTH

lesser of (i) actual flange width $= 3750/2$

$= 1875$ mm

or (ii) $350 + \dfrac{0.7 \times 7000}{10}$

$= 840$ mm

$\therefore\ b = 840$ mm

Output: $b = 840$ mm

MAIN REINFORCEMENT

POSITION	$\frac{M}{bd^2}$ N/mm²	$\frac{100A_s}{bd}$	A_s mm²	BARS
Near middle of end span	0·76	0·2	752	2T25 – (982)
1st int. suppt.	2·22	0·61	955	2T25 – (982)
Mid. int. spans	0·59	0·15	564	2T20 – (628)

The usual checks for :-

i) the position of the neutral axis;

ii) shear;

iii) deflection;

should be carried out as illustrated previously. For example see pages 13/11, 13/12 and 13/13
However when only two bars are provided cracking can give rise to problems.

CURPE CONSULTANTS
46 Orburn Road
Dunfield

Project: | Job ref:
Part of structure: CONTINUOUS BEAMS & SLABS – ONE WAY | Calc sheet no 15/13 rev 1
Drawing ref: | Calc by BC. | Date: | Check by: | Date:

Ref	Calculations	Output

3.4.7 CHECK CRACKING

table 3.30 Maximum clear distance between bars

$= 160$ mm

Actual clear distances :

i) near middle of end span

$= 350 - 2 \times (30 + 10 + 25)$

$= 220$ mm

ii) 1st interior support

as (i) $= 220$ mm

iii) middle of interior spans

$= 350 - 2 \times (30 + 10 + 20)$

$= 230$ mm

All these distances are too large. A third bar must be introduced to satisfy the requirements of cracking.

Near middle of end span

USE – 3T20 mm ϕ bars = 943 mm² — Output: Mid end span 3T20

1st interior support

USE – 2T25 + 1T20 = 1296 mm² — Output: 1st int. suppt. 2T25 + 1T20

Middle of interior spans

USE – 3T16 mm ϕ bars = 603 mm² — Output: Mid. int. span 3T16

Since this check can alter the main reinforcement and so affect other checks it is prudent, in a lightly loaded beam, to check cracking first.

CURPE CONSULTANTS 46 Orburn Road Dunfield	Project			Job ref	
	Part of structure CONTINUOUS BEAMS & SLABS - ONE WAY			Calc sheet no 15/14	rev 1
	Drawing ref	Calc by BC	Date	Check by	Date

Ref	Calculations	Output
	2. Beam on grid line B	
	LOADING - beam carries 3.75 m width of slab	
	Imposed loading on beam	
	$= 3.00 \times 3.75 = 11.25$ kN/m	
	Dead loading from slab to beam	
	$= 6.03 \times 3.75 = 22.61$ kN/m	
	Self weight of beam downstand, from page	
	$= 2.88$ kN/m	
	DESIGN LOAD F $= 1.4 G_k + 1.6 Q_k$	
	$= [1.4(22.61 + 2.88) + 1.6 \times 11.25]\,7$	
	$= 375.80$ kN	F = 375.8 kN
3.4.3 table 3.6.	ULTIMATE BENDING MOMENTS & SHEAR FORCES	

Position	B.M (kN.m)	S.F. (kN)
At outer suppt.	0	0.45F = 169.11
Near mid. end span	0.09Fl = 236.75	—
At 1st int. suppt.	−0.11Fl = −289.37	0.6F = 225.48
Mid int. spans	0.07Fl = 184.14	—

CURPE CONSULTANTS 46 Orburn Road Dunfield	Project			Job ref	
	Part of structure CONTINUOUS BEAMS & SLABS - ONE WAY			Calc sheet no 15/15	rev 1
	Drawing ref	Calc by B.C.	Date	Check by	Date

Ref	Calculations	Output
	EFFECTIVE DEPTH	
	$d = 447.5$ mm as calculated on page	d = 447.5
3.4.1.5	EFFECTIVE FLANGE WIDTH	
	lesser of (i) actual flange width	
	$= 3750$ mm	
	or (ii) $350 + \frac{0.7 \times 7000}{5}$	
	$= 1330$ mm	
	$\therefore b = 1330$ mm	b = 1330 mm

MAIN REINFORCEMENT

POSITION	$\frac{M}{bd^2}$ N/mm²	$\frac{100A_s}{bd}$	A_s mm²	BARS
Near middle of end span	0.89	0.22	1309.39	2T25 + 2T20 (1610)
1st int. suppt.	4.13	1.24	1942.15	4T25 - (1960)
Mid. int. spans	0.69	0.18	1071.3	4T20 - (1260)

The checks for :-

i) the position of the neutral axis;

ii) shear;

iii) deflection;

iv) cracking;

CURPE CONSULTANTS 46 Orburn Road Dunfield	Project			Job ref	
	Part of structure CONTINUOUS BEAMS & SLABS - ONE WAY			Calc sheet no 15/16	rev 1
	Drawing ref	Calc by B.C.	Date	Check by	Date

Ref	Calculations	Output
	v) minimum areas of reinforcement; should be carried out in the manner previously illustrated, which should be now familiar to the reader.	

16

Two-way slab: worked example

CURPE CONSULTANTS 46 Orburn Road Dunfield	Project	Job ref
		Sheet no 16/1
	Sub-section TWO-WAY SLAB	Made by BC
		Date
	Structural Summary Sheet	Checked by
		Date

A REINFORCED CONCRETE SLAB CONTINUOUS IN TWO DIRECTIONS AT RIGHT ANGLES

A reinforced concrete slab which is 175mm thick is supported on a rectangular grid of beams and columns as shown below. The slab is to be designed for mild conditions of exposure and is to satisfy BS 8110. The design data is as follows:

i) imposed load = 5.0 kN/m^2;

ii) dead load, including finishes and partitions but excluding self weight = 2.25 kN/m^2;

iii) f_{cu} (the characteristic strength of the concrete) = 40 N/mm^2;

iv) f_y (the characteristic strength of the reinforcement) = 460 N/mm^2;

v) the floor is one in a seven storey building;

vi) all beams and columns are 350 mm wide.

CURPE CONSULTANTS 46 Orburn Road Dunfield	Project	Job ref
	Sub-section TWO-WAY SLAB	Sheet no 16/2
	Structural Summary Sheet	Made by BC
		Date
		Checked by
		Date

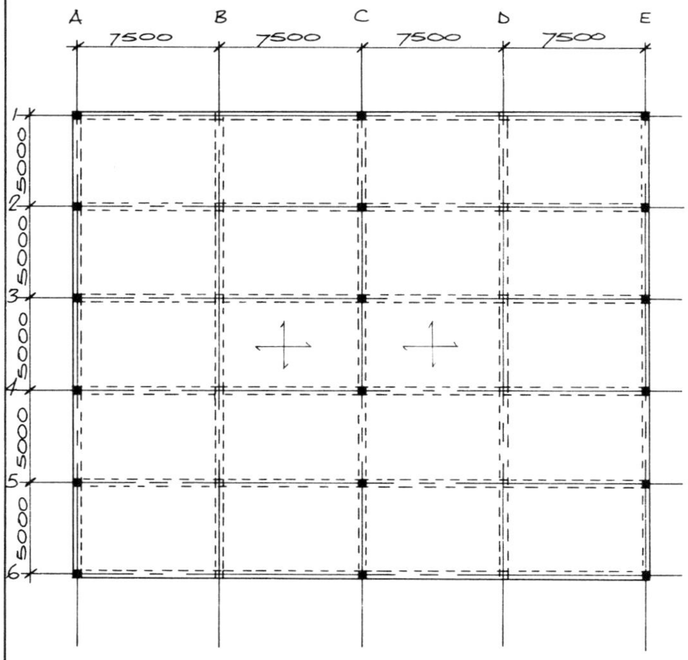

PLAN OF SLAB

CURPE CONSULTANTS 46 Orburn Road Dunfield	Project			Job ref	
	Part of structure TWO WAY SLAB			Calc sheet no 16/3	rev 1
	Drawing ref	Calc by B.C.	Date	Check by	Date

Ref	Calculations	Output
	LOADING	
BS8110	Dead	
	Self weight $= 0.175 \times 2400 \times 9.81 \times 10^{-3} = 4.12$ kN/m^2	
	Finishes and partitions $= 2.25$	
	Total $g_k = 6.37$	
	Imposed $q_k = 5.0$ kN/m^2	
3.5.3.2	Design Load	
	$n = 1.4 g_k + 1.6 q_k$	
	$= 1.4 \times 6.37 + 1.6 \times 5.0$	
	$= 16.92$ kN/m^2	$n = 16.92$ kN/m^2
3.5.3.4 3.5.3.5	BENDING MOMENTS	
	The bending moments are found from equations 14 and 15 which depend upon values of β_{sx} and β_{sy}. These in turn depend upon the span ratio l_y/l_x.	
	It is also required to divide the floor slab into various panels to accord with the description given in table 3.15.	
	Thus the design sequence is as follows:	
Fig. 3.9	1. Divide the floor into panels 2. Divide the panels into middle and edge in each 3. Calculate l_y/l_x	

CURPE CONSULTANTS 46 Orburn Road Dunfield	Project	Job ref
	Part of structure: TWO WAY SLAB	Calc sheet no: 16/4 rev: 1
	Drawing ref / Calc by: BC / Date	Check by / Date

Ref	Calculations	Output
table 3.15	4. Calculate values of β_{sx} and β_{sy} for each panel and each position	
eqn 14 & 15.	5. Calculate bending moments	

LOCATION	Short Span Coeffs.		Short span B.M's. kN.m		Long Span Coeffs.		Long Span B.M's kN.m	
	−ve	+ve	−ve	+ve	−ve	+ve	−ve	+ve
Interior panels	0.053	0.040	22.42	16.92	0.032	0.024	13.54	10.15
One short edge discontinuous	0.058	0.043	24.53	18.19	0.037	0.028	15.65	11.84
One long edge discontinuous	0.073	0.055	30.88	23.27	0.037	0.028	15.65	11.84
Two adjacent edges discontinuous	0.078	0.059	32.99	24.96	0.045	0.034	19.04	14.38

CURPE CONSULTANTS 46 Orburn Road Dunfield	Project	Job ref
	Part of structure: TWO WAY SLAB	Calc sheet no: 16/5 rev: 1
	Drawing ref / Calc by: BC / Date	Check by / Date

Ref	Calculations	Output

EFFECTIVE DEPTH

There are two effective depths – one for each span direction. The majority of the bending moment will be taken by the shorter, i.e the stiffer, span. Thus this should have the larger effective depth.

table 3.4 — cover, for mild conditions of exposure, = 20mm.

Assuming 12 mm dia. bars

d(short span) = 175 − 20 − 12/2

= 149 mm — **Output:** d(short) = 149mm

d(long span) = 175 − 20 − 12 − 12/2

= 137 mm — **Output:** d(long) = 137mm

MAIN REINFORCEMENT

1. Middle Strips.

The main reinforcement for the middle strips is calculated in the normal way, using chart 2. This is shown in tabular form on the following page

CURPE CONSULTANTS
46 Orburn Road
Dunfield

Project		Job ref
Part of structure: TWO WAY SLAB		Calc sheet no 16/6 rev 1
Drawing ref	Calc by BC / Date	Check by / Date

Ref	Calculations	Output

LOCATION	B.M (kN.m)	M/bd^2 (N/mm^2)	$100A_s/bd$	A_s (mm^2/m)	BARS
INTERIOR PANEL					
S.S. −ve	22·42	1·01	0·27	402·3	T12 @ 250 c/c
S.S +ve	16·92	0·76	0·20	298·0	T12 @ 350 c/c
L.S −ve	13·54	0·72	0·19	260·3	T12 @ 350 c/c
L.S +ve	10·15	0·54	0·14	191·8	T10 @ 400 c/c
ONE SHORT EDGE DISCONT.					
SS −ve	24·53	1·10	0·29	432·1	T12 @ 250 c/c
SS +ve	18·19	0·82	0·22	327·8	T12 @ 300 c/c
LS −ve	15·65	0·83	0·22	301·4	T10 @ 250 c/c
LS +ve	11·84	0·63	0·17	232·9	T10 @ 300 c/c
ONE LONG EDGE DISCONT.					
SS −ve	30·88	1·39	0·37	551·3	T12 @ 200 c/c
SS +ve	23·27	1·05	0·28	417·2	T12 @ 250 c/c
LS −ve	15·65	0·83	0·22	301·4	T10 @ 250 c/c
LS +ve	11·84	0·63	0·17	232·9	T10 @ 300 c/c
TWO ADJACENT EDGES DISCONT.					
SS −ve	32·99	1·49	0·40	596	T12 @ 175 c/c
SS +ve	24·96	1·12	0·29	432·1	T12 @ 250 c/c
LS −ve	19·04	1·01	0·27	369·9	T12 @ 300 c/c
LS +ve	14·38	0·77	0·20	274	T10 @ 250 c/c

Note:
i) SS denotes short span
ii) LS denotes long span
iii) the bars chosen may be adjusted later in the design or to suit ease of detailing.

CURPE CONSULTANTS
46 Orburn Road
Dunfield

Project		Job ref
Part of structure: TWO WAY SLAB		Calc sheet no 16/7 rev 1
Drawing ref	Calc by BC / Date	Check by / Date

Ref	Calculations	Output
3.5.3.5 (4)	2. Edge strips	
table 3.27	minimum $\frac{100A_s}{A_c} = 0{\cdot}13$	
	$A_s = \frac{0{\cdot}13 \times 1000 \times 175}{100}$	
	$= 227{\cdot}5\ mm^2/m$	
	T10 mm ϕ bars at 300 mm c/c ($262\ mm^2/m$)	T10 @ 300 c/c
3.5.3.5 (5)	3. Torsion reinforcement	
	This is required where the slab is simply supported on two edges meeting at the corner ie A1, A6, E1 and E6.	
	The maximum mid span moment occurs in the S.S. +ve moment in the panel where two adjacent edges are discontinuous. From table on page 16/6 area = $432{\cdot}1\ mm^2/m$	
	Area of torsional reinforcement	
	$= 3/4 \times 432{\cdot}1$	
	$= 324{\cdot}08\ mm^2/m$	
	T12 mm ϕ bars at 300 mm c/c ($377\ mm^2/m$)	Outer corners T12 @ 300 c/c
(6)	This is required at outer edges eg B1, C1, D1, 2A, 3A, 4A, 5A etc.	
	$1/2 \times 324{\cdot}08 = 162{\cdot}04\ mm^2/m$	
	T10 mm ϕ bars at 400 mm c/c ($196{\cdot}5\ mm^2/m$)	edges T10 @ 400 c/c

CURPE CONSULTANTS
46 Orburn Road
Dunfield

Project			Job ref	
Part of structure: TWO WAY SLAB			Calc sheet no: 16/8	rev: 1
Drawing ref	Calc by: B. C.	Date	Check by	Date

Ref	Calculations	Output
3.5.7	CHECK DEFLECTION	
	The maximum +ve moment is the short span direction in the panel where two adjacent edges are discontinuous	
3.4.6.3	basic $l/d = 26$	
eqn 8	$f_s = \frac{5 \times 460}{8} \times \frac{432.1}{452} \times 1$	
	$= 274.84\ N/mm^2$	
eqn 7	Modification factor for tension reinforcement	
	$= 0.55 + \frac{(477 - 274.84)}{120(0.9 + 1.12)}$	
	$= 1.38$	
	Permissible $l/d = 26 \times 1.38$	
	$= 35.88$	
	Actual $l/d = \frac{5000}{149} = 33.56$	
	< 35.88	
	∴ SATISFACTORY	Deflection O.K.
3.12.11.2.7	CHECK CRACKING	
	maximum clear distance $= 3d$	
	ie either $3 \times 149 = 447\ mm$	
	or $3 \times 137 = 411\ mm$	
	Satisfied in every case.	
	∴ SATISFACTORY	Cracking O.K.

CURPE CONSULTANTS
46 Orburn Road
Dunfield

Project			Job ref	
Part of structure: TWO WAY SLAB			Calc sheet no: 16/9	rev: 1
Drawing ref	Calc by: B. C.	Date	Check by	Date

Ref	Calculations	Output
3.12.3	CHECK STABILITY - TIES	
3.12.3.4	1) Internal ties.	
	F_t is lesser of $(20 + 4n_s)$ or 60	
	ie $(20 + 4 \times 7) = 48$ or 60	$F_t = 48\ kN/m$
	a) For the 5.0m span, the tie force	
	$= \frac{(g_k + q_k)}{7.5} \cdot \frac{l_r}{5} \cdot F_t$	
	$= \frac{(6.37 + 5.0)}{7.5} \cdot \frac{5}{5} \cdot 48$	
	$= 72.77\ kN/m > F_t$	
	∴ Design for $72.77\ kN/m$	
	Area of reinforcement required	
	$= \frac{72.77 \times 10^3}{460}$	
	$= 158.19\ mm^2/m$	
	This is satisfied in all short span cases by the designed reinforcement.	
	b) For the 7.50m span, the tie force	
	$= \frac{(6.37 + 5.00)}{7.5} \cdot \frac{15.0}{5} \cdot 48$	
	$= 218.30\ kN/m > F_t$	
	∴ Design for $218.30\ kN/m$	

CURPE CONSULTANTS 46 Orburn Road Dunfield	Project			Job ref	
	Part of structure TWO WAY SLAB			Calc sheet no 16/10	rev 1
	Drawing ref	Calc by B. C.	Date	Check by	Date

Ref	Calculations	Output
	Area of reinforcement required	
	$= \frac{218.30 \times 10^3}{460}$	
	$= 474.57\ mm^2/m$	
	This is NOT satisfied in any long span case by the designed reinforcement.	
	All long span +ve reinforcement must be increased to satisfy the tie requirement.	All L.S +ve
	ie T12mm ϕ bars at 200mm c/c	T12 @ 200 c/c
	$- 566 mm^2/m$	
3.12.3.5	2) Peripheral ties	
	- see page 15/9	
	SECONDARY REINFORCEMENT	
	-as edge strip reinforcement on page 16/7 - 179	Distribution
	T10 mm bars at 300 mm c/c	T10@ 300 c/c

17

Pad foundation: worked examples

CURPE CONSULTANTS 46 Orburn Road Dunfield	Project	Job ref
		Sheet no 17/1
	Sub-section PAD FOUNDATION	Made by BC
		Date
	Structural Summary Sheet	Checked by
		Date

A PAD FOUNDATION SUBJECTED TO AXIAL LOAD ONLY

A reinforced concrete pad foundation is required to satisfy BS 8110 and the following:

i) axial imposed load = 500 kN ;
ii) axial dead load = 1000 kN ;
iii) the loads are transferred to the base through a centrally positioned column which is 400 mm square ;
iv) the pad is square
v) the safe ground bearing pressure = 150 kN/m^2
vi) f_{cu} (the characteristic strength of the concrete) = 35 N/mm^2
vii) f_y (the characteristic strength of the reinforcement) = 460 N/mm^2

CURPE CONSULTANTS
46 Orburn Road
Dunfield

Project			Job ref	
Part of structure: PAD FOUNDATION			Calc sheet no: 17/2	rev: /
Drawing ref	Calc by: BC	Date	Check by	Date

Ref | Calculations | Output

LOADING

Service loading 'N' $= G_k + Q_k$

$= 1000 + 500$

$= 1500$ kN — Output: **N = 1500 kN**

Design loading 'F' $= 1.4G_k + 1.6Q_k$

$= 1.4 \times 1000 + 1.6 \times 500$

$= 2200$ kN — Output: **F = 2200 kN**

CALCULATE THE PAD SIZE

Since the ground has to withstand the base itself an allowance must be made for self weight. Where soil has to be removed to construct the base, the only allowance to be made is for the difference between the weight of the soil removed and the weight of the concrete imposed. This can be done by subtracting the densities of the two materials.

Density of concrete = 2400 kg/m^3
Density of soil, say = 1700 "

Difference = 700 "

This is approximately equivalent to 7 kN/m^3

For a base say 700 thick, the difference $= 7 \times 0.7 = 4.9\ kN/m^2$

The loading from the ground floor slab will also be supported on the ground. Thus an allowance must be made for this loading - say 5.0 kN/m^2

Total $= 4.9 + 5.0 \simeq 10.0\ kN/m^2$

CURPE CONSULTANTS
46 Orburn Road
Dunfield

Project			Job ref	
Part of structure: PAD FOUNDATION			Calc sheet no: 17/3	rev: /
Drawing ref	Calc by: BC	Date	Check by	Date

Ref | Calculations | Output

The safe ground bearing pressure is reduced by this amount.

ie $150 - 10 = 140\ kN/m^2$

Area required $= \text{Load}/\text{SGBP}$

$= 1500/140$

$= 10.71\ m^2$

For a square pad try:

3.40 m × 3.40 m × 0.70 m — Output: **Pad size 3.4 × 3.4 × 0.7**

DESIGN PRESSURE

$= \text{design load}/\text{area}$

$= 2200/3.4^2$

$= 190.31\ kN/m^2$ — Output: **Design press 190.31 kN/m^2**

EFFECTIVE DEPTH

Since the base is square the bending moments can act in either direction, thus the effective depth should be taken as the average of the two directions.

table 3.4. — cover, for moderate exposure (table 3.2)

$= 35$mm

3.3.1.4. — minimum cover for concrete cast against blinding $= 40$mm

CURPE CONSULTANTS 46 Orburn Road Dunfield	Project			Job ref	
	Part of structure PAD FOUNDATION			Calc sheet no 17/4	rev 1
	Drawing ref	Calc by BC	Date	Check by	Date

Ref	Calculations	Output
	Assuming 25 mm dia. bars	
	effective depth	
	$d = 700 - (40+25)$	
	$= 635mm$	$d = 635mm$
	DESIGN FOR BENDING	
3.11.2.2.	The critical section for bending is at the face of the column.	
	The moment is caused by the upward reactive pressure from the ground producing a cantilever moment about the face of the column. This pressure is assumed to be uniformly distributed under the base	

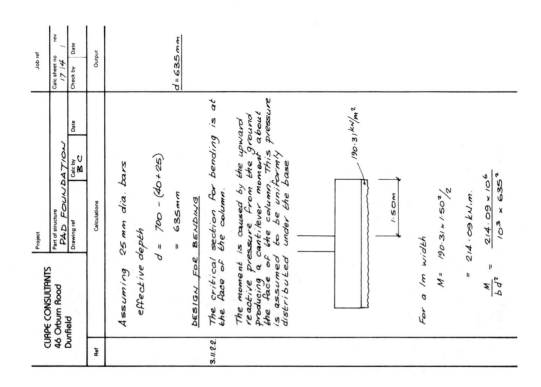

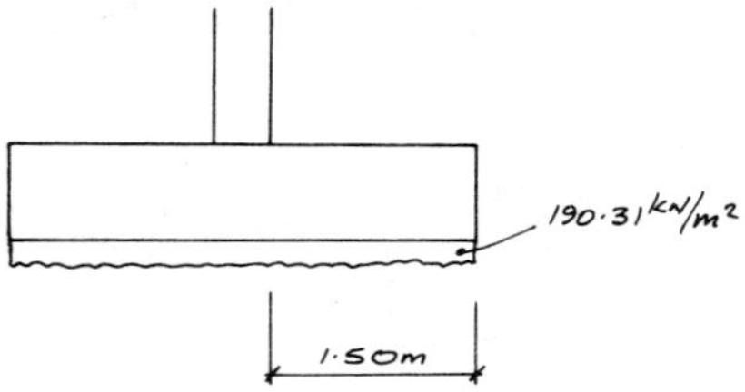

Ref	Calculations	Output
	For a 1m width	
	$M = 190{\cdot}31 \times 1{\cdot}50^2/2$	
	$= 214{\cdot}09 \, kN.m.$	
	$\dfrac{M}{bd^2} = \dfrac{214{\cdot}09 \times 10^6}{10^3 \times 635^2}$	

CURPE CONSULTANTS 46 Orburn Road Dunfield	Project			Job ref	
	Part of structure PAD FOUNDATION			Calc sheet no 17/5	rev 1
	Drawing ref	Calc by BC	Date	Check by	Date

Ref	Calculations	Output
	$= 0{\cdot}53$	
chart 2	$\dfrac{100A_s}{bd} = 0{\cdot}15$	
	$A_s = \dfrac{0{\cdot}15 \times 10^3 \times 635}{10^2}$	
	$= 952{\cdot}5 \, mm^2/m$	
	T20 mm ϕ bars at 300 mm c/c (1050 mm²/m)	T20@300c/c
3.11.3.4	CHECK SHEAR	
	There are two conditions of shear which must be checked.	
	i) shear along a vertical section	
	ii) punching shear	
	i) shear along a vertical section	
	The design ultimate shear stress is obtained from table 3.9	
	$\dfrac{100A_s}{bd} = \dfrac{100 \times 1050}{10^3 \times 635}$	
	$= 0{\cdot}165$	
table 3.9	v_c, by linear interpolation, $= 0{\cdot}35 \, N/mm^2$	
	enhancement factor $= (f_{cu}/25)^{1/3} = (35/25)^{1/3}$	
	$= 1{\cdot}12$	
	$\therefore \; v_c = 1{\cdot}12 \times 0{\cdot}35 = 0{\cdot}39 \, N/mm^2$	$v_c = 0{\cdot}39 \, N/mm^2$
	Since the column face is looked upon as being a support, the shear strength can be enhanced.	

CURPE CONSULTANTS
46 Orburn Road
Dunfield

Project				Job ref	
Part of structure: PAD FOUNDATION				Calc sheet no: 17/6	rev: 1
Drawing ref	Calc by: BC	Date		Check by	Date

Ref	Calculations	Output
3.4.5.8	The enhancement factor = $2d/a_v$	
	where a_v is the distance from the column face to the section under consideration	
	At $a_v = d$, enhancement factor = 2.0 ie v_c = 0.78	
	at $a_v = 1\tfrac{1}{2}d$, enhancement factor = 1.33 ie v_c = 0.52	
	at $a_v = 2d$, enhancement factor = 1.0 ie v_c = 0.39	
eq^n 21	$v = \dfrac{V}{bd}$ ie $V = v.b.d$	
	Allowable shear force across the base	
	At 2d = $0.39 \times 3400 \times 635 \times 10^{-3}$ = 842.01 kN	
	at $1\tfrac{1}{2}$d = $0.52 \times 3400 \times 635 \times 10^{-3}$ = 1122.68 kN	
	at d = $0.78 \times 3400 \times 635 \times 10^{-3}$ = 1684.02 kN	

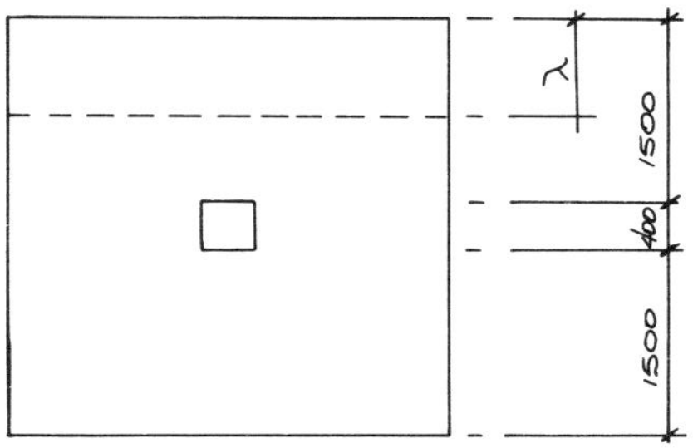

PLAN OF BASE

CURPE CONSULTANTS
46 Orburn Road
Dunfield

Project				Job ref	
Part of structure: PAD FOUNDATION				Calc sheet no: 17/7	rev: 1
Drawing ref	Calc by: BC	Date		Check by	Date

Ref	Calculations	Output
	Actual shear force	
	= design pressure × area outside the perimeter	
	= $190.31 \times 3400 \times \lambda \times 10^{-3}$	
	= $647.05\,\lambda$ kN	
	At 'd' from column λ = 0.865 m	
	at '$1\tfrac{1}{2}$d' from column λ = 0.548 m	
	at '2d' from column λ = 0.230 m	
	Actual shear forces:	
	At 'd' from column = 647.05×0.865 = 559.70 kN	
	at '$1\tfrac{1}{2}$d' from column = 647.05×0.548 = 354.59 kN	
	at '2d' from column = 647.05×0.230 = 148.82 kN	
	Since these are all less than the respective allowable values calculated on page 17/6 shear along a vertical section is	
	SATISFACTORY	vertical shear OK
3.7.7	ii) Punching shear	
eq^n 27	$v_{max} = \dfrac{V}{u_o\, d}$	
3.7.1.1.	where u_o is the effective length of the perimeter which touches a loaded area i.e. the column sides	
	= 4 × 400 = 1600 mm	

CURPE CONSULTANTS 46 Orburn Road Dunfield	Project	Job ref
	Part of structure: PAD FOUNDATION	Calc sheet no: 17/8 rev: 1
	Drawing ref: — Calc by: BC Date: —	Check by: — Date: —

Ref	Calculations	Output
	$\therefore v_{max} = \dfrac{2200 \times 10^3}{1600 \times 635}$	
	$= 2.17\ N/mm^2$	
	$< 0.8\sqrt{f_{cu}}$ or $5\ N/mm^2$	
3.7.7.6.	The first perimeter to be checked is 1.5d from the loaded area. ie $1.5 \times 635 = 952$ mm	
	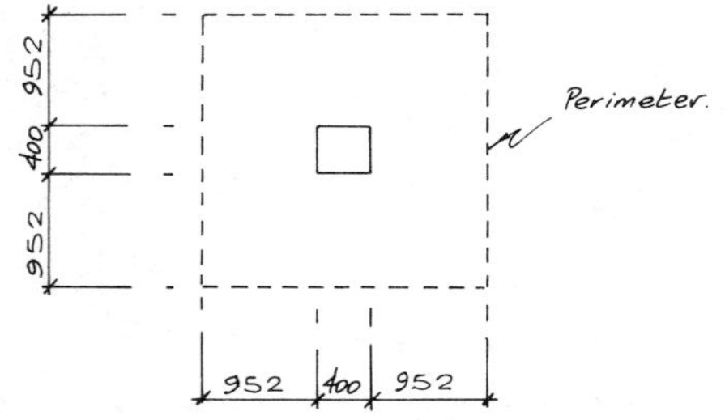	
	length of perimeter	
	$= 4(0.952 + 0.4 + 0.952)$	
	$= 9.216$ m.	
	Area within the perimeter	
	$= (0.4 + 2 \times 0.952)^2$	
	$= 5.308\ m^2$	
	Area outside the perimeter	
	$= 3.4^2 - 5.308$	

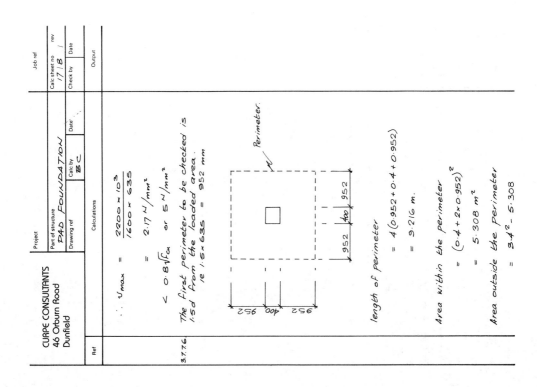

CURPE CONSULTANTS 46 Orburn Road Dunfield	Project	Job ref
	Part of structure: PAD FOUNDATION	Calc sheet no: 17/9 rev: 1
	Drawing ref: — Calc by: BC Date: —	Check by: — Date: —

Ref	Calculations	Output
	$= 6.25\ m^2$	
	Shear force on the perimeter	
	= design pressure × area outside	
	$= 190.31 \times 6.25$	
	$= 1189.44$ kN	
3.7.7.3. eqn 28	Punching shear stress	
	$= \dfrac{V}{u.d}$	
	$= \dfrac{1189.44 \times 10^3}{9216 \times 635}$	
	$= 0.20\ N/mm^2$	
	$< v_c$ of $0.39\ N/mm^2$	
	ie punching shear is SATISFACTORY	Punching shear O.K.
	NO SHEAR REINF. IS REQUIRED	
3.12.11.2.7	CHECK CRACKING	
	Actual clear space $= 300 - 20$ $= 280$ mm	
	Permissible clear space ≯ 750mm or	
	3d ie $3 \times 635 = 1905$mm	
	Since the reinforcement percentage is 0.165 ie < 0.3%, no further check is required.	
	∴ SATISFACTORY	Cracking O.K.

CURPE CONSULTANTS
46 Orburn Road
Dunfield

Project			Job ref	
Part of structure: PAD FOUNDATION			Calc sheet no: 17 / 10	rev: 1
Drawing ref	Calc by: BC	Date	Check by	Date

Ref	Calculations	Output
3.11.3.2	DISTRIBUTION OF REINFORCEMENT	
	$l_c = 0.5 \times 3.40 = 1.70m$	
	$\frac{1}{4}(3c + 9d) = \frac{1}{4}(3 \times 0.40 + 9 \times 0.635)$	
	$= 1.73m$	
	Since $l_c < \frac{1}{4}(3c + 9d)$ the reinforcement should be uniformly distributed over the width of the base.	

18

Reinforced concrete columns

Definition: A reinforced concrete column may be defined as a vertical load-bearing member whose greater overall cross-sectional dimension does not exceed four times its smaller dimension.

Braced and unbraced columns (3.8.1.5)

It is important that the difference between braced and unbraced columns is appreciated, since if an unbraced column is designed as a braced column serious consequences and even failure could result. This is illustrated as follows:

The basic formula for short-column design is that of Euler, which states that:

$$\text{The load } P = \frac{\pi^2 EI}{l^2}$$

where E = Young's modulus of elasticity;
I = the second moment of area;
l = the effective height.

For a given column, E, I and π are constant. However, l for a braced column, which is properly restrained in direction at both ends, is equal to $0.75l_0$ (Table 3.21, BS 8110). For a similar unbraced column, the value given in Table 3.22 of BS 8110 is $1.2l_0$. Since the Euler formula has l^2, the difference in load-bearing ability of an unbraced column designed as a braced column is the difference of 0.75^2 and 1.2^2, i.e. 0.56 and 1.44, i.e. 2.57. Thus the column will have about 0.39 of the strength it should have and collapse is likely.

A *braced column* is one which is not required to resist lateral forces. These forces are resisted by shear walls, cores, bracing or buttressing.

An *unbraced column* is one which is required to resist lateral forces. It may be part of a frame which is free to deflect laterally under load and resistance is supplied by the frame action of the structure.

Effective height (3.8.1.6)

The effective height or effective length of a column is different from the clear distance between end restraints. This difference depends on the restraints of the column ends. The sketches in Fig. 18.1 show the theoretical values rather than the real values. For example, it is almost impossible to achieve either a pure pin joint or a fully fixed joint in practice. BS 8110, in Tables 3.21

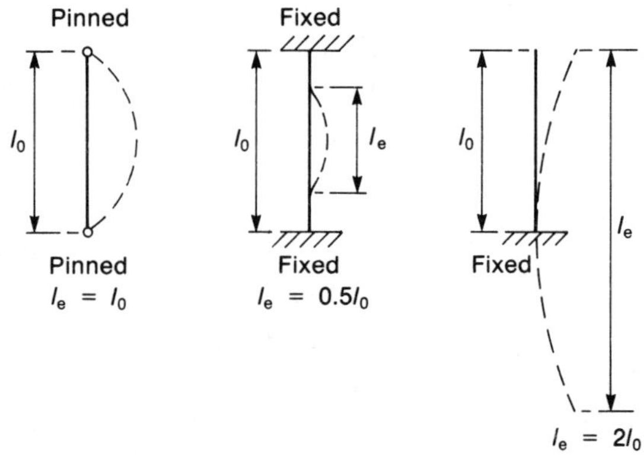

Fig. 18.1

and 3.22, gives values for the relationship between l_0 and l_e depending on the end conditions.

Short or slender columns (3.8.1.3)

The difference between a short and a slender column is based on the slenderness ratio of the column. This may be defined as the effective height of the member divided by the breadth (Fig. 18.2). In practice it is necessary to consider *two* slenderness ratios, namely those which correspond to the principal axis of cross-section, i.e.

$$\frac{l_{ex}}{h} \text{ and } \frac{l_{ey}}{b}$$

A column is said to be short when the ratios $\frac{l_{ex}}{h}$ and $\frac{l_{ey}}{b}$ are *both* less than 15 for a braced column, and less than 10 for an unbraced column.

Short columns

There are two methods given in BS 8110 for designing short columns. These are:

1) by design charts;
2) by design formula.

1) By design charts

This will be the most common method of design for short columns. However, the charts deal only with symmetrically reinforced columns, (normally of rectangular section).

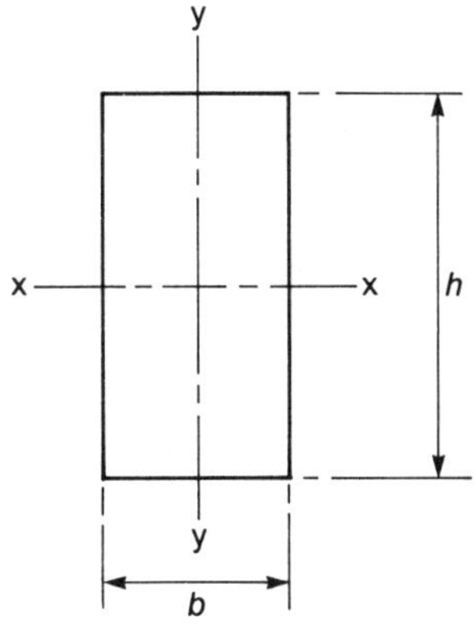

Fig. 18.2

In order to select the appropriate chart, f_{cu}, f_y and d/h must be known. Since the d/h ratio is given in increments of 0.05 only there will be occasions when the exact value required is unavailable. In this case, where a required d/h lies between two charts, both may be read and the exact value for $100A_{sc}/bh$ found by linear interpolation. However, a conservative answer may be found by using the chart below the exact value necessary, e.g. if $d/h = 0.87$ use the chart where $d/h = 0.85$.

A column according to 3.8.2.4, BS 8110 should not be designed for a moment less than the ultimate axial load and at a minimum eccentricity equal to 0.05 times the overall dimension of the column in the plane of bending considered, but not more than 20 mm.

2) By design formula

Consider a column having a cross-sectional area of concrete A_c with a total area of symmetrically placed reinforcement A_{sc} (Fig. 18.3). BS 8110 gives the design stress of concrete in compression as $0.67f_{cu}/\gamma_m$ and with $\gamma_m = 1.5$ for concrete, this stress value becomes $0.45f_{cu}$.

The design stress for the reinforcement in compression is f_y/γ_m, and with $\gamma_m = 1.15$, for reinforcement, this stress value becomes $0.87f_y$. The ultimate axial load N which can be carried by the column is the load carried by the concrete plus the load carried by the reinforcement, i.e.

$$N = 0.45f_{cu}A_c + 0.87f_yA_{sc}$$

However, to cater for constructional inaccuracies BS 8110 has reduced the stresses above by about 10%, giving

$$N = 0.4f_{cu}A_c + 0.75A_{sc}f_y \quad \text{[eqn. 38, BS 8110]}$$

It should be noted that such restrictions as are necessary to allow this formula to be used are rarely found and could only be expected where exceptional control and supervision are available.

Where such accuracy is not obtainable, but the following two parameters are achieved,

1) the beams are designed for uniformly distributed imposed loads,

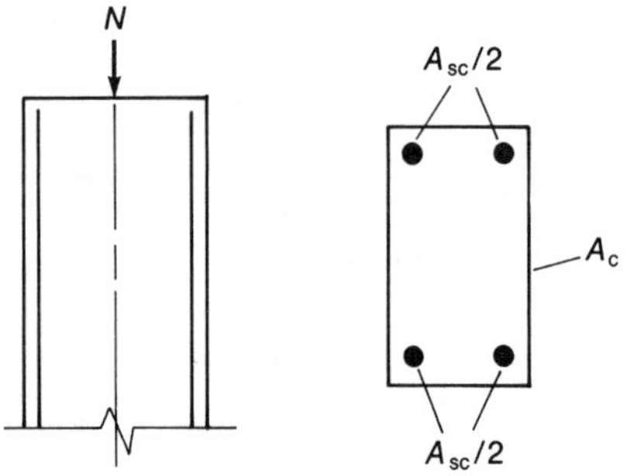

Fig. 18.3

2) the beam spans do not differ by more than 15% of the longer,

then, by reducing the stresses in equation 38 by a further 10% a fair approximation for N can be made. Thus:

$$N = 0.35f_{cu}A_c + 0.67A_{sc}f_y \quad \text{[eqn. 39, BS 8110]}$$

Eccentrically loaded columns

Formulae allowing both N and M to be calculated can be derived for eccentrically loaded columns. However, in a design using such formulae an answer is obtained by a process of trial and error which is usually very tedious. For these eccentrically loaded columns the design charts make the calculations much simpler.

Design examples for short columns

Example 1

Calculate the ultimate axial load of a short, braced 400 mm square reinforced concrete column where the only eccentricity is due to constructional inaccuracies. The longitudinal reinforcement consists of 6 no. 32 mm diameter bars. Also,

$$f_{cu} = 35 \text{ N/mm}^2 \text{ and } f_y = 460 \text{ N/mm}^2$$

Using equation [38, BS 8110],

$$N = 0.4f_{cu}A_c + 0.75A_{sc}f_y$$

A_c can be taken as the nominal cross-sectional area of the column.

$$N = [(0.4 \times 35 \times 400^2) + (0.75 \times 4830 \times 460)] \times 10^{-3}$$

$$= 3906.35 \text{ kN}$$

Example 2

Calculate the ultimate axial load for the column in Example 1, if it were an internal column in a multi-storey, equal bay, structure.

Using equation [39, BS 8110]

$$N = 0.35f_{cu}A_c + 0.67A_{sc}f_y$$

$$= [(0.35 \times 35 \times 400^2) + (0.67 \times 4830 \times 460)] \times 10^{-3}$$

$$= 3448.61 \text{ kN}$$

Example 3

Design a square, short, braced reinforced concrete column to support an ultimate axial load of 2500 kN, if $f_{cu} = 35 \text{ N/mm}^2$ and $f_y = 460 \text{ N/mm}^2$. It can be assumed that control and supervision are excellent.

Using equation [38, BS 8110]

$$N = 0.4f_{cu}A_c + 0.75A_{sc}f_y$$

$$2500 \times 10^3 = (0.4 \times 35 \times A_c) + (0.75 \times A_{sc} \times 460)$$

There are two unknowns, namely A_c and A_{sc}. However, BS 8110 gives minimum and maximum areas of reinforcement. Table 3.27 gives the minimum area as 0.4% while clause 3.12.6.2 gives the maximum areas as 6%, 8% and 10% for vertically cast, horizontally cast and at laps in columns, respectively.

It is common to have 2–3% reinforcement, which usually gives an amount of reinforcement which can easily be handled on site. We shall assume therefore, an amount of reinforcement equal to, say 2.7%, i.e.

$$A_{sc} = 0.027A_c$$

Now substitute this in the equation above.

$$2500 \times 10^3 = (0.4 \times 35 \times A_c) + (0.75 \times 0.027A_c \times 460)$$

i.e. $A_c = 107\,227.11 \text{ mm}^2$

For a square column, the dimension of each side

$$= \sqrt{107\,227.11}$$

$$= 327.46 \text{ mm}$$

$\therefore$ Use a column which is 350 mm square.

Now from equation [38, BS],

$$2500 \times 10^3 = (0.4 \times 35 \times 350^2) + (0.75 \times A_{sc} \times 460)$$

$$A_{sc} = 2275.36 \text{ mm}^2$$

$$6T25 \text{ mm } \varnothing = 2950 \text{ mm}^2$$

Links

In order to prevent buckling of the longitudinal bars, links (or binders) are necessary. These links are designed in accordance with 3.12.7.1, i.e.

Size $= \frac{1}{4} \times 25 = 6.25$ mm, i.e. 8 mm bars

Spacing $= 12 \times 25 = 300$ mm

R8 @ 300 mm c/c.

Summary

Size	350 mm square
Main reinf.	6T25 bars
Links	R8 @ 300 c/c.

Example 4

An internal column in a braced, three-storey building having equal bays, carries an ultimate load of 3000 kN. It has cross-sectional dimensions of 400 mm × 300 mm and a height between end restraints of 4.20 m. These restraints are such that, from Table 3.21:

$$l_{ex} = 0.9 l_0, \text{ and } l_{ey} = 0.95 l_0$$

Calculate the reinforcement required if $f_{cu} = 40$ N/mm^2: $f_y = 460$ N/mm^2.

Check if the column is short or slender:

$$\frac{l_{ex}}{h} = \frac{0.9 \times 4200}{400} = 9.45$$

$$\frac{l_{ey}}{b} = \frac{0.95 \times 4200}{300} = 13.3$$

Since both are less than 15 the column is short. Using equation [39, BS 8110],

$$N = 0.35 f_{cu} A_c + 0.67 A_{sc} f_y$$

$$3000 \times 10^3 = (0.35 \times 40 \times 400 \times 300) + (0.67 A_{sc} \times 460)$$

$$A_{sc} = 4282.93 \text{ mm}^2$$

$4T25 + 2T40 = 4470$ mm^2 or $6T32 = 4830$ mm^2

Links size

$= \frac{1}{4} \times 40 = 10$ mm or $\frac{1}{4} \times 32 = 8$ mm

Links spacing

$= 12 \times 25 = 300$ mm or $12 \times 32 = 384$ mm

R10 @ 300 mm c/c or R8 @ 375 mm c/c

The column designs illustrated in examples 1–4 have not included bending moments. As such they would not be common. When bending moments are considered they can bend the column either about one axis or about both axes. These columns are designed most conveniently using design charts rather than design formulae.

Example 5

An internal column of a braced frame in a single-storey building carries an ultimate axial load of 300 kN and an ultimate bending moment of 175 kN m about the major axis. The column has cross-sectional dimensions of 450 mm × 300 mm and a height of 4.50 m between the end restraints. These restraints are such that $l_{ex} = 0.9 l_0$ and $l_{ey} = 0.9 l_0$.

Assume:
1) $f_{cu} = 40$ N/mm^2
2) $f_y = 460$ N/mm^2
3) $\varnothing$ of main reinforcement $= 25$ mm
4) $\varnothing$ of the links $= 8$ mm

Design the column in accordance with BS 8110.

Check if the column is short or slender:

$$\frac{l_{ex}}{h} = \frac{0.9 \times 4500}{450} = 9$$

$$\frac{l_{ey}}{b} = \frac{0.9 \times 4500}{300} = 13.5$$

Since both are less than 15 the column is short.

Check minimum moment (3.8.2.4, BS 8110):

$$e_{min} = 0.05 \times 450$$

$$= 22.5 \text{ mm} > 20 \text{ mm}$$

$\therefore$ Use 20 mm

$$M_{min} = 300 \times 0.02$$

$$= 6 \text{ kN m} < 175 \text{ kN m}$$

$\therefore$ Design for 175 kN m.

Choose the appropriate design chart.

Cover to the links, with moderate conditions of exposure since the bottom of the column will be below ground level, = 30 mm (T.3.4, BS 8110).

$$d = 450 - \left(30 + 8 + \frac{25}{2}\right) = 399.50 \text{ mm}$$

$$\frac{d}{h} = 399.50/450 = 0.888$$

Round down to 0.85 and use Chart No. 38, BS 8110.

Calculate the main reinforcement.

$$\frac{N}{bh} = \frac{300 \times 10^3}{300 \times 450} = 2.22 \text{ N/mm}^2$$

$$\frac{M}{bh^2} = \frac{175 \times 10^6}{300 \times 450^2} = 2.88 \text{ N/mm}^2$$

$$\frac{100A_{sc}}{bh} = 1.4$$

$$\therefore \quad A_{sc} = \frac{1.4 \times 300 \times 450}{100} = 1890 \text{ mm}^2$$

$\therefore$ Use 4T25 (1960 mm^2)

Links

Size $= \frac{1}{4} \times 25 \doteqdot 8$ mm

Spacing = 12 × 25 = 300 mm

$\therefore$ Use R8 @ 300 mm c/c.

Max. and min. areas of reinforcement
Min. areas: These are built into charts and are therefore satisfactory.
Max. areas: (3.12.6.2) – none exceeded.

Example 6 – Biaxial bending.
A short column carries an ultimate axial load of 1000 kN with ultimate bending moments of 280 kN m about the major axis and 120 kN m about the minor axis.

Assume:
1) $f_{cu} = 30 \text{ N/mm}^2$
2) $f_y = 460 \text{ N/mm}^2$
3) $d/h = 0.90$
4) $h = 450$ mm
5) $b = 300$ mm

Design the column in accordance with BS 8110.

Calculate the increased moments (3.8.4.5, BS 8110):

$$h' = 450 \times 0.9 \left(\text{since } \frac{d}{h} = 0.9\right)$$

$$= 405 \text{ mm}$$

$$b' = 300 - (450 - 405)$$

$$= 255 \text{ mm}$$

$$\frac{M_x}{h'} = \frac{280 \times 10^6}{405} = 691 \times 10^3 \text{ N}$$

$$\frac{M_y}{b'} = \frac{120 \times 10^6}{255} = 470.59 \times 10^3 \text{ N}$$

i.e. $\frac{M_x}{h'} > \frac{M_y}{b'}$

$\therefore$ Use equation [40, BS].

From Table 3.24, BS 8110,

$$\frac{N}{bhf_{cu}} = \frac{1000 \times 10^3}{300 \times 450 \times 30} = 0.247$$

$\therefore$ $\beta = 0.69$ by interpolation.

Using eqn. [40, BS],

$$M'_x = M_x + \beta \frac{h'}{b'} M_y$$

$$= 280 + 0.69 \times \frac{405}{255} \times 120$$

$$= 411.51 \text{ kN m}$$

Design for N and M'_x.

$$\frac{N}{bh} = \frac{1000 \times 10^3}{300 \times 450} = 7.41 \text{ N/mm}^2$$

$$\frac{M'_x}{bh^2} = \frac{411.51 \times 10^6}{300 \times 450^2} = 6.77 \text{ N/mm}^2$$

From Chart No. 29, BS 8110,

$$\frac{100A_{sc}}{bh} = 3.20$$

$$A_{sc} = 3.20 \times \frac{300}{100} \times 450$$

$$= 4320 \text{ mm}^2$$

6T32 mm bars = 4830 mm^2

Links

Size $= \frac{1}{4} \times 32 = 8$ mm

Spacing $= 12 \times 32 = 384$ mm

Use R8 @ 375 mm c/c.

Slender columns

In the design of slender columns the fact that the columns deflect must be taken into account. This is done by designing, not only for the calculated moment, but also for additional moments induced by this deflection. A slender column design can fall into one of three basic categories:

1) Bending about a minor axis.
2) Bending about a major axis.
3) Biaxial bending.

How each is designed is illustrated in the following sections:

1) Bending about a minor axis

A reinforced concrete column has cross-sectional dimensions of 600 mm × 450 mm and is part of a braced frame. The effective height of the column in relation to both axes is 7.250 m. Use BS 8110 to design the column for the following conditions:

1) the axial load $N = 3500$ kN,
2) the moment about the Y-axis at the bottom = 110 kN m,
3) the moment about the Y-axis at the top = 175 kN m,
4) $f_{cu} = 40 \text{ N/mm}^2$,
5) $f_y = 460 \text{ N/mm}^2$,
6) conditions of exposure are moderate.

Check if the column is short or slender (3.8.1.3, BS 8110):

$$\frac{l_{ex}}{h} = \frac{7250}{600} = 12.08$$

$$\frac{l_{ey}}{b} = \frac{7250}{450} = 16.11$$

Since both are not less than 15 the column is slender.

Calculate the design moment (3.8.3.2, BS 8110):

$$M_i = 0.4M_1 + 0.6M_2 \geqslant 0.4M_2 \text{ (Eqn. [36, BS])}$$

$$= (0.4 \times 110) + (0.6 \times 175) \geqslant 0.4 \times 175$$

$$= 149 \text{ kN m} \geqslant 70 \text{ kN m}$$

Maximum design moment will be the greatest of:

1) M_2
2) $M_i + M_{add}$
3) $M_1 + M_{add}/2$
4) $e_{min}N$

i.e. 1) $M_2 = 175$ kN m

2) Combining equations [32, BS 8110], [34, BS 8110] and [35, BS 8110],

$$M_{add} = \frac{NKh}{2000}\left(\frac{l_e}{b'}\right)^2$$

An initial, conservative, value for K is 1.00.

$$\text{Thus} \quad M_{add} = \frac{3500 \times 1 \times 0.450}{2000}\left(\frac{7.250}{0.450}\right)^2$$

$$= 204.41 \text{ kN m}$$

$$\therefore \quad M_i + M_{add} = 149 + 204.41$$

$$= 353.41 \text{ kN m}$$

3) $M_1 + M_{add}/2$

$$= 110 + 204.41/2 = 212.21 \text{ kN m}$$

4) $e_{min}N = 0.05 \times 0.450 \times 3500$

$$= 67.5 \text{ kN m}$$

The greatest of these is (2), i.e. 353.41 kN m.

Choose the appropriate design chart.

T.3.4, BS 8110 cover = 30 mm

Assuming 32 mm main bars and 8 mm links,

$$d = 450 - (30 + 8 + 32/2)$$

$$= 396 \text{ mm}$$

$$d/h = 396/450 = 0.88$$

∴ Use Chart No. 38, BS 8110.

$$N/bh = 3500 \times 10^3/(600 \times 450)$$

$$= 12.96 \text{ N/mm}^2$$

$$M/bh^2 = 353.41 \times 10^6/(600 \times 450^2)$$

$$= 2.91 \text{ N/mm}^2$$

$$100A_{sc}/bh = 0.8$$

$$A_{sc} = 0.8 \times \frac{450}{100} \times 600$$

$$= 2160 \text{ mm}^2$$

4T32 Bars - 3320 mm²

It is possible now to reduce M_{add} by the appropriate 'K' factor from Chart 38 and substitute this value in equation [32, BS].

i.e. from Chart 38, with $N/bh = 12.96 \text{ N/mm}^2$ and $M/bh^2 = 2.91 \text{ N/mm}^2$

$$K = 0.68 \text{ approximately.}$$

Thus $$M_{add} = 204.41 \times 0.68$$

$$= 139.0 \text{ kN m}$$

∴ $$M_c + M_{add} = 149 + 139$$

$$= 288 \text{ kN m}$$

Now $$M/bh^2 = 288 \times 10^6/(600 \times 450^2)$$

$$= 2.37 \text{ N/mm}^2$$

From Chart 38, BS 8110, $100A_{sc}/bh = 0.4$ (i.e. the minimum).

$$A_{sc} = \frac{0.4 \times 600 \times 450}{100}$$

$$= 1080 \text{ mm}^2$$

4T25 Bars - 1960 mm²

Note 1: By comparing this area of reinforcement with that calculated with $K = 1$, it can be appreciated why it was earlier stated that $K = 1$ gives a conservative estimate.

Note 2: Chart 38 of BS 8110 now gives $K = 0.61$ approximately. This value could be taken and the calculation repeated on an iterative basis until the optimum is reached.

Links $\frac{1}{4} \times 25 \doteqdot 8$ mm

$12 \times 25 = 300$ mm

8 mm links @ 300 mm c/c

2) Bending about a major axis

Redesign the column in the previous example, assuming zero moments about the Y–Y axis but a moment of 300 kN m at the bottom and 425 kN m at the top, about the X–X axis. As calculated previously, the column is slender.

Check clauses 3.8.3.4 and 3.8.3.5

$$\frac{l_e}{h} = \frac{7250}{600} = 12.08 < 20$$

$$\frac{600}{450} = 1.33 < 3.00$$

Note: Both of the above values are less than the critical values stated, hence the column can be designed following the principles illustrated in the previous example. If either critical value had been exceeded, the column would have had to be designed as biaxially bent, as illustrated in the next example.

Calculate the design moment:

$$M_i = 0.4M_1 + 0.6M_2 \geqslant 0.4M_2 \qquad \text{[eqn. 36, BS]}$$

$$= (0.4 \times 300) + (0.6 \times 425) \geqslant 0.4 \times 425$$

$$= 375 \text{ kN m} \geqslant 170 \text{ kN m}$$

Max design moment will be the greatest of:

1) M_2
2) $M_i + M_{add}$
3) $M_1 + M_{add}/2$
4) $e_{min}N$

i.e. 1) $M_2 = 425$ kN m

2) Assuming for initial calculation $K = 1$,

$$M_{add} = \frac{NKh}{2000}(l_e/b')^2$$

(Eqns. [32, 34, 35, BS 8110])

$$= \frac{3500 \times 1 \times 0.600}{2000}\left(\frac{7.250}{0.45}\right)^2$$

$= 272.55$ kN m

$M_i + M_{add} = 375 + 272.55$

$= 647.55$ kN m

3) $M_1 + M_{add}/2$

$= 300 + 272.55/2 = 436.28$ kN m

4) $e_{min}N$

$e_{min} = 0.05 \times 600 = 30$ mm,

but max $= 20$ mm (3.8.2.4)

$\therefore e_{min}N = 0.020 \times 3500$

$= 70$ kN m

The greatest of these is (2), i.e. 647.55 kN m.

Choose the appropriate design chart.

T.3.4, BS 8110 cover $= 30$ mm

Assuming 32 mm main bars and 8 mm links,

$d = 600 - (30 + 8 + 32/2)$

$= 546$ mm

$d/h = 546/600 = 0.91$

$\therefore$ Use Chart No. 39, BS 8110.

$N/bh = 3500 \times 10^3/(450 \times 600)$

$= 12.96$ N/mm^2

$M/bh^2 = 647.55 \times 10^6/(450 \times 600^2)$

$= 4.00$ N/mm^2

From Chart 39, BS 8110, $K = 0.76$ approx.

Thus $M_{add} = 272.55 \times 0.76$

$= 207.14$ kN m

$\therefore M_i + M_{add} = 375 + 207.14$

$= 582.14$ kN m

$M/bh^2 = 582.14 \times 10^6/(450 \times 600^2)$

$= 3.59$ N/mm^2

From Chart 39, BS 8110,

$$\frac{100A_{sc}}{bh} = 1.25$$

$A_{sc} = 1.25 \times 450 \times 600/100$

$= 3375$ mm^2

4T32 + 2T25 $= 3220 + 982 = 4202$ mm^2

Links $\frac{1}{4} \times 32 = 8$ mm

Spacing $12 \times 25 = 300$ mm

$\therefore$ Use R8 @ 300 mm c/c.

3) Biaxial bending

Redesign the column in the previous examples assuming that the moments about the X–X and Y–Y axes coexist. As calculated previously the column is slender.

Calculate the design moments (3.8.3.6, BS 8110):

Design moment in the Y-direction
$= 353.41$ kN (see page 193).
Design moment in the X-direction.
When this was calculated previously it was based on b' being the smaller dimension of the column, i.e. 450 mm. However for biaxial bending b' is to be taken as the dimension of the column in the plane of bending.
Thus additional moment will be the greatest of:

1) $M_2 = 425$ kN m

2) $M_i + M_{add}$ with $K = 1$ and $b' = 600$ mm

$$M_{add} = \frac{3500 \times 1 \times 0.6}{2000}\left(\frac{7.25}{0.6}\right)^2$$

$= 153.31$ kN m

$\therefore M_i + M_{add} = 375 + 153.31$

$= 528.31$ kN m

3) $M_1 + M_{add}/2 = 300 + 153.31/2$

$= 376.66$ kN m

4) $e_{min}N = 70$ kN m

The greatest of these is 2), i.e. 528.31 kN m. Equations [40, BS] and [41, BS], as appropriate, give the increased design moments.

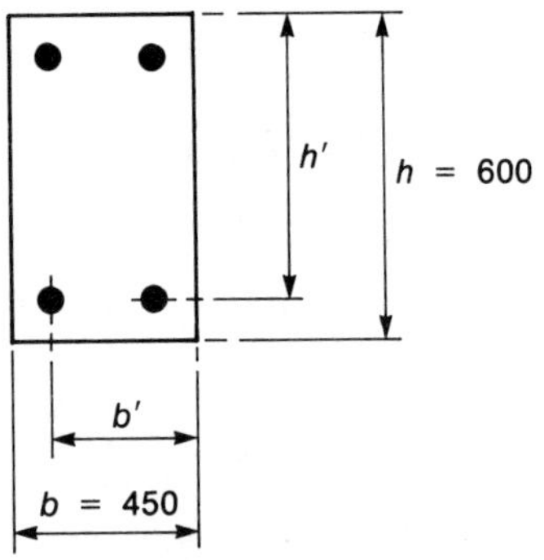

Fig. 18.4 *Plan of column*

Assuming moderate conditions of exposure, 32 mm diameter main bars and 8 mm links,

$$b' = 450 - 30 - 8 - 32/2$$
$$= 396 \text{ mm}$$
$$h' = 600 - 30 - 8 - 32/2$$
$$= 546 \text{ mm}$$

$$M_x/h' = \frac{528.31 \times 10^6}{546} = 967.6 \times 10^3$$

$$M_y/b' = \frac{353.41 \times 10^6}{396} = 892.45 \times 10^3$$

i.e. $M_x/h' > M_y/b'$

$\therefore$ Use equation [40, BS 8110].

$$M'_x = M_x + \beta \frac{h'}{b'} M_y$$

β is obtained from Table 3.24, BS 8110.

$$\frac{N}{bhf_{cu}} = \frac{3500 \times 10^3}{600 \times 459 \times 40}$$
$$= 0.324$$

$\therefore \quad \beta = 0.62$

$$\therefore \quad M'_x = 528.31 + 0.62 \times \frac{546}{396} \times 353.41$$
$$= 830.42 \text{ kN m}$$

Choose design chart.

$$\frac{d}{h} = \frac{h'}{h} = \frac{546}{600} = 0.91$$

$\therefore$ Use Chart No. 39, BS 8110.

$$N/bh = \frac{3500 \times 10^3}{450 \times 600} = 12.96 \text{ N/mm}^2$$

$$\frac{M'_x}{bh^2} = \frac{830.42 \times 10^6}{450 \times 600^2} = 5.13 \text{ N/mm}^2$$

$K = 0.81$

Thus M_x and M_y and hence M'_x can be reduced.

In Y-direction, $M_{add} = 204.41 \times 0.81$

$= 165.57$ kN m

$\therefore \quad M_i + M_{add} = 149 + 165.57$

$= 314.57$ kN m

In X-direction, $M_{add} = 153.31 \times 0.81$

$= 124.18$ kN m

$\therefore \quad M_i + M_{add} = 375 + 124.18$

$= 499.18$ kN m

$$M'_x = 499.18 + 0.62 \times \frac{546}{396} \times 314.57$$
$$= 768.09 \text{ kN m}$$

$$\frac{M'_x}{bh^2} = \frac{768.09 \times 10^6}{450 \times 600^2} = 4.74 \text{ N/mm}^2$$

From Chart 39, BS 8110, with $\frac{N}{bh} = 12.96$ N/mm^2,

$$\frac{100A_{sc}}{bh} = 2.05$$

$$A_{sc} = \frac{2.05 \times 450 \times 600}{100}$$
$$= 5535 \text{ mm}^2$$

Because the column is bent about both axes, $A_{sc}/2$ must be provided on each face, i.e. 5535/2 = 2767.50 mm^2.

∴ Use 4T32 (3220 mm^2) in each face.

Links $\frac{1}{4} \times 32 = 8$ mm

$12 \times 32 = 384$ mm

R8 @ 375 mm c/c

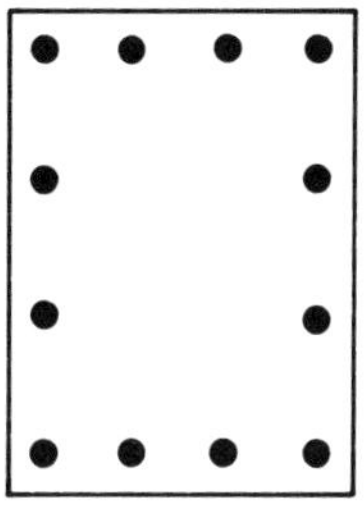
i.e. total reinforcement
= 12T32 bars.

Fig. 18.5 *Layout of designed reinforcement*

Appendix 1

Bibliography

The following publications have been found helpful by the authors in the preparation of this book, and will be of some assistance to students who may require further or explanatory information.

1) Used by both authors:

Extracts from British Standards for Students of Structural Design, The British Standards Institution.

BS 648: 1969 *Schedule of weights of building materials.*

BS 6399: Part 1: 1984 *Code of practice for dead and imposed loads.*

2) Used by R.A. Sharpe in the preparation of Chapters 1–11:

BS 5268 *Structural use of timber:* Part 2: 1984.

BS 5628 *Code of practice for use of masonry:* Part 1: *Unreinforced masonry,* 1978.

BS 5950 *Structural use of steelwork in building:* Part 1: 1985.

CP 117: Part 1 *Composite steel/concrete beams,* 1965.

Steelwork Design Guide to BS 5950: Part 1: 1985. The Steel Construction Institute, Croydon.

MacGinley, T.J. and Ang, T.C., *Structural Steelwork Design to Limit State Theory,* Butterworths, London, 1987.

Pask, J.W., *Manual on Connections,* BCSA Ltd, London.

Davies, C., *Steel–Concrete Composite Beams for Buildings,* Goodwin, 1975.

Knowles, P.R., *Composite Steel and Concrete Construction,* Butterworths, London, 1973.

Structural Masonry Designers' Manual, Granada, 1982.

Formwork – A Guide to Good Practice, The Concrete Society, 1986.

Baird, J.A. and Ozelton, E.C., *Timber Designers' Manual,* Crosby Lockwood Staples, London, 1976.

3) Used by B. Currie in the preparation of Chapters 12–18:

BS 8110: *Structural use of concrete:* 1985.

Kong, F.K. and Evans, R.H., *Reinforced and Prestressed Concrete,* 3rd edn, Van Nostrand Reinhold (UK), Wokingham, 1987.

Allen, A.H., *Reinforced Concrete Design to BS 8110 Simply Explained,* Spon, London, 1987.

Handbook to BS 8110: 1985, Spon, London.

Manual for the Design of Reinforced Concrete Building Structures, The Institution of Structural Engineers, London, 1985.

All British Standards are available from:
British Standards Institution Sales, Linford Wood, Milton Keynes, Bucks, MK14 6LE.

Appendix 2

Area chart for reinforcement

Steel area chart for number of bars

Bar diameter (mm)	*Areas in mm² for numbers of bars*									
	1	*2*	*3*	*4*	*5*	*6*	*7*	*8*	*9*	*10*
6	28	57	85	113	142	170	198	226	255	283
8	50	101	151	201	252	302	352	402	453	503
10	79	157	236	314	392	471	550	628	707	785
12	113	226	339	452	566	679	792	905	1,020	1,130
16	201	402	603	804	1,010	1,210	1,410	1,610	1,810	2,010
20	314	628	943	1,260	1,570	1,890	2,200	2,510	2,830	3,140
25	491	982	1,470	1,960	2,450	2,950	3,440	3,930	4,420	4,910
32	804	1,610	2,410	3,220	4,020	4,830	5,630	6,430	7,240	8,040
40	1,260	2,510	3,770	5,030	6,280	7,540	8,800	10,100	11,300	12,600

Steel area chart for spacing of bars

Bar diameter (mm)	*Areas in mm² for spacings in mm*								
	50	*75*	*100*	*125*	*150*	*175*	*200*	*250*	*300*
6	566	377	283	226	189	162	142	113	94
8	1,010	671	503	402	335	287	252	201	168
10	1,570	1,050	785	628	523	449	393	314	262
12	2,260	1,510	1,130	905	745	646	566	452	377
16	4,020	2,680	2,010	1,610	1,340	1,150	1,010	804	670
20	6,280	4,190	3,140	2,510	2,090	1,800	1,570	1,260	1,050
25	9,820	6,550	4,910	3,930	3,270	2,810	2,450	1,960	1,640
32	16,100	10,700	8,040	6,430	5,360	4,600	4,020	3,220	2,680
40	25,100	16,800	12,600	10,100	8,380	7,180	6,280	5,030	4,190

Wireweld metric fabric - British Standard preferred types

	British Standard reference	*Mesh size Nominal pitch of wires* Main (mm)	Cross (mm)	*Size of wires* Main (mm)	Cross (mm)	*Cross sectional area per metre width* Main (mm)	Cross (mm)	*Nominal mass per square metre (kg)*
Square mesh fabric:	A 393	200	200	10	10	393	393	6.16
	A 252	200	200	8	8	252	252	3.95
	A 193	200	200	7	7	193	193	3.02
	A 142	200	200	6	6	142	142	2.22
	A 98	200	200	5	5	98	98	1.54
Structural fabric:	B 1.131	100	200	12	8	1,131	252	10.9
	B 785	100	200	10	8	785	252	8.14
	B 503	100	200	8	8	503	252	5.93
	B 385	100	200	7	7	385	193	4.53
	B 283	100	200	6	7	283	193	3.73
	B 196	100	200	5	7	196	193	3.05
Long mesh fabric:	C 785	100	400	10	6	785	70.8	6.72
	C 636	100	400	9	6	636	70.8	5.55
	C 503	100	400	8	5	503	49.0	4.34
	C 385	100	400	7	5	385	49.0	4.41
	C 283	100	400	6	5	283	49.0	2.61

Sheet size and availability: square and long meshes are normally available from stock in sheets of 4.8 m x 2.4 m.

Index